Emissions Trading
and Business

Ralf Antes
Bernd Hansjürgens
Peter Letmathe (Editors)

Emissions Trading and Business

With 64 Figures and 52 Tables

sponsored by
Stiftungsfonds Dresdner Bank
im Stifterverband für die Deutsche Wissenschaft

Physica-Verlag
A Springer Company

Editors

PD Dr. Ralf Antes
University of Halle-Wittenberg
Chair of Corporate Environmental Management
Große Steinstraße 73
06099 Halle
Germany
antes@wiwi.uni-halle.de

Prof. Dr. Bernd Hansjürgens
UFZ – Centre for Environmental Research
Department of Economics
Permoserstraße 15
04318 Leipzig
Germany
bernd.hansjuergens@ufz.de

Prof. Dr. Peter Letmathe
University of Siegen
Chair of Value Chain Management
Hölderlinstraße 3
57068 Siegen
Germany
peter.letmathe@uni-siegen.de

Library of Congress Control Number: 2006931849

ISBN-10 3-7908-1747-3 Physica-Verlag Heidelberg New York
ISBN-13 978-3-7908-1747-8 Physica-Verlag Heidelberg New York

Physica-Verlag is a part of Springer Science+Business Media

springer.com

Typesetting: Camera ready by the author
Cover-design: Erich Kirchner, Heidelberg
Production: LE-TEX Jelonek, Schmidt & Vöckler GbR, Leipzig

SPIN 11799672 Printed on acid-free paper 88/3100YL 5 4 3 2 1 0

Foreword

Since 2003 CO_2 emissions are no longer a free good. This is the result of the European Union's Emission Trading Scheme (EU ETS), which has created the world's largest market in emission allowances, involving around 1200 companies or an equivalent of 12.000 installations across the 25 EU member states.

The major risk affecting companies in the EU ETS essentially come from possible movements in price and the availability of a sufficient number of allowances at a specific point in time. In order to manage these risks efficiently and profitably appropriate physical and financial instruments are expected to enter the market soon.

Spot transactions, Forwards and Swaps, Options, Repos as well as Hybrid or Basket products are offered or are under development. CO_2 is bridging the gap between gas, power and commodity prices worldwide. Bourses and trading platforms are using the opportunities to catch a share of the rapidly growing carbon market.

A future challenge will be the integration of other regional activities or country trading schemes, for example in Asia or North-America, into the on-going European carbon market. Strengthen liquidity is one of the key factors for optimizing the market framework.

Analysing the economic background and market drivers is a huge challenge for academic research and a real demand for the business practice.

Armin Sandhövel
Head Carbon Risk, Dresdner Bank AG, Germany

Preface and acknowledgements

This book forms part of our work on the project Corporate Sustainability, which is funded by the Dresdner Bank Stiftungsfonds im Stifterverband für die Deutsche Wissenschaft. It is part of the sponsorship on the occasion of the 500-year anniversary of the University Halle-Wittenberg. We would like to express our gratitude for the financial and administrative support, particularly to Dr. Armin Sandhövel (Dresdner Bank AG), Dr. Gabriele Jachmich and Karin Heyl (Stiftungsfonds Dresdner Bank), Dr. Heinz-Rudi Spiegel and Sigrid Westermann (Stifterverband für die Deutsche Wissenschaft).

In the discourse on sustainable development, interdisciplinary cooperation is often and appropriately demanded from both economists and management scholars. Quite to the contrary and remarkably rare, intradisciplinary cooperation between the neighbouring disciplines economics and management science is demanded and undertaken. Bridging both disciplines by discussing problems and questions at their interface is therefore an additional aim of the book. Accordingly, this volume is based on cooperation between the German Society for Operations Research and the Faculty of Economics of the University Halle-Wittenberg, in particular the Chair of Environmental Economics and the Chair of Corporate Environmental Management.

We are pleased to have brought together young and established scholars from both disciplines and from different countries to discuss with us the topic of emissions trading. We had 30 submissions from research institutions of eleven countries. After a review process we accepted the papers of 17 young scholars from Finland, Germany, Italy, the Netherlands, Spain, the United Kingdom, and the United States of America. Additional Denny Ellerman (MIT Boston), Katie Begg (University of Edingburgh), Katja Barzantny (EUtech Energie & Management GmbH, Aachen), Edeltraud Günther (Dresden University of Technology), Joachim Schleich (Fraunhofer Institute for Systems and Innovation Research, Germany), and Wolf Fichtner (University of Technology Cottbus) publish key papers in the volume.

The book is inspired by the contributions presented at the workshop "Business and Emissions Trading" in November 2003 at the Leucorea, the former University of Wittenberg/Germany. For the book the contributions were revised and updated.

In its early phase emissions trading was often criticised as a kind of letter of indulgence. However, it is a pure coincidence of history that the place from where Martin Luther in the 16th century fought against the 'original' letters of indulgence was Wittenberg (Luther was Professor for Theology at the Leucorea from 1508 on). Founded as a University in 1502, resettled and united with the University Halle in 1817 and re-established as a foundation in 1994, the old buildings of the Leucorea were completely renovated, i. e. as a venue for the University Halle-Wittenberg, a few years ago.

We would like to thank all those individuals who contributed to this book for sharing their work and ideas with us, as we learned a great deal from them. We would like to thank Professor Dr. Hans-Ulrich Zabel for his constant support. We would like to thank Paul Ronning, who did a great job in improving the English texts for the non-native speakers, and Katharina Wetzel-Vandai, Gabriele Keidel and Ilse Wittig from Springer Publishers for their great patience and helpfulness at all times. And finally, we would like to give special thanks to Kay Fiedler who provided, with his meticulous sense of commitment, the layout for the book.

Halle (Saale) / Leipzig / Siegen July 2006

Ralf Antes, Bernd Hansjürgens and Peter Letmathe

Contents

Foreword V

Preface and acknowledgements VII

Introduction 1
Ralf Antes, Bernd Hansjürgens, Peter Letmathe

Part A - Institutional design, decision making and innovation

Abatement costs vs. compliance costs in multi-period emissions trading – the firms' perspective 11
Sven Bode

1 Introduction 12
2 Emissions trading and allocation of allowances 13
3 The model 16
4 Numerical analysis 20
5 Conclusion 23

Generous allocation and a ban on banking – implications of a simulation game for EU emissions trading 27
Joachim Schleich, Karl-Martin Ehrhart, Christian Hoppe, Stefan Seifert

1 Introduction 28
2 Banking in the EU ETS 29
3 The emissions trading simulation SET UP 30
4 Results of SET UP 33
5 Conclusions 36

Emissions trading and innovation in the German electricity industry – impact of possible design options for an emissions trading scheme on innovation strategies in the German electricity industry 39
Martin Cames, Anke Weidlich

1 Introduction 40
2 Innovation and windows of opportunity 40

3 Time windows in the German electricity market ... 41
4 Innovation incentives of different allocation methods ... 45
5 Conclusions ... 49

A dynamic game of technology diffusion under emissions trading: an experiment ... 53
Ivana Capozza

1 Introduction ... 54
2 The model ... 55
3 The game ... 58
4 The experiment ... 65
5 Concluding remarks ... 69

Sustainability entrepreneurship in the context of emissions trading ... 73
Anne Gerlach

1 Introduction ... 74
2 Conceptual framework ... 74
3 Sustainability intra- and interpreneurship in the context of emissions trading ... 80
4 Conclusions ... 84

Part B - Investment and management strategies under emissions trading

Optimal strategies for emissions trading in a Putty-Clay Vintage Model ... 91
Peter Letmathe, Sandra Wagner

1 Introduction ... 92
2 Emissions trading from the firm's perspective ... 92
3 Short-term production planning ... 94
4 Long-term production planning with the Vintage Production Functions and the Putty-Clay Model ... 97
5 Integrated investment and production planning ... 100
6 Conclusions ... 102

Strategic production management of companies participating in the European greenhouse gas emission allowance trading scheme ... 105
Wolf Fichtner

1 Introduction ... 106
2 Characterisation of the new production factor emission allowances ... 106
3 A model for investment and production planning within electric utilities considering the framework of the European CO_2 emission allowance trading ... 107

4 Long term planning of energy supply concepts in energy-intensive production companies 112
5 A model to analyse the efficiency of international cooperation in mitigating climate change 115
6 Summary 116

Decision making in the emissions-market under uncertainty 119
Gorden Spangardt, Michael Lucht, Christian Wolf, Christian Horn

1 Background 120
2 Modelling the decision process 120
3 Modelling the stochastic variables 122
4 Optimization model 124
5 Exemplary results 127
6 Summary and outlook 131

The impact of climate policy on heat and power capacity investment decisions 133
Harri Laurikka

1 Introduction 134
2 Investment decision process 135
3 Quantitative investment appraisal 138
4 Discussion and conclusions 145

Implications of the European emissions trading scheme for strategic energy management in small and medium enterprises 151
Anja Pauksztat, Martin Kruska

1 Introduction 152
2 Relevance of the European emissions trading scheme for small and medium enterprises 152
3 Implications of the community scheme for small and medium enterprises 154
4 Scenario analysis involving dependence on the emissions allowance price 159
5 Conclusion 163

Management and optimization of environmental data within emissions trading markets – VEREGISTER and TEMPI 165
Bernhard Grimm, Stefan Pickl, Alan Reed

1 Introduction 166
2 Management of greenhouse gas emissions 166
3 Forecasting and econo-mathematics 169
4 Emissions trading market 173
5 Conclusion 176

Emissions trading with changing future commitments – some initial thoughts 177
Marcus Stronzik

1 Introduction 178
2 Application of the discounted cash flow approach 179
3 Emissions trading and real options 181
4 Conclusions 185

Part C - Emissions trading and business administration

Emission Trading North – important findings from a business perspective 189
Katja Barzantny

1 Introduction 190
2 The pilot project 190
3 Results 192
4 Final remarks 196

Corporate greenhouse gas management in the context of emissions trading regimes 199
Ralf Antes

1 Introduction 200
2 How GHGs affect companies 200
3 Emerging organizational fields of GHG management 205
4 Internal effects of GHG emissions trading regimes on companies 208

Accounting for emission rights 219
Edeltraud Günther

1 Introduction 220
2 Evolution of the climate policy up until the introduction of the scheme for greenhouse gas emission allowance trading 221
3 Greenhouse gas emission reporting 223
4 Treatment of emission rights in the annual accounts 229
5 Evolution of accounting at the international level 235
6 Possible development of the recognition of environmental issues in annual accounts 237
7 Summary 239

The role of stakeholder driven corporate governance – the example of BP's climate change strategy 241
Thomas Langrock

1 Introduction 242
2 Background: evaluation research and policy network analysis 243

3 Evaluation and assessment of the GHG commitment 244
4 The policy network approach applied to the BP plc. policy making process 248
5 Conclusions from the case study 252

Emissions trading and effects on financial markets 257
Timo Busch

1 Introduction 258
2 Interactions of financial markets and sustainability 258
3 Emissions trading and effects at a company level 259
4 New business opportunities in financial markets 262
5 Current developments and a framework for pro-active involvement 266
6 Conclusions 270

Part D - Effects of emissions trading schemes existing and being implemented

The EU emissions trading scheme and its competitiveness effects upon European business – results from the CGE model DART 275
Sonja Peterson

1 Introduction 276
2 Emissions trading and competitiveness in a globalizing world 277
3 Simulation of competitiveness effects of the EU emissions trading scheme 278
4 Simulation results 282
5 Summary and conclusions 287

Implementing the EU emissions trading directive in Spain: a comparative study of corporate concerns and strategies in different industrial sectors ... 293
Pablo del Río González

1 Introduction: aim, scope and methodology 294
2 The theoretical approach: public choice and climate policy 295
3 Firms, emissions trading and allocation: the views and strategies of the sectors covered by the EU ETS 296
4 Concluding remarks 311

UK's climate change levy and emissions trading scheme: implications for businesses' productivity and economic efficiency 313
Adarsh Varma

1 Introduction 313
2 Overview of UK's climate change levy (CCL) and emissions trading scheme (ETS) 314
3 The stochastic translog frontier cost function model 316

4 Dynamics of the model 319
5 The experience so far 320
6 Conclusion 323

The sources of emission reductions: evidence from U.S. SO_2 emissions from 1985 through 2002 327
A. Denny Ellerman, Florence Dubroeucq

1 Introduction 328
2 Data and methodology 330
3 Decomposition results 334
4 Conclusion 340

Policy-business interaction in emissions trading between multiple regions 353
Jürgen Scheffran, Marian Leimbach

1 Emissions trading on multiple levels 354
2 Links between economic growth, emission reductions and mitigation costs 357
3 Choice between technical options 360
4 Non-linear ratios 361
5 Model-based data and computations for the world's regions and selected cases 364

The changing role of the project mechanisms in emissions trading 369
Katherine G. Begg

1 Introduction 370
2 Development of the project mechanisms 370
3 What can be done to reduce costs in the CDM and where is progress being made? 377
4 Conclusions 383

Prevailing technologies and locations of CDM projects: the current situation compared with expectations 387
Jobert Winkel

1 Introduction 388
2 Overview of CDM projects at present 388
3 Determinants of the IRR of a landfill project 395
4 Conclusions 399

Introduction

Ralf Antes[I, II], Bernd Hansjürgens[III, IV], Peter Letmathe[V]

[I] Martin Luther-University Halle-Wittenberg
Faculty of Economics
Department of Corporate Environmental Management
Große Steinstraße 73, 06099 Halle, Germany
ralf.antes@wiwi.uni-halle.de

[II] Carl von Ossietzky-University Oldenburg
Faculty of Economics
CENTOS – Oldenburg Center for Sustainability Economics and Management
Uhlhornsweg, 26129 Oldenburg, Germany
ralf.antes@uni-oldenburg.de

[III] Martin Luther-University Halle-Wittenberg,
Faculty of Economics
Professorship of Environmental Economics
Universitätsring 3, 06099 Halle, Germany

[IV] UFZ – Centre for Environmental Research
Department of Economics
Permoserstraße 15, 04318 Leipzig, Germany
bernd.hansjuergens@ufz.de

[V] University of Siegen
School of Economic Disciplines
Chair of Value Chain Management in Small and Medium Sized Enterprises
Hoelderlinstraße 3, 57068 Siegen, Germany
peter.letmathe@uni-siegen.de

Keywords: Emissions trading, milestones, preconditions and criticisms, state-of-the-art in economics and business studies

1 State-of-the-art of emissions trading (ET) in theory and practice

Since the preliminary works of Coase (1960), upon which Dales (1968) built his concept and the formal efficiency-proof of Montgomery (1972), a well-formed methodological framework for theory has been long established. Furthermore, environmental permits have found on-going application within the context of practical environmental policies, in the course of which there exists particular emphasis upon the areas of clean-air policies and climate protection (see Hansjürgens 2005, for an overview). Figure 1 illustrates this development.

Tradable emission permits are, in this regard, a relatively new and innovative instrument in the world of environmental policies. Its special appeal can be attributed, however, not only to the fact that environmental protection is made possible at minimal cost to national economies.[1] This has long been recognised by economists. The special appeal of tradable emission permits can also be attributed to the fact that this instrument can play a strategic role for the implementation of sustainable environmental policies. Tradable environmental permits are namely the only (!) environmental-policy tool that is directly targeted at the steering of overall emission quantities. This is an advantage that is often overlooked, and not only by economists. If the politics to protect the environment is successful in its attempt to define the stock of natural assets, and on this basis is able to purport quantity targets for allowable exploitation of said natural assets, then tradable permits are, in this fashion, ideally suited to both instrumental implementation of these performance targets and assured preservation of (natural) stocks.

Introspection is most certainly required in this context, since the effectiveness of this instrument is inherently linked to further prerequisites. To avoid local and regional distribution effects (hot spots) the local concentration of emissions, which can occur as a consequence of trading permits must be ecologically irrelevant. Ecologists (as opposed to environmental activists!) say that this is indeed the case with green house gases. Furthermore, the homogeneity of the traded goods is necessary. Again, this is the case, per se, with green house gases. But there are serious problems with other natural resources for which permit trading schemes might be implemented, for example water, land or habitats. These goods differ in quality, and therefore are not homogeneous goods per se. Moreover, methodological problems concerning the monetary valuation of the social and ecological consequences of economic activities can hardly be denied. Market prices (e.g. for permits) need not reflect ecological scarcities, but follow another logic. More fundamentally, it is argued that monetary calculus is part of the problem itself, and therefore could not serve as the exclusive solution. There is a need, so it is concluded, for addi-

[1] As the EU Commission argues in their proposal for an EU-wide emissions trading scheme: "The key economic rationale behind emissions trading is to ensure that emissions reductions required to achieve a pre-determined environmental outcome take place where the cost of reduction is the lowest." (COM(2001) 581 final; Brussels, 23.10.2001)

tional non-monetary coordination, such as ethics. Finally, as a consequence of weak institutions, problems involving hot and tropical air could arise.

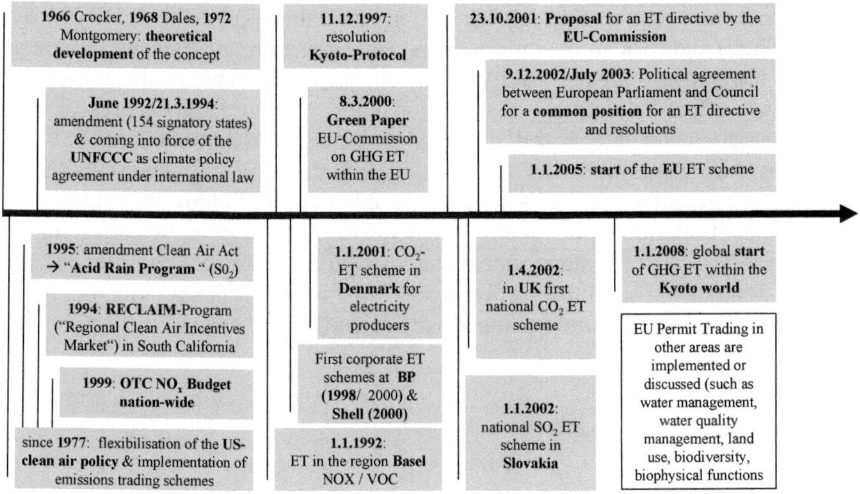

Fig. 1. Milestones of emissions trading

Notwithstanding all these considerations, the EU-CO_2 emissions trading, the worldwide largest emissions trading scheme, started on January 1st, 2005 and became reality for companies sited within the EU.

Denoted as remarkable are hence the extreme differences which have emerged between theory and practice regarding the approach to this topic. Economics has been developing considerable expertise in this regard for over 30 years. But it was, oddly enough, the practical application capacities of enterprises which were, for the most part, neither pro-active nor anticipatory in their implementation of emissions trading. This was even the case for large and only-indirectly-affected enterprises, such as the financial services industry. Nevertheless, enterprises will have to adjust in one form or another by the time, at the latest, that emissions trading schemes are implemented. On the other hand, business/management studies – i.e. the scientific discipline whose study topic (companies) and selection criterion (the economic calculation of internal company decision-making) have been essentially designated as the consignee of emissions trading and, therefore, would be assumed to be able to produce significant material regarding clarification and formation of ET schemes – have so far, surprisingly, not dealt seriously with the topic. A few individual publications have been brought forward on issues of risk and production management, and various advisory market studies have been specifically made for financial products – these dominate the sparse body of literature.

The reticence of the majority of companies, i.e. the empirical field of management research may be one reason, of course. But is it not possible that a more profound justification prevails, namely that market coordination for the primarily hierarchically-construed environmental politics and environmental law represents an "alien"? And consequently, would not the same then apply for environmental management, which reflects these institutional arrangements on an internal (within the enterprise) basis? Not to mention the fact that an adjustment, in the context of an internal emissions-trading scheme (i.e. in internal markets such as BP or Royal Dutch Shell) signifies an oxymoron. This is the situation, at all events, if one advocates the analysis of Coase and Williamson, according to whom the existence of enterprises offers hierarchical coordination under certain circumstances, even in free-enterprise economies based on higher efficiency of transaction costs.

These short expositions show that, as regards the effects of emissions trading upon businesses, there still exists a considerable deficit of knowledge. In order to reduce this, it appears that an intellectual exchange between the two disciplines of economics and business/management studies would constitute a considerably productive approach. The following four parts in 24 contributions provide the results for such an introductory and internationally scientific exchange.

2 Overview of the book

Part A includes contributions regarding the institutional design, decision making and innovation aspects of emissions trading. In the first contributions the allocation of allowances is analysed. *Sven Bode* compares the allocation based on historical emissions and based on benchmarks by using a two-player, two-period model. He finds that, depending on their marginal abatement cost, participants have different preferences with regard to the allocation method over time, as individual compliance costs can change as well. Based on a simulation with companies, *Joachim Schleich et al.* analyse the effects of banning banking while allowances are allocated fairly generously. Their findings suggest that an EU-wide ban on banking would lead to efficiency losses in addition to those losses which arise from the lack of inter-temporal flexibility. *Martin Cames* and *Anke Weidlich* are focussing on different allocation methods, too. In an analysis of the German electricity industry they find that alternative designs for emissions trading schemes have an impact on technological change, in particular for the fuel switch away from carbon-intensive lignite to natural gas. Due to long investment cycles the capacity to broaden techno-economic windows is crucial. The next two contributions again discuss innovation effects: *Ivana Capozza* models a dynamic duopoly. She shows that technology diffusion is one of the possible stationary equilibria of the game. Replication of the game in a laboratory experiment leads her to suggest that firms' behaviour will eventually lead to innovation diffusion. *Anne Gerlach* puts emphasis on sustainability entrepreneurs, particularly on entrepreneurs within organisations (intrapreneurs). She firstly develops a general conceptual framework for sustainability entrepreneurship and then applies it to the context of emissions trading. The findings suggest that organisations that want to gain advantage from innovative emissions reductions indeed have to foster sustainability entrepreneurship.

In **Part B** the authors focus the broad perspective of decision-making on decisions about investment and management strategies. Combining vintage production functions with the Putty-Clay Model, *Peter Letmathe* and *Sandra Wagner* show that technical progress rates, timing of investment and prices of allowances play an important role in defining the optimal strategy for firms in order to cope with emissions trading. *Wolf Fichtner* takes a look at investment and production planning issues, too. He develops a European energy model to quantify various impacts of emissions trading. Furthermore, a model for the economic assessment of CO_2 emission reduction technologies and strategies on a detailed company level is presented, which integrates inter-company energy supply concepts. Finally he shows how the flexible mechanisms (JI/CDM) can be considered in investment strategy decisions. *Gordon Spangard et al.* then present a stochastic optimisation model for decision-making in the emissions market under uncertain boundary conditions. Using the scenario approach, the model aims at finding a strategy for profit-optimal emissions trading and reduction. The paper by *Harri Laurikka* explores how different instruments of climate policy, such as emissions trading and taxes, affect heat and power capacity investment decisions and how flexibility can help to cope with those impacts. He shows that there are structural differences in flexibility between heat and power generation technologies. Make-or-buy decisions in the strategic energy management of those enterprises covered by the scheme are the subject of the analyses of *Anja Pauksztat* and *Martin Kruska*. While building different scenarios, they analyse the financial burden for an industrial boiler and for a cement production plant. How information technology can support optimal energy management decisions under an emissions trading regime is demonstrated by *Bernhard Grimm*, *Stefan Pickl*, and *Alan Reed*. They describe and discuss two software tools, TEMPI and VEREGISTER. Again, the investment behaviour of companies covered by the EU Directive is the subject of the last contribution to this part. Using real option theory *Marcus Stronzik* looks at two different approaches to design an allocation free of charge, grandfathering without updating the base year and a rolling base year.

The contributions of **Part C** pick out as central theme the impact of emissions trading on business administration. To prepare companies in Hamburg and Schleswig-Holstein, Germany for the EU scheme the pilot "Emissions Trading North" was carried out. *Katja Barzantny* presents the main findings, which were on information, reporting, decision-making processes, and on an optimal emission management strategy. Based on the theoretical concepts of ecological involvement, the stakeholder approach and the organizational field, *Ralf Antes* analyses the impact greenhouse gas emissions trading regimes have on companies in general. Afterwards, the main consequences of the EU emissions trading scheme are discussed, in particular strategy options for procurement management for greenhouse gas certificates and the scheme's impact on corporate environmental management. One impact is that companies have to account for emissions rights. *Edeltraud Günther* shows alternatives for greenhouse gases to be reported. The focus of the article is based on the influence of climate policy on accounting. Moreover, the interdependencies of the development of accounting and the treatment of ecological resources are shown. *Thomas Langrock* analyses why BP has adopted a commitment to reduce the greenhouse gas emissions from its operations. This commitment has been and will be implemented using a variety of measures, including emissions trading and emission credits. The research approach rests on using policy network analysis and evaluation. In the last article of this part *Timo Busch* sheds light on the financial sector and financial institutions. Due to the financial links of emissions trading to companies, there is also a strong linkage to financial markets and new business opportunities. Furthermore, Busch shows that financial institutions can contribute to establishing and fostering emission trading as a business case in general through proactive involvement.

In the final **Part D** papers are put together, which look at existing emissions trading schemes or at the implementation of the EU scheme. *Sonja Peterson* assesses the range of possible implications of the EU trading scheme. Using the computable general equilibrium model DART, she discusses the likely competitiveness effects for European business, taking into account the effects that occur due to European and international linkages. A very conflicting issue is the distribution of allowances between polluting sources. By adopting a public choice perspective, *Pablo del Río González* takes a closer look at the Spanish allocation process, analysing the interests and strategies of the different actors and their interactions. The next two contributions deal with trading schemes, which already existed before the EU scheme was set into force. Using a stochastic translog frontier cost function with a cost inefficiency component, *Adarsh Varma* sets out a theoretical model to analyse the effect of UK's climate change levy and the UK emission trading scheme on the productivity and overall economic efficiency of businesses. *Denny Ellerman* and *Florence Dubroeucq* study the SO2 reductions in the US using a unit-level data base of emissions from almost 2000 electric generating units from 1985 through 2002. The results indicate that cleaning up the old plants has made by far the greatest contribution (in contrast to replacing the old with new 'clean' plants), and that this contribution has been especially large since the introduction of the SO2 cap-and-trade program in 1995. The policy-business interactions in emissions trading between multiple regions are investigated by *Jürgen Scheffran and Marian Leimbach.* They apply a multi-region model with stylised data for 11 world regions and four cases (business as usual, equal per capita, 10% reductions of baseline, stabilization). The project based mechanisms JI and CDM then, are the subject of the last two papers. Firstly, *Katherine G. Begg* looks at criticisms aimed at both mechanisms, in particular the conflict between environmental integrity and the increasing transaction costs due to the complexity of the system to ensure that integrity. Her paper therefore traces the evolution of the project mechanisms and some of the latest developments in carbon accounting for reductions in large and small-scale projects, and then takes a look at the implications for projects under the EU and UK emissions trading schemes. *Jobert Winkel*, secondly, analyses prevailing types of technology and the locations of CDM projects.

The distribution of technology types can be generally explained by the values for the internal rates of return (including the revenues from CERs), except for hydropower, which is the prevailing technology. Factors that may have substantial impact on the popularity of certain technologies or project locations are examined, such as the internal rate of return of a project. Furthermore, the sensitivity of the project's internal rate of return is tested in a case study.

References

Coase R H (1960) The problem of social cost, In: Journal of Law and Economics, vol 3, pp. 1-44

Commission of the European Communities: Proposal for a directive of the European Parliament and of the Council establishing a scheme for greenhouse gas emission allowance trading within the Community and amending Council Directive 96/61/EC, COM(2001) 581 final, Brussels, 23.10.2001

Dales J H (1968) Land, water, and ownership, In: Canadian Journal of Economics, vol 1, no 4/1968, pp. 791-804

Hansjürgens B (2005) Introduction, In: Hansjürgens B (ed) Emissions Trading for climate policy. US and European views, Cambridge: Cambridge University Press, pp. 1-14

Montgomery D (1972) Markets in licenses and efficient pollution control programs, In: Journal of Economic Theory, vol 5, 3, pp. 395-418

Part A

Institutional design, decision making and innovation

Abatement costs vs. compliance costs in multi-period emissions trading – the firms' perspective

Sven Bode

Hamburg Institute of International Economics
Neuer Jungfernstieg 21, 20347 Hamburg, Germany
sven.bode@hwwa.de

Abstract

Greenhouse gas emissions trading has become increasingly important in the context of climate change. Recently, a discussion on trading at the entity, i.e. company, level has started. Emitters obliged to participate have argued for an initial allocation of the emission rights free of charge. In this paper, the implication of such an allocation based on historical emissions and on benchmarks in multi-period emissions trading is analysed. Different allocation rules for successive periods are applied, namely allocations with reference figures that are either constant or that change over time. The analysis is carried out using a two-player, two-period model. I find that, depending on their marginal abatement cost, participants have different preferences with regard to the allocation method over time, as individual compliance costs can change as well. Total costs remain, however, unaffected by the individual allocations, as emissions are reduced where abatement is cheapest.

Keywords: Abatement costs, compliance costs, allocation of emission rights, multi period emissions trading

Acknowledgement: I would like to thank Richard Tol and Katrin Rehdanz (Centre for Marine and Climate Research, Hamburg University, Hamburg, Germany) for valuable discussions in an early stage of the paper.

1 Introduction

In the context of increasing awareness of human-induced climate change, emissions trading as one possible instrument for reducing greenhouse gas (GHG) emissions has moved into the centre of the discussion in the last few years. In 1997, the Kyoto-Protocol was agreed upon to set the framework for GHG emissions trading at the state level. The so-called Annex B countries can either make use of the so-called flexible mechanisms under the Protocol by buying emission rights abroad to meet their emission targets, or alternatively, they can reduce emissions by employing domestic policies and measures (P&M). Among these P&Ms, market based instruments have attracted a lot of attention, especially by economists. Recently emissions trading at the entity level has become the focal point of the discussion. Numerous "governmental – industry working groups" were set up to discuss what a potential trading scheme involving private actors could look like (see for example AGE 2002; AGO 1999; MIES 2000; NZME 1998). The most important step may be the directive for a Europe-wide emissions trading scheme proposed by the European Commission in 2001 (Com 2001a).

When creating national trading schemes, many design features must be decided. An important point from the firms' perspective is the question of how the emission rights[1] will be allocated. This issue has been discussed in numerous studies (AGE 2002; AGO 1999; CCAP 2002; Field 2000; MIES 2000; NZME 1998). However, they only address general issues or, in case they do go into detail, they only consider a one period game and do not discuss the impact of different allocation options in subsequent commitment periods. This paper tries to shed some light on this issue: The implications of different allocation options for multi-period emissions trading focussing on the total costs (i.e. efficiency of the instrument) as well as on the individual firms compliance costs for an allowance selling and a buying firm, respectively, will be analysed.

In the following chapter general options for allocating emission rights are discussed before focussing on an allocation based on historical emissions and on benchmarks using the two-player two-period model. As the analytical solution is difficult to interpret, numerical examples are provided. Chapter five concludes.

[1] The term emission right and allowance are used equivalently. Other authors talk about permits in this context. Since the EU Commission has proposed a directive on emissions trading this might be a bit misleading as the term "permit" is used in another sense in this directive (like a permission or approval).

2 Emissions trading and allocation of allowances

Emissions trading[2] allows for the possibility of meeting an absolute emission target cost-efficiently. However, the answer to the question of who is bearing the costs is not incorporated in the instrument itself and has to be treated separately, also known as "question of burden sharing". Participants in the trading scheme can decide whether they want to abate emissions internally or to buy emission rights on the market. At the end of a period each emitter has to have at least as many allowances as he actually emitted into the atmosphere. The decision to buy allowances is driven by the question of whether or not in-house marginal abatement costs are lower than the allowance price. As each player faces this problem, marginal abatement costs are equalised among the sources at the end of the period.

Before trading can start the participants have to be allocated a certain quantity of emission rights. Allowances may either be provided for free or the participants may be charged. For trading at the company level, economists have argued in favour of a charged allocation or, more precisely, an auction, as giving away the allowances for free would result in extra revenue for the receivers of the allowances (Cramton and Kerr 2002; Field 2000, p. 31; Woerdman 2002, p. 620). However, others have argued that this question can only be answered when comparing the concrete design of an auction[3] and a charge scheme respectively (for example Bohm 2002). On the other hand, those who would be obliged to participate ask for an allocation free of charge (Com 2001b, p. 2). The EU Directive prescribes on allocation almost free of charge.[4]

A very important point for the allocation of allowances in GHG trading schemes as it is understood in this paper, is that the total budget is more or less given by the commitments made under the Kyoto-Protocol.[5] Consequently, a simple bottom-up approach, that is to say, wherein the individual allocation is made without reflecting the total budget constraint, does not seem appropriate.[6] An adjustment with the national budget has to be made, what is also known as the top-down approach.[7]

---

[2] In this paper emissions trading is used in the sense of a cap-and-trade system.

[3] For example "How is the revenue from the auction recycled?"

[4] More precisely: At least 95% of allowances have to be allocated free of charge for the initial period 2005-2008 and at least 90% for the subsequent period (EU 2003).

[5] This is not really true for any trading before the entry into force of the Protocol. For a discussion see (Bode 2003; Rehdanz and Tol 2002)

[6] Such an approach would indeed be possible if different sectors, other than the ones participating, would get a smaller share of available national budget. This could generate a lot of resistance from the other sectors, as they would have to bear a stronger burden. This is why such an approach is not considered further.

[7] In the context of the Kyoto-Protocol, the budget can be enlarged by the purchase of CERs/ERUs from CDM and JI-projects respectively. However, since this induces additional costs which have to be borne by someone, it is not considered as an option: from the participants' point of view the total costs are most relevant regardless of exactly where they accrue.

2.1 Determining the share of emission allowances (static analysis)

The available quantity for all participants has to be distributed among participants in some manner. One may think of the following:

- Negotiations (for example: AGE 2001, p. 63)
- Reference figure
  - Production (for example: NERA 2002, p. 31)
  - Emission related in a certain year/period (for example: CCAP 2002, p. 17)
  - Benchmarks (for example: CCAP 2002, p. 17)
- Abatement Costs / Potential (Com 2001a, Annex III, 3)
- Others (e.g. turn over, employees etc.)
- Any combination of the above

Below, the focus is on an allocation based on historical emissions, which is often referred to as grandfathering[8], and on benchmarks that are both by far the most frequently mentioned in the reports cited above.

A straightforward approach for an allocation based on historical emissions could be that emissions are allocated proportionately to the single emitters' share of the total emissions, formally:

$$A_i = \frac{I_i q_i}{\sum_i I_i q_i} A \quad i = 1, 2, \ldots N \tag{1}$$

where A_i = emissions allowances allocated to company i; I_i = emission intensity of company i; q_i = output of company i; A = total quantity of emission rights to be distributed to the participants.

For a general benchmark[9] the allocation could appear as shown in equation (2).[10]

$$A_i = s\, q_i \tag{2}$$

[8] However, as emission reductions are still necessary for almost all Annex B countries, grandfathering can *generally* not mean distribution of emission rights equal to historical emissions (bottom-up approach) An adjustment according to the total budget available is necessary - or in other words: grandfathering is an allocation method based on historical emissions (top-down approach).

[9] Theoretically, individual benchmarks for each emitter are conceivable. However, as there are already more than 11000 installations to participate in the initial phase of the EU-trading scheme, the use of such company specific benchmarks seems at least administratively infeasible.

[10] When creating real world systems one should be aware of the fact that in case different sectors, such as heat and power, pulp and paper etc., participate, different "sub-budgets" for each benchmark must be determined since specific emission factors are only applicable if the reference figure is used by all members of the subset.

$$\text{s.t. } \sum_i A_i \leq A \quad i = 1, 2, \ldots N \tag{3}$$

where s = general specific emission factor (t CO_2/unit), q_i = output of company i (unit)

2.2 Determining the share of emission rights in successive periods

Trading schemes can be designed as either one-period or multi-period schemes, whereby the latter option is much more likely to be realised. The allocation in successive periods can either be carried out on the basis of the initial allocation, which is referred to as *constant allocation* in this paper, or on the basis of data gathered in a period later than the initial one, which is referred to as *rolling allocation*. One could think of a constant base period as the intuitive approach. In this case the allocation for all periods is determined at the very beginning of the trading scheme. Thus, emitters who reduce emissions by more than their initial allocations are directly awarded by this approach, since they have longer-term revenue from the sale of the allowances. However, simple reduction of output for whatever reason can also be beneficial with regard to the allowance balance. One might question whether it is desirable that a company, which changes its business portfolio, receives emission rights in 2050 on the basis of emissions it had released sometime in the 1990s. An approach in which the allocation in subsequent periods is based on more recent emissions, or output, rather than on the initial emissions, i.e. a rolling allocation, may take this aspect into account. On the other hand, this could reduce the incentive to invest in in-house reductions as today's decision could influence the next period allocation: "The more you reduce now the less emission rights you get in the future". Edwards and Hutton (2001, p. 375) argue similarly, stating: "However, if companies expect that, by emitting more now, they get a larger allowance allocation in the future, there would be a counterbalancing incentive to continue emitting."

So far, the discussion of the allocation of emission rights has more or less neglected the question of future commitment periods. Boemare and Quirion (2002) for example, analyse ten trading schemes in order to give some recommendations for the design of the European scheme. The issue of allocation over time is, however, not addressed. The same goes for Holmes and Friedmann (2000) who discuss design alternatives for a carbon trading scheme in the US, and for Svendsen (1997) who focuses on Denmark. This issue is neither addressed in the "governmental-industry reports" mentioned above, nor discussed in a recent paper on "how to develop a national allocation plan" by the European Commission (Com 2003). CCAP (1999) focuses on the allocation of greenhouse gas reduction responsibilities, but does not consider dynamic aspects. Finally, UNEP et al. (2002, p. 14) states that the allocation rule can change over time without providing a more detailed analysis.

On the other hand, rolling allocations have been used in general equilibrium analyses on the impact of different allocation options on the large scale economy.

However, due to this intention, they neglect the individual firm's perspective. For example, Jensen and Rasmussen (2000) allocate emission rights to the firms in a certain period, inter alia, according to the market share in the previous period. Edwards and Hutton (2001) apply a regular updated benchmark (a variant of the "best-practice" approach) and find that this approach would even improve GDP and welfare compared to other approaches that were analysed, i.e. auction and grandfathering. Unfortunately, it remains unclear whether the top-down alignment was employed. Nera analysed a so-called updating approach whereby the allocation is based on the output in different periods (Nera 2002, pp. 25-30). Unfortunately, they do not consider changing emissions as a basis for allocation, nor do they analyse the individual company's point of view. In summation, a systematic analysis of the different options and the resulting impacts on seller and buyer is missing. I attempt to increase the understanding of this issue as follows.

When analysing successive periods one must add a time index and rewrite equations (1) - (3) as follows:

$$A_{i,t} = \frac{I_{i,t-j} q_{i,t-j}}{\sum_i I_{i,t-j} q_{i,t-j}} * A_t \quad \text{i = 1, 2, ...N ; t = 1,2,...T ; j = 1,2,...T} \tag{4}$$

$$A_{i,t} = s_{t-j} * q_{i,t-j} \tag{5}$$

$$\text{s.t.} \ \sum_i A_{i,t} \le A_t \quad \text{i = 1, 2, ...N ; t = 1,2,...T} \tag{6}$$

In the following section a general model for an analytical analysis of the different allocation options is introduced before a numerical approach is used to study the impacts.

3 The model

The underlying model is taken from Rhedanz and Tol (2002). Consider a two-player, two-period market. Emitters face different abatement costs which are assumed to be quadratic in the reduction technologies implemented. Denote the buying firm with the index *a* and the firm with the lower abatement costs (seller) with the index *b*. No banking between the periods is allowed. The lifetime of the investment is one period[11], so that each firm can decide[12] on the abatement level in

[11] As long as the total reduction obligation (i.e. total expected emissions minus total budget of allowances) is equal or less in the following periods, each emitter reduces at least as much as he did in the preceding period so that the affect could be interpreted as a lifetime of two periods for those reductions undertaken in the first period and additional reductions in the second one.

each period independently. Emissions are reduced by an end-of-pipe technology[13], that is to say specific emissions during production do not change. Emissions are rather removed at the end of the production chain. In the interest of brevity, the mathematical solution for only special case is presented, namely emission based allocation. For the benchmark, an introductory comment is made. [14]

Especially with the rolling allocation approach, the number of permits allocated in the second period may depend on the emissions, and thus on the emission reductions, in the first period. The two players then minimise costs over both periods already at the start of the trading scheme. Transaction costs are neglected. They face the following optimisation problem:

$$\min_{R_{a1},R_{a2},P_1,P_2} \quad C_a = \alpha_{a1}R_{a1}^2 + \pi_1 P_1 + \frac{\alpha_{a2}R_{a2}^2}{1+\delta} + \frac{\pi_2 P_2}{1+\delta} \; ; \tag{7}$$

$$\min_{R_{b1},R_{b2},P_1,P_2} \quad C_b = \alpha_{b1}R_{b1}^2 - \pi_1 P_1 + \frac{\alpha_{b2}R_{b2}^2}{1+\delta} - \frac{\pi_2 P_2}{1+\delta}$$

where, C = Costs, α = parameter, R = investment in in-house abatement technologies (t $CO_{2\text{-eq}}$), π = allowance price (assuming a perfect market where each participant faces the same price), P = quantity of allowances bought or sold, δ = discount factor.

3.1 Abatement by end-of-pipe technologies

When using an end-of-pipe technology the emitters have to consider the constraints

$$R_{at} + P_{at} \geq I_a q_{at} - A_{at} \;\; ; \;\; R_{bt} - P_{bt} \geq I_b q_{bt} - A_{bt} \;\; \text{t} = 1{,}2 \tag{8}$$

index 1; 2 = period one and two respectively, I = Emission, q = output (product), A = allocation

The allocation changes according to the approach applied.

3.1.1 Allocation based on historical emissions

3.1.1.1 Constant allocation

In the case of a constant allocation based on emissions, the individual allocation would be:

---

[12] Indeed, emitters do not really make decisions regarding their own about their reduction level. In a perfect market, as is assumed here, reductions are rather prescribed by the total reduction obligation and the marginal cost curves of the different players.

[13] As for example, removal of carbon dioxide removal from flue gases.

[14] Other solutions can be provided upon request.

$$A_{i,t} = \frac{I_{i,0}q_{i,0}}{\sum_i I_{i,0}q_{i,0}} * A_t \quad \text{i = a, b ; t = 1, 2} \tag{4a}$$

Minimising eq. (7), (8) and (4a) gives

$$R_{a1} = \frac{\alpha_b(I_a q_{a1} + I_b q_{b1} - A_1)}{\alpha_b + \alpha_a}\ , \ R_{b1} = \frac{\alpha_a(I_a q_{a1} + I_b q_{b1} - A_1)}{\alpha_b + \alpha_a}\ , \tag{9}$$

$$P_1 = \frac{I_a^2 q_a \alpha_{a*} + I_a q_{a1} I_b q_{b0} \alpha_a - I_a q_{a0} A_1 \alpha_a - I_a q_{a0} I_b q_{b1} \alpha_b - I_b^2 q_{b*} \alpha_b + I_b q_{b0} A_1 \alpha_b}{(I_a q_{a0} + I_b q_{b0})(\alpha_a + \alpha_b)}\ ,$$

$$\pi_1 = \lambda_{b1} = \lambda_{b1} = 2\frac{\alpha_*(I_a q_{a1} + I_b q_{b1} - A_1)}{\alpha_b + \alpha_a}\ ,$$

$$R_{a2} = \frac{\alpha_b(I_a q_{a2} + I_b q_{b2} - A_2)}{\alpha_b + \alpha_a}\ , \ R_{b2} = \frac{\alpha_a(I_a q_{a2} + I_b q_{b2} - A_2)}{\alpha_b + \alpha_a}\ ,$$

$$P_2 = \frac{I_a^2 q_{a0} q_{a2} \alpha_a + I_a q_{a2} I_b q_{b0} \alpha_a - I_a q_{a0} A_2 \alpha_a - I_a q_{a0} I_b q_{b2} \alpha_b - I_b^2 q_{b0} q_{a2} \alpha_b + I_b q_{b0} A_2 \alpha_b}{(I_a q_{a0} + I_b q_{b0})(\alpha_a + \alpha_b)}$$

$$\lambda_{a2} = \lambda_{b2} = 2\frac{\alpha_a \alpha_b (I_a q_{a2} + I_b q_{b2} - A_2)}{(\alpha_b + \alpha_a)(1+\delta)}\ ,$$

$$\pi_2 = 2\frac{\alpha_a \alpha_b (I_a q_{a2} + I_b q_{b2} - A_2)}{\alpha_b + \alpha_a}$$

We see that the problem is nearly symmetrical for both periods. Marginal abatement costs in period one and two differ by the discount factor. Other differences may occur in case the output and thus emissions changes over time.

3.1.1.2 Rolling base period

A rolling period means that allocation is done on the basis of a period later than the initial one (i.e. j<t). For example, one may allocate emission rights for the period 2005-2008 based on emissions in 1998-2000 and for the subsequent period 2008-2012 on emissions in 2003-2005.

In the case of a rolling allocation based on (historical) emissions, the individual allocation would be:

$$A_{i,t} = \frac{I_{i,t-1}q_{i,t-1}}{\sum_i I_{i,t-1}q_{i,t-1}} * A_t \quad \text{i = a, b ; t = 1, 2} \tag{4b}$$

With a rolling allocation (7), (8) and (4b) minimise as (9) except for P_2 which is:

$$P_2 = \frac{I_a q_{a2} \alpha_a A_1 + I_b q_{b1} \alpha_b A_2 - I_b q_{b2} \alpha_b A_1 - I_a q_{a1} \alpha_a A_2}{A_1(\alpha_b + \alpha_a)} \tag{9b}$$

We can see that, unlike in the case with constant allocation, the quantity traded in period two is dependent on the total budget of period one (A_1). This seems to be sound as the total budget in period one determines the players' reductions and thus emissions in that period which can in turn form the basis for the allocation in period two.

3.1.2 Allocation based on benchmarks

In this analysis the term "benchmark" is used in the sense of a specific emission factor, i.e. $\frac{emissions}{output}$.[15]

As mentioned above, the allocation based on a general benchmark could be calculated as

$$A_{i,t} = s_{t-j} * q_{i,t-j} \tag{5}$$

As already pointed out above, an alignment between a bottom-up (benchmark) approach and a top-down constraint would be necessary to consider the constraint given in inequality (6). A straightforward approach would be the introduction of a correction factor as discussed, for example in PwC (2003) and AGE (2002)

$$c_t = \frac{A_t}{\sum_i A_{i,t}} \quad t = 1, 2$$

Taking into account this factor, (5) changes to

$$A_{i,t} = s_{t-j} * q_{i,t-j} * c_t \qquad \Leftrightarrow A_{i,t} = s_{t-j} q_{i,t-j} \frac{A_t}{\sum_i A_{i,t}} \tag{10}$$

$$\Leftrightarrow A_{i,t} = s_{t-j} q_{i,t-j} \frac{A_t}{\sum_i s_{t-j} q_{i,t-j}}$$

$$\Leftrightarrow A_{i,t} = q_{i,t-j} \frac{A_t}{\sum_i q_{i,t-j}}$$

As we can see by transforming equation (10), a benchmark based allocation, that takes into account the national budget (e.g. the Kyoto Commitment), results in an

[15] Theoretically, any benchmark, such as labour productivity or turnover, could be used for allocation. For an emission benchmark, other reference figures than the output could also be used (e.g. input see for example PwC 2003).

individual allocation that is only proportional to a participant's output and not at all related to emission intensities. This might be somewhat surprising as the intention of the use of a benchmark is generally to consider the specific emissions.[16]

The amount of allowances for the case of constant and rolling allocation are given in equation (10a) and (10b) respectively.

$$A_{i,t} = q_{i,0} \frac{A_t}{\sum_i q_{i,0}} \quad t = 1, 2 \tag{10a}$$

$$A_{i,t} = q_{i,t-1} \frac{A_t}{\sum_i q_{i,t-1}} \quad t = 1, 2 \tag{10b}$$

As mentioned above, the solution for the benchmark allocation can be provided upon request.

4 Numerical analysis

As the difference between the approaches is not obvious from the analytical solution, a numerical example is provided. A specific case is analysed by presenting the solution in detail in order to understand the players' actions, assuming that they minimise costs as a function of the output which is exogenously given, i.e. non-strategic behaviour.

The analysis is based on the values given in Table 1. The corresponding results of the model are presented in Table 2, Table 3 summarises the preferences of the two players when the output changes.

Table 1. Values used for scenario analysis

Name	Symbol	Period 0 (base period)	Period 1	Period 2
Total Budget	Λ	/	38	36
Emission Intensity (a)	I_a	2	2	2
Emission Intensity (b)	I_a	2	2	2
Parameter	α_a	/	4	4
Parameter	α_b	/	2	2

Values have been chosen more or less arbitrarily. The only condition is $\alpha > \beta$ as long as player a is supposed to have the higher abatement costs.

[16] Compliance costs can of course differ, but not as a consequence of different allocations, but rather as a consequence of different abatement costs (see also numerical examples below).

Discussion

For the discussion of the result shown in Table 2, the second period is concentrated upon, since the different allocation options start to be effective at that time. Total emissions for both players are equal to the total budget as given in Table 1 (see point 1). Whereby emissions trading emissions are reduced where it is cheapest, reductions by both players are the same for both a constant and a rolling allocation (see point 2). Consequently, total costs do not change by choosing different allocation options for the same total reduction obligation (see point 3), since the abatement cost curves are not affected by the initial allocation.

What can indeed be affected are the compliance costs for the players. In the first period, there is no change in the individual compliance costs with varying approaches. In the second period, i.e. when the rolling approach starts being effective, compliance costs may however change considerably. Only when the output is constant is there no difference in the second period for the benchmark allocation. This is due to the fact that the allocation is only a function of the output (see section on general benchmark), which is kept constant.

With an emission-based allocation the situation is different. As player *a* reduces less emissions in the first period due to the higher abatement costs compared to *b* (see point A), he receives more allowances in the second period (see point B), which in turn reduces his compliance costs. The opposite is true for player *b*. Driven by the market, he reduces more in the first period than *a* – resulting in a reduced allocation and higher compliance costs in period two (see point C).

Table 2. Impact of different allocation options over time with constant output for both players and reductions by end-of-pipe technology

		Period 0	Period 1			Period 2 (emission based)			Period 2 (benchmark)		
Name	Symbol		Constant	Rolling	Change	Constant	Rolling	Change	Constant	Rolling	Change
Output	/										
Output (a)	q_a	10	10	10	/	10	10	/	10	10	/
Output (b)	q_b	10	10	10	/	10	10	/	10	10	/
Emissions	/										
Emissions (a)	$I_a q_a$	20	19.33	19.33	0	18.67	18.67	0.00	18.67	18.67	0.00
Emissions (b)	$I_b q_b$	20	18.67	18.67	0	17.33	17.33	0.00	17.33	17.33	0.00
Total (a+b)	/	40	38.00	38.00	0	36.00	36.00	0.00	36.00	36.00	0.00
Allocation											
Allocation (a)	A_a	/	19.00	19.00	0	18.00	18.32	0.32	18.00	18.00	0.00
Allocation (b)	A_b	/	19.00	19.00	0	18.00	17.68	-0.32	18.00	18.00	0.00
The Market											
Allowance price	π	/	5.33	5.33	0	10.67	10.67	0.00	10.67	10.67	0.00
Allowances traded	P	/	0.33	0.33	0	0.67	0.35	-0.32	0.67	0.67	0.00
Reduction investments (a)	R_a	/	0.67	0.67	0	1.33	1.33	0.00	1.33	1.33	0.00
Reductions investments (b)	R_b	/	1.33	1.33	0	2.67	2.67	0.00	2.67	2.67	0.00
Costs (a)		/									
Reductions	$\alpha_a R_a^2$	/	1.78	1.78	0	7.11	7.11	0.00	7.11	7.11	0.00
Allowances	πP	/	1.78	1.78	0	7.11	3.74	-3.37	7.11	7.11	0.00
Total (a)	C_a	/	3.56	3.56	0	14.22	10.85	-3.37	14.22	14.22	0.00
Costs (b)		/									
Reductions	$\alpha_b R_b^2$	/	3.56	3.56	0	14.22	14.22	0.00	14.22	14.22	0.00
Allowances	πP	/	-1.78	-1.78	0	-7.11	-3.74	3.37	-7.11	-7.11	0.00
Total (b)	C_b	/	1.78	1.78	0	7.11	10.48	3.37	7.11	7.11	0.00
Total (a+b)	/	/	5.33	5.33	0	21.33	21.33	0.00	21.33	21.33	0.00

Figures and letters are explained below.

The situation is different when output changes. Then, the benchmark approach delivers another outcome in the second period as well. With the buyer expending business, he prefers the rolling benchmark allocation to a constant one. For the scenarios studied[17], however, player *a*, or the buyer, prefers the rolling emission based allocation to the rolling benchmark in absolute terms. Obviously, with the model set-up as chosen, there is no incentive for the player with the lower abatement costs to renounce investments in the first period in order to avoid reduced allocation in the second period with a rolling allocation. His cost minimising strategy is to invest regardless.

In summation, seller and buyer do have different preferences with regard to the design of the scheme which are summarised in Table 3.

It can be seen that with equal emission intensities the net buyer always prefers the emission based rolling allocation that in turn, as it is a zero sum game, ranks lowest in the seller's priority.

Table 3. Preferences with reductions by end-of-pipe technology and varying output and equal initial emission intensities

	Player a			Player b		
	Output (a/b)			Output (a/b)		
	constant / constant	growing / constant	Constant / growing	constant / constant	growing / constant	constant / growing
1. Priority	emi / rol	emi / rol	emi / rol	all other	emi / const & bm / const	bm / rol
2. Priority	all other	bm / rol	emi / const & bm / const	emi / rol	bm / rol	emi / const & bm / const
3. Priority		emi / const & bm / const	bm / rol		emi / rol	emi / rol

emi = emission based allocation, bm = benchmark allocation, const = constant allocation, rol = rolling allocation, k = allocation, const = constant allocation, rol = rolling allocation

5 Conclusion

GHG-Emissions trading has become more and more important in the context of climate change in recent years. After trading on state level has been agreed in 1997 in the Kyoto-Protocol, it has been implemented in the EU. Before implementing any scheme, many design features have to be decided upon, especially the question of how to allocate the allowances, given that this is a very important one

[17] A number of scenarios with different parameters than those given in Table 1 have been studied.

from the participants' point of view. The latter have argued for an allocation free of charge.

For this, the implications of an allocation free of charge based on historical emissions and on benchmarks in multi-period emissions trading have been analysed. Different allocation methods in the successive periods have been studied in this paper using a two-player two-period model. Even though all participants are likely to agree on an allocation free of charge, they have different preferences with regard to the allocation method over time. With equal emission intensities, the net-buying emitter prefers a rolling, emission-based allocation to any other option as the number of allowances received increases in the second period. Thus, compliance costs are reduced. The net-seller, on the other hand, is likely to reject this option as it entails the highest compliance costs. With different emission intensities, preferences may change depending on the intensity. Furthermore, it became evident that a benchmark-based allocation which takes into account a total budget constraint, as for example the Kyoto targets of Annex B countries, results in an individual allocation that only depends on the emitters' share of the total output and that is thus no longer dependent on the individual emission intensities. Consequently, there would be no need to try to determine any such benchmarks as it is currently done in the context of the upcoming EU directive on emissions trading.

The total costs for meeting the overall reduction target are, however, not affected by the individual allocation as emissions are reduced where abatement is cheapest. This would imply that national authorities developing a national allocation plan could choose whatever option they wish or simply allocate emission rights to those emitters "who are shouting the loudest".

All results are derived from the specific model and the assumptions made. Further analyses of different emission intensities, changing output, the role of banking and different life-times of the investments are necessary. All errors are mine.

References

AGE (2001) Materialienband zum Zwischenbericht 2001. Arbeitsgruppe Emissionshandel zur Bekämpfung des Treibhauseffektes (German Emissions Trading Group), http://www.ag-emissionshandel.de

AGO (1999) National Emission Trading. Discussion paper 1-4, Australian Greenhouse Office, Canberra

Bode S (2003) Implications of Linking National Emission Trading Schemes prior to the Start of the First Commitment Period of the Kyoto Protocol. HWWA Discussion Paper, January 2003, retrievable on: http://www.hwwa.de

Boemare C, Quirion Ph (2002) Implementing greenhouse gas trading in Europe: Lessons from the economic literature and international experience. Ecological Economics 43: 213-230

Bohm P (2002) Comparing Permit Allocation Options: The Main Points. Working Paper in Economics 2002: 11, Department of Economics, University of Stockholm

CCAP (2002) Design of a Practical Approach to Greenhouse Gas Emissions Trading Combined with Policies and Measures in the EC, Center for Clean Air Policy, retrievable on: http://www.ccap.org

CCAP (1999) Allocation of Greenhouse Gas Reduction Responsibilities Among and Within the Countries of the European Union, Center for Clean Air Policy, retrievable on: http://www.ccap.org

COM (2003) The EU Emission trading scheme: How to develop a National Allocation Plan. Non-Paper of the 2nd meeting of Working 3, Monitoring Working Committee, April 2003

COM (2001a) Proposal for a DIRECTIVE OF THE EUROPEAN PARLIAMENT AND OF THE COUNCIL establishing a framework for greenhouse gas emissions trading within the European Community and amending Council Directive 96/61/EC; Com (2001) 581, 23.10.2001

Com (2001b) Greenpaper on Greenhouse Gas Emission Trading within the European Union, Summary of Submission, 14 May 2001, Retrievable on http://europa.eu.int/comm/environment/docum/0087_summary.pdf

Cramton P, Kerr S (2002) Tradable permit auctions How and why to auction not grandfather. Energy Policy 30: 333-345

Edwards TH, Hutton JP (2001) Allocation of carbon permits within a country: a general equilibrium analysis of the United Kingdom. Energy Economics 23: 371-386

EU (2003) Directive 2003/2003/87/EG of the European Parliament and of the Council establishing a scheme for greenhouse gas emission allowance trading within the Community and amending Council Directive 96/61/ EC, Brussels

Field (2000) Designing Options for Implementing an Emissions Trading Regime for Greenhouse Gases in the EC, Final Report, Foundation for International Environmental Law and Development, retrievable on: http://www.field.org.uk

Holmes KJ, Friedmann RM (2000) Design alternatives for a domestic carbon trading scheme in the United States. Global Environmental Change 10: 273-288

Jensen J, Rasmussen T (2000) Allocation of CO_2 Emission Permits: A General Equilibrium Analysis of Policy Instruments. Journal of Environmental Economics and Management 40: 111-136

Matsuo N (1997) Key elements related to the emissons trading for the Kyoto Protocol in: Energy Policy 26: 263-273

MIES (2000) Implementing an Emissions Credits Trading System in France to Optimize Industry's Contribution to Reducing Greenhouse Gases (Final), MIES - Industry Working Group

Nera (2002) Evaluation Of Alternative Initial Allocation Mechanisms In A European Union Greenhouse Gas Emissions Allowance Trading Scheme, National Economic Research Associates

NZME (1998) Technical Design Issues for a Domestic Emissions Trading Regime for Greenhouse Gases: A Working Paper, Ministry for the Environment (New Zealand), http://www.mfe.govt.nz/about/publications/climate/climatechange.htm; accessed: January 17, 2002

PwC (2003) Allowance allocation within the Community-wide emissions allowance trading scheme, PriceWaterhouseCoopers, Utrecht

Rehdanz K, Tol RSJ (2002) On National and International Trade in Greenhouse Gas Emission Permits, Research Unit Sustainability and Global Change FNU-11 (revised), Centre for Marine and Climate Research, Hamburg University, Hamburg

Svendsen GT (1997) A general model for CO_2 regulation: the case of Denmark. Energy Policy 26: 33-44

UNEP, UNCTAD (2002) An emerging market for the environment: A Guide to Emissions Trading. United Nations Publication, first edition 2002

Woerdman E (2000) Organizing emissions trading: The barrier of domestic permit allocation. Energy Policy 28: 613-623

Generous allocation and a ban on banking – implications of a simulation game for EU emissions trading*

Joachim Schleich[I,II], Karl-Martin Ehrhart[III], Christian Hoppe[III], Stefan Seifert[III,IV]

[I] Fraunhofer Institute for Systems and Innovation Research (ISI)
Breslauer Strasse 48, 76139 Karlsruhe, Germany
j.schleich@isi.fraunhofer.de

[II] Virginia Polytechnic Institute and State University
Blacksburg, VA 24061, USA

[III] University of Karlsruhe, WIOR, Zirkel 2, 76128 Karlsruhe, Germany
erhart@wiwi.uni-karlsruhe.de; hoppe@wiwi.uni-karlsruhe.de

[IV] Takon GmbH, Ritterstraße 7, 76133 Karlsruhe, Germany
seifert@takon.de

Abstract

Admitting banking in emissions trading systems reduces overall compliance costs by allowing for intertemporal flexibility: cost savings can be traded over time. However most, EU Member States prohibit the transfer of unused allowances from the period of 2005-2007 into the first commitment period under the Kyoto Protocol, i.e. 2008-2012. At the same time, allowances appear to be allocated fairly generously to the emissions trading sector. In this paper, we first explore the implications of such a ban on banking when initial emission targets are lenient. This analysis is based on a simulation which was recently carried out in Germany with companies and with a student control group. The findings suggest that an EU-wide ban on banking would lead to efficiency losses in addition to those losses which arise from the lack of intertemporal flexibility.

Keywords: Emissions trading, climate policy, banking, simulation

Acknowledgement: The authors are grateful to A. Denny Ellerman, MIT, for his critical comments and valuable suggestions.

---

*This paper is a shorter and slightly modified version of Schleich et al. (2006).

1 Introduction

According to the EU Directive on Emissions Trading (CEC 2003), large installations of the energy industry and most other carbon-intensive industries are part of an EU-wide CO_2 trading system (EU ETS) since 2005. As the heart of the European Climate Change Programme, the EU ETS is expected to result in the world's largest emissions trading system and is supposed to help achieve the EU's obligations under the United Nations Framework Convention on Climate Change and the Kyoto Protocol in a cost-effective way (CEC 2000).

The EU ETS is a cap-and-trade allowance trading system and requires companies to submit a number of allowances for cancellation corresponding to their actual annual CO_2 emissions. At the beginning of each year, participants receive allowances in the so-called primary allocation either for free or they have to buy them through an auction. According to the Directive, at least 95% of allowances will be allocated free of charge for the years 2005-2007. For the first Kyoto commitment period, i.e. for 2008-2012, not more than 10% may be auctioned. Every company can sell its surplus allowances or, if permitted, save them for future years (*banking*). Failure to submit a sufficient amount of allowances results in severe sanction payments and, in addition, companies have to surrender the missing allowances in the following year.

From an economic point of view, emissions trading is expected to achieve efficiency gains in reaching the emissions target: companies which can abate their emissions at low cost have an incentive to do this to an increasing extent, since they can sell their surplus allowances at a profit to companies with high abatement costs. Since abatement measures will be realised where they are cheapest, environmental targets can – under ideal conditions – be met at minimum costs. This property is also called *static efficiency* property. Estimated cost savings for the Acid Rain Program, which is the most well-known and most intensively-studied trading system, range around 50% compared to command-and-control regulation (Ellerman et al. 2000; Carlson et al. 2000).

In most existing emissions trading programs, banking has been permitted (Boemare and Quirion 2002). Theoretical and empirical analyses from existing programs suggest that banking (and borrowing, i.e. the use of future allowances to cover actual emissions) reduces overall compliance costs by allowing intertemporal flexibility: cost savings can be traded over time (Kling and Rubin 1997; Tietenberg 1999; Ellerman et al. 2003; Ellerman 2002; Ellerman and Montero, 2002). Likewise, since it provides a safety cushion for unexpected high demand in the future, banking tends to dampen price fluctuations.

However, a first analysis of the National Allocation Plans[1] (NAPs) suggests that the banking of allowances from the first trading period (2005-2007) to the second trading period (2008-2012) in the EU ETS will be prohibited in most are

[1] In the National Allocation Plan, which each Member State develops individually, Member States have to state (i) the total quantity of allowances that will be allocated in each period and (ii) how these allowances will be allocated to individual installations.

Member States.[2] The only exceptions are France and Poland. At the same time, a quantitative analysis of the NAPs shows that Member States allow for a generous allocation of allowances to the emissions trading sector compared to historic emissions (Betz et al. 2004).[3]

In this paper, we explore the implications of a ban on banking from the first commitment period to the second commitment period in the EU ETS when initial emission targets are lenient. The analysis is based on a simulation game with a company group and a student control group which was recently carried out in Germany (Schleich et al. 2002). The results of the simulation suggest that a ban on banking in the EU ETS would be a major source of inefficiency and may prevent cost savings potentials from being realised. The remainder of the paper is organised as follows. In Section 2, we discuss possible reasons why banking from 2007 to 2008 may be prohibited in the EU ETS. In Section 3, the design of the emissions trading simulation is described. Results of the simulation game are presented in Section 4, Section 5 concludes.

2 Banking in the EU ETS

The Directive for the EU ETS generally allows the unrestricted transfer of surplus allowances into future years – with one exception: banking from 2007 into the first commitment period under the Kyoto Protocol starting in 2008 is prohibited unless it is explicitly included in individual Member States' national implementations of the Directive. Borrowing is allowed within commitment periods but prohibited between 2007 and 2008 as well as between all future commitment periods. Because of the homogeneous nature of allowances – without harmonised banking rules across EU Member States – all surplus allowances can be expected to move into the (registries of the) Member States that allow unlimited banking from 2007 into 2008 due to arbitrage reasons. In a worst case scenario, individual Member States permitting unlimited banking of surplus allowances from 2007 to 2008 may fail to meet the national emission targets that they are committed to under the Burden-Sharing Agreement beginning in 2008. Alternatively, allocation to the ET sector would have to be lower or other domestic sectors outside the EU ETS would have to make up for an unexpectedly large amount of banked allowances. Moreover, from a practical point of view, it would be very difficult to project the amount of banked allowances when Member States draw up their national allocation plans in early 2006 for the Kyoto commitment period 2008-2012. Theoretically, a small Member State may even encounter a situation in which the number of banked allowances exceeds the number of Assigned Amount Units (AAUs) it has available for the five-year Kyoto commitment period.

2 Based on simple game-theoretic considerations, Schleich et al. (2006) that an EU-wide ban on banking would be the outcome which is likely to be a prisoners' dilemma situation.

3 Schleich et al. (2006) and del Rio (this volume) provide a discussion of possible underlying reasons for lenient emission targets.

The following two-country example illustrates the problems which may arise if some Member States permit banking in the national implementation of the directive, but others do not. Suppose France permits banking between 2007 and 2008 but Germany does not. If, in 2008, a German company sells allowances of 100.000 t of CO_2 to a French company, 100.000 Assigned Amount Units (AAUs) will be subtracted from the German account and added to the French account. Thus, France would be allowed to emit an additional 100.000 t of CO_2 in 2008 or later. However, if that same trade took place in 2007, there would be no adjustment of AAUs across Member States. Since – in the example – banking is not restricted in France, the French company (but not France!) would be entitled to emit the additional 100.000 t of CO_2 in 2008 or later. Thus, to meet the target under the Burden-Sharing Agreement, other sectors in France would have to cut emissions to make up for the transferred allowances. To add another twist to the example, assume that the French company decided to sell the originally banked 100.000 t of CO_2 back to the German company in 2008. In this case, the allocation of allowances would be the same as at the outset, but the German account of AAUs would increase by 100.000 t of CO_2 at the expense of the French account.

Recently, concerns have been raised by the Director General of the Competition Directorate at the European Commission that banking may involve some kind of state aid because banked allowances need to be backed by Assigned Amount Units (AAUs) from the national account which could otherwise be sold on the market.[4] By the same logic, however, allowances would have to be auctioned off entirely. Finally, at least in theory, large quantities of banked allowances may trigger the Commitment Period Reserve (CPR) rule[5]. If a national government has a reserve of AAUs which is less than 90% of its Assigned Amount of greenhouse gases, all transfers of allowances between the national registries will be stopped until the national reserve is replenished.

To explore the potential implications of a ban on banking between 2007 and 2008 in the EU ETS when the primary allocation is generous and the cap is tightened over time, we simulated emissions trading under such boundary conditions in a quasi-controlled field experiment, which was recently carried out in Germany under the name SET UP.[6]

3 The emissions trading simulation SET UP

For some of the aforementioned emissions trading systems in the US, experiments were carried out to test selected properties before the actual implementation of the systems (for example, Muller 1999). This so-called *testbedding methodology*

[4] Pointcarbon, 22 July 2004, "EC questions banking of allowances" (http://www.pointcarbon.com).

[5] Without a CPR, countries Kyoto-targets may have an incentive to sell all their AAUs right at the beginning and then pull out of the system.

[6] SET UP was sponsored by the Ministry for the Environment and Transport of the German Federal State of Baden-Württemberg.

makes it possible to assess the validity of results from theoretical models. This explorative approach may also lead to new findings which – because of the complexity of the environment – could not have been derived from theory.[7] In addition to controllable laboratory experiments, more realistic field experiments may also be carried out, but they tend to suffer from two major drawbacks: lack of controllability and high costs. The simulation game SET UP combined the advantages of a field experiment and a controlled lab experiment, because the companies that participated in the simulation will actually take part in such a system.[8] The artificial testing environment developed for SET UP allowed the controlled analyses of the participants' actual decision processes and strategies.

In SET UP, twelve companies, including producers of power, steel, cement, glass, paper, drive systems, printing presses, and automobiles, participated in the trading simulation using their own data. The rules of the game were based on the Directive Proposal for an EU ETS from October 2001 (CEC 2001). Parallel to the group of companies, a student control group carried out the simulation with identical rules and identical company data. Thus, the EU ETS could be tested in a realistic environment and compared to a theoretical point of reference, that is, the theoretical optimum. This set-up allowed conclusions to be drawn with respect to the actual design of the trading system. In particular, the impact of a one-time ban on banking was analysed.

General setting, trading structure and abatement measures

- The system included direct energy- and process-based CO_2 emissions from installations operated by the participating companies.
- The simulated period from 2005 to 2013 was structured – similar to the actual scheme – into commitment periods, each of them subdivided into three trading periods (years). The transition into the first Kyoto commitment period, beginning in 2008, was also modelled, especially the potential ban on banking from 2007 to 2008. The first commitment period from 2005–2007 is supposed to mirror the three-year period of the planned EU ETS prior to the first five-year commitment period of the Kyoto Protocol. In the simulation game, subsequent

7 For overviews, see Sturm and Weimann (2001) or Muller and Mestelmann (1998). Loewenstein (1999) discusses the extent to which findings from experiments may be transferred (external validity). For the general role of experimental economics in research on regulation, see Eckel and Lutz (2003).

8 Other CO_2 trading simulations conducted include Baron (2000, 2002) and Eurelectric et al. (1999). Baron (2000, 2002) focuses on the role of national governments' policy-making in the context of international emissions trading for countries as set up under the Kyoto Protocol. Baron (2002) also includes several power generating companies in order to analyse the interaction of international emissions trading with the electricity market. Eurelectric et al. (1999) includes only (virtual) power generating companies as well as electricity trading. In these trading simulations, participants failed to achieve the optimal solution. However, since the set-up of these simulations was such that company and government players did not necessarily have to minimise costs, the optimal solution is not a valid reference point. Similarly, these simulations did not include control groups.

time intervals of three years were chosen for practical reasons: there was only enough time to simulate three trading periods within one afternoon session.

- In each trading period, there were two trading dates when allowances could be sold or purchased by the participants. The market can be described as a *closed system*: market prices and volumes of trade exclusively resulted from the participants' market interaction.
- The companies participated with their real installations and abatement measures. The set of available abatement measures was fixed at the beginning of the simulation and could not be changed afterwards. The participants, on average, had five abatement measures available. At the beginning of each period they could decide on the implementation. When making their decisions on abatement and trading, individual participants only knew the costs of their own measures and emissions.
- Participants still in possession of allowances at the end of the third commitment period in 2013 received a credit in order to avoid endgame behaviour. The credit was a function of the average market prices in the preceding trading period.
- The participants had to maximise profits. All payments for investments, trading, and (possibly) sanctions for each participant were tracked through individual accounts.

Primary allocation and overall emission target

- Existing installations received allowances free of charge based on historic emissions (grandfathering). Thought has been given to early action. For further details, see Schleich et al. (2002).
- Allowances for new installations had to be purchased on the spot market.
- Allowances for phased-out installations were withdrawn in the next period.
- The overall emission target (cap) was tightened over time. Figure 1 shows the primary allocation with the projected actual emissions (without abatements) for the simulation period (2005 to 2013). The resulting annual need for abatement, also cumulated, is displayed as well. For reasons given in Section 2, the targets for the first three years of the simulation were rather lenient, so that no net emission reductions were required during that time period (excess allocation).
- Participants were informed about their allocation for the current and following commitment period. Since the Directive for the EU ETS requires that, at the start of a commitment period, participants know their individual allocation only for that actual commitment period, participants of SET UP were better informed about future allocations than they will be under the EU ETS.

Simulation environment

- An Internet platform was developed so that the trading simulation could be carried out in a decentralised manner.

- To allow a meaningful analysis and evaluation of the simulation results to be made, some important parameters like interest rates, and fuel prices were set as defaults.
- The participants had access to a decision-support tool through the Internet platform which made it possible to assess the profitability of abatement measures, the so-called indicator price (see Ehrhart et al. 2003).
- Parallel to the company group, a *student control group* at Karlsruhe University carried out the simulation with identical rules and data. The data was anonymised, and no information was shared with the companies during the simulations. Both groups were given identical information at every point in time.

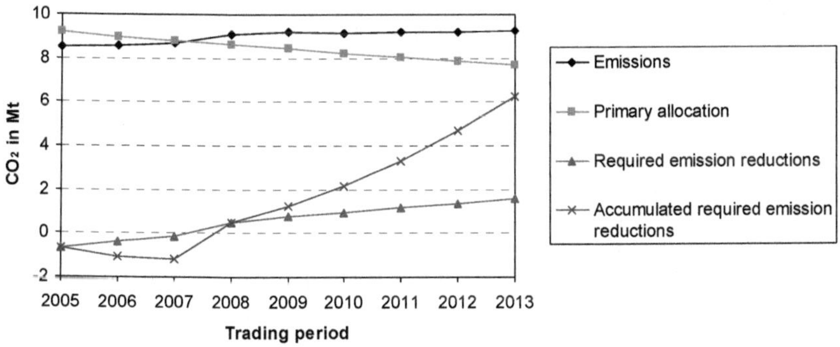

Fig. 1. Projected emissions (without abatement), primary allocation (caps) and required annual and cumulated emission reductions for the entire simulation period.

4 Results of SET UP

The simulation results were examined for both groups individually and compared with the calculated *theoretical cost-efficient optimum*. Figure 2 and Figure 3 show the time path of the observed market prices for the company group and the student control group respectively, together with the optimum. To calculate the cost-efficient optimum, all abatement measures and emissions were considered and the required global emission reduction is realised through the globally cheapest measures. Minimum marginal abatement costs increase over time because targets become tighter, because borrowing is prohibited, and because of the opportunity costs of investing in abatement measures, i.e. interest payments. Furthermore, since the set of available abatement measures was fixed at the beginning of the simulation, companies could not search for new, potentially cheaper abatement measures and add them to their options.

4.1 Company group

The market price development for the company group is shown in Figure 2. In the first three trading periods, market prices were generally low and a complete price collapse could be observed in 2007. The low market prices are primarily a result of the surplus allocation of allowances in the first commitment period of the simulation (2005-2007). Together with the ban on banking, surplus allocation resulted, as expected, in a drop in allowance prices towards the end of the commitment period. Since unused allowances could not be transferred into the year 2008, their economic value was zero. Also, companies which based their abatement decisions on current allowance prices initiated only low-cost measures. In the second commitment period (2008-2010), a dramatic increase in the prices of allowances could be observed. Starting in 2008, when emission targets started to become more stringent, demand for allowances increased, while supply dropped. Many participants with surplus allowances saved these for security reasons instead of selling them on the market and in addition failed to initiate low-cost measures. Therefore, allowance market prices started to rise continuously. At the same time, some companies initiated more costly abatement measures in response to the observed high market prices. Due to implementation lags, many of these measures did not become effective until the third commitment period (2011-2013). The combination of surplus allocation of allowances in the beginning and the ban on banking resulted in a price path in the shape of a bubble for the first two commitment periods. In the third commitment period, market prices came back down to the expected level.

4.2 Student group

The results for the student control group were astonishingly similar to those of the company group: initiated measures, trading volumes and market price development were qualitatively completely analogous. The observed time path for market prices for the student control group, which also takes on the shape of a bubble (with a dip) is displayed in Figure 3. Compared to the company group, the student group played “better”: overall reduction costs were slightly lower than for the company group.

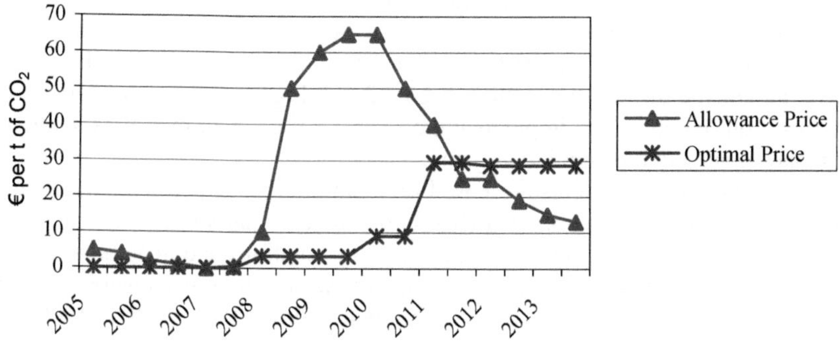

Fig. 2. Observed price path in the company group and optimal price path

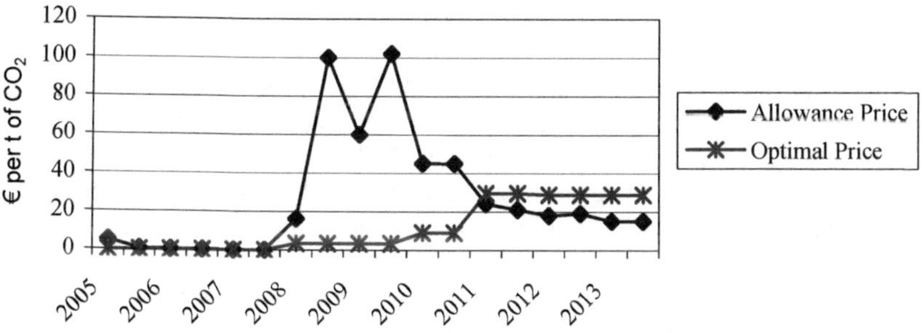

Fig. 3. Observed price path in the student group and optimal price path

4.3 Comparison of simulation results with theoretical optimum

Company and student participants made decisions on abatement measures which were individually rational: they carried out abatement measures in the ascending order of their costs. When regarded across all participants, the choice of measures was not always by minimum cost: some cost-effective measures were not realised, whereas more expensive measures were. In Figure 2 and Figure 3 there is a gap between observed prices and optimal prices, particularly in the second commitment period. Misinterpreting the current prices as future scarcity signals, the participants activated many expensive abatement measures that presumably would not have been activated following a 'correct' price development. In both groups, this led to overall abatement costs that were much higher by far than the expected costs, considering the overall optimal solution. The similar inefficient results of the control group strengthen the supposition that there is a systematic reason behind these observations. Possible reasons for this development include incomplete information about the scarcity of allowances in the future and other participants' abatement costs, as well as risk aversion.[9] It should be noted that the observed level and shape of the price path in Figure 2 and Figure 3 depends – among other things – on the costs and implementation lags of abatement measures available to the participants in the simulation game.

Of course, these findings are based on a simplifying experimental setting. But there are at least two reasons which support the external validity of our findings: first, companies participated using their real installations and abatement measures and they were represented by decision makers who will be responsible for the future emissions trading. Our second argument follows Plott (1982, 1989): if the theoretical prediction of a high degree of efficiency gains fails in a simplifying experiment, then there is little reason to expect these gains to occur in the more complicated real world.

5 Conclusions

Banning the transfer of allowances from 2007 to 2008 increases overall compliance costs because cost savings cannot be traded over time. In addition, for lenient emission targets in the first phase of the EU ETS, the results of the simulation game for both participant groups suggest that market prices will not reflect the "true" opportunity costs, if banking is prohibited, which will then lead to an inefficient choice of abatement measures. More specifically, it is to be expected that a generous allocation of allowances in the first phase results in the collapse of market prices towards the end of the first commitment period and a sharp increase afterwards when targets become stricter. In addition, a generous allocation tends to reduce market transactions in the first phase so that the expected gains in experience in emissions trading between 2005 and 2007 may only be small. In the actual EU ETS there are various mechanisms in place which prevent a price bubble after a ban on banking which was observed in SET UP. First, according to the so-called "Linking Directive", companies will be able to use relatively low-cost credits from project-based mechanisms (Clean Development Mechanism) as early as 2005. Thus, if there is sufficient supply of these relatively low-cost credits, their price may set an upper limit for EU allowance prices. Secondly, in SET UP the set of available measures was, for technical reasons, assumed to be fixed. However, if there is increasing scarcity in the market because emission targets become tighter and because the potential for existing measures becomes exploited, increasing allowance prices will provide incentives for companies to develop and invest in new abatement measures. Finally, as Ehrhart et al. (2004) argue in the context of the

[9] Ideally, to test these hypotheses, additional experiments should be run which differ in the banking rules as well as in the type of information given to the participants, such as other participants' abatement costs.

potential ban on banking in the EU ETS, if forward markets exist or if part of the allowances are auctioned off prior to the start of the periods, reliable price signals for the future scarcity of allowances may be generated which would lead to lower overall abatement costs. However, in the actual EU ETS these mechanisms may not be readily available: (i) the EU ETS only allows for a small portion of allowances to be auctioned off by Member States, and (ii) a generous allocation of allowances may lead to a thin spot market which may prevent derivative markets from developing sufficiently enough to allow for robust price signals and for hedging.

References

Baron R (2000) Emission trading: a real time simulation. International Energy Agency Information Paper for COP6, Paris. http://www.iea.org/envissu/cop6/emistr.pdf

Baron R (2002) Trading CO_2 and electricity in the Baltic Sea region. Report on the simulation of the Baltic Sea Region Energy Cooperation BASREC, IEA, Paris

Betz R, Eichhammer W, Schleich J (2004) Designing national allocation plans for EU emissions trading - a first analysis of the outcomes. Energy & Environment 15: 375-425

Boemare C, Quirion P (2002) Implementing greenhouse gas trading in Europe: lessons from economic theory and international experiences. Ecological Economics 43: 213-230

Carlson C, Burtraw D, Cropper M, Palmer KL (2000) Sulphur dioxide control by electric utilities: what are the gains from trade? Journal of Political Economy 108: 1292-1326

CEC (2000) Communication from the Commission to the Council and the European Parliament - on EU Policies and Measures to Reduce Greenhouse Gas Emissions: Towards a European Climate Change Programme (ECCP). COM (2000) 88 Final. Brussels, 8 March 2000

CEC (2001) Proposal for a Directive of the European Parliament and of the Council Establishing a Framework for Greenhouse Gas Emissions Trading within the European Community and Amending Council Decision 96/61/EC. COM (2001) 581 Final. Brussels, 21 October 2001

CEC (2003) Directive 2003/87/EC of the European Parliament and the Council of 13 October 2003 Establishing a Scheme for Greenhouse Gas Emission Allowance Trading within the Community and Amending Council Directive 96/61/EC, OJ L275, 25.10.2003. Brussels, pp. 32-46

Eckel C, Lutz N (2003) Introduction: what role can experiments play in research on regulation? Journal of Regulatory Economics 23: 103-107

Ehrhart KM, Hoppe C, Schleich J, Seifert S (2003) Strategic aspects of CO_2 emissions trading: theoretical concepts and empirical findings. Energy & Environment 14: 579-595

Ehrhart KM, Hoppe C, Schleich J, Seifert S (2004) EU Emissions Trading: Design Parameters and Efficiency. Working Paper, University of Karlsruhe, Fraunhofer Institute for Systems and Innovation Research, Karlsruhe, and Takon GmbH

Ellerman AD, Harrison D, Joskow PL (2003) Emissions trading: experience, lessons, and considerations for greenhouse gases. Pew Center for Global Climate Change, Arlington, VA

Ellerman AD, Schmalensee R, Joskow PL, Montero JP, Baily E (2000) Markets for Clean Air: the U.S. Acid Rain Program. Cambridge University Press, Cambridge

Ellerman AD (2002) U.S. Experience with emissions trading: lessons for CO_2 emissions trading. Paper presented at the 10th Symposium of the Egon-Sohmen-Foundation on Climate Policy: US and European Views, Dresden, 25-26 October 2002

Ellerman AD, Montero JP (2002) The temporal efficiency of SO_2 emissions trading. MIT - Center for Energy and Environmental Policy Research Working Paper 2002-003 Cambridge

Eurelectric, IEA, and ParisBourse (1999) Greenhouse gas and electricity trading simulation 2. Reference 1999-420-0013, http://www.iea.org/clim/cop5/pubs/report.pdf

Kling C, Rubin J (1997) Bankable permits for the control of environmental pollution. Journal of Public Economics 64: 101-115

Loewenstein G (1999) Experimental economics from the vantage-point of behavioural economics. Economic Journal 109: 25-34

Muller RA (1999) Experimental methods for research into trading of greenhouse gas emissions. Working Paper, Department of Economics, McMaster University, Hamilton

Muller RA, Mestelmann S (1998) What have we learned from emissions trading experiments? Managerial and Decision Economics 19: 225-238

Plott CA (1982) Industrial organization theory and experimental economics. Journal of Economic Literature 20: 1485-1527

Plott CA (1989) An updated review of industrial organization: applications of experimental methods, In: Schmalensee R., Willig, RD. (eds), Handbook of Industrial Organization, vol. 2, North-Holland, Amsterdam, pp. 1109-1176

Schleich J, Betz R, Wartmann S, Ehrhart KM, Hoppe C, Seifert S (2002) Simulation eines Emissionshandels für Treibhausgase in der baden-württembergischen Unternehmenspraxis (SET UP), Endbericht an das Ministerium für Umwelt und Verkehr Baden-Württemberg, Karlsruhe. http://www.isi.fhg.de/u/planspiel/endber.pdf. Summary available in english under http://www.isi.fhg.de/u/e-projekte/e-bawueplan.htm

Schleich J, Ehrhart KM, Hoppe C, Seifert S (2006) Banning Banking in EU emissions trading, Energy Policy 34: 112-120

Sturm B, Weimann J (2001) Experimente in der Umweltökonomik. FEMM Working Paper No. 7/2001, November 2001, University of Magdeburg

Tietenberg T (1999) Tradable permit approaches to pollution control: faustian bargain or paradise regained? In: Kaplowitz MD, Property Rights, Economics, and the Environment, JAI Press Inc., Stamford, pp. 175-199

Emissions trading and innovation in the German electricity industry – impact of possible design options for an emissions trading scheme on innovation strategies in the German electricity industry

Martin Cames[I], Anke Weidlich[II]

[I] Öko-Institut – Institute for Applied Ecology
Novalisstraße 10, 10115 Berlin, Germany
m.cames@oeko.de

[II] Universität Karlsruhe (TH)
Lehrstuhl für Informationsbetriebswirtschaftslehre
Englerstraße 14, 76131 Karlsruhe, Germany
anke.weidlich@iw.uni-karlsruhe.de

Abstract

The paper examines what impact different design options of emissions trading have on the innovation process in the electric power industry. Recent concepts of innovation research in evolutionary economics are reviewed and investment cycles in the German power sector are examined before taking a closer look at different emissions trading design options and their respective impact on power generation costs.

Keywords: Climate change, emissions trading, innovation, electricity, industry, new entrants, plant closure, windows of opportunity

1 Introduction

The European emissions trading scheme has been introduced in order to stimulate innovation in the direction of a more efficient and less CO_2-intensive means of production, particularly in the electricity sector. The hypothesis of this paper is that the impact of this new instrument on the innovation process depends on the way in which it is designed.

In this paper, emissions trading is put into the context of findings of modern innovation research. Section two provides a brief introduction to selected elements of innovation research. The current situation in the German power sector is reviewed in section three: investment cycles and the stages of technological development of two important power technologies – advanced lignite-fired plants and combined-cycle gas turbines – are examined.

Section four describes possible design options for an emissions trading system and their respective innovation effects. The focus of this paper will be on the examination of different regulations concerning the method of allowance allocation (auctioning or free allocation) and the treatment of plant closures and new entrants. This enables the likely general innovation effect that emissions trading has on the German power sector to be deduced, in addition to an estimate of the importance that the cited technologies will have for power production under the new conditions of CO_2 emissions trading. These aspects are dealt with in section five.

2 Innovation and windows of opportunity

Innovation can be of a technical, social or socio-technical nature and is always the result of planned human action. Innovation includes the invention, development, adaptation and diffusion of new products or product components, new materials, new production processes and new organizational set-ups. Changes in lifestyle, institutions, values and other aspects of social life can also be regarded as innovation.

In neo-classical economic theory, technology and innovation are largely seen as an exogenous factor; the causes of and incentives for innovation are not further examined. In evolutionary theory, however, technological innovation is seen as a fundamental source or endogenous factor of economic growth, and the driving forces and dynamics of technological change are placed at the centre of examination. According to evolutionary theory, economic development is characterized by a sequence of rather stable phases with only incremental innovations, and by unstable phases that represent opportunities for basic technological changes, where already implemented technologies can be replaced by new, alternative technologies. These unstable phases are considered *windows of opportunity*. In unstable phases, insignificant initial advantages in favour of a particular technology can become a self-reinforcing factor due to learning effects (learning by doing, learning by using), network externalities and economies of scale. This can, in certain conditions, push development along a particular path and lead to the success of a technology that may be technologically inferior to other solutions (David 1985).

Windows of opportunity summarize a set of favourable conditions for innovation. These conditions can be observed at different levels. Most important is the techno-economic level of competing technologies. Zundel et al. (2003) differentiate between two different constellations of techno-economic windows:

- Innovation competition between new technologies that are at a similar stage of maturity. This kind of time window is particularly important at an early stage of competition in a newly emerging market.
- Competition between an old, dominant technology and new technologies. This competition becomes topical when the investment cycles of the dominant solutions come to an end and new technologies are potentially competitive. Unstable phases thus mainly depend on the dynamics of the old technology.

In capital-intensive sectors such as the power sector, sunk costs and the duration of development periods and investment cycles are especially sensitive to the time factor. Investing in a new technology means high sunk costs for a company that cannot be recovered for a long period of time. At the end of an investment cycle, when investments are written off, sunk costs are zero and a switch to a new technology is easier. When investment cycles are synchronous in an industrial sector, this idea can be applied to the entire sector (Zundel et al. 2003). The end of investment cycles in an industrial sector thus opens techno-economic windows of opportunity.

Besides the techno-economic factors, other, especially political and institutional factors determine the emergence of time windows. In the case of competition between an old dominant technology and new technologies, political and institutional factors are highly relevant for the emergence of time windows. Dependence on the institutional path (determined by norms and standards, lobbying by syndicates, "regulatory capture" etc.) linked with dependence on the technological path, stabilizes the dominant path and may hinder the use of a techno-economic time window (Nill 2002).

3 Time windows in the German electricity market

In large-scale technological systems with long investment cycles, several attempts – in the sense of good opportunities – are usually required to give new direction to a system's development (Zundel et al. 2003). Evolutionary economics show that far-reaching innovation beyond dominant development paths requires, at the same time, favourable conditions at different levels. It is highly likely that a technological regime change can only be successful if favourable social, political and economic factors come together at one period of time and reinforce each other.

In order to get an indication as to how emissions trading might affect investment decisions and the direction of innovation, these timing aspects will therefore be further examined in the following sections. Investment cycles in the German power sector will be identified and a closer look taken at the development stages of two competing energy technologies – combined-cycle gas turbines and advanced lignite-fired power plants.

3.1 Investment cycles in the German electricity sector

As mentioned above, decisions in favour of certain technologies create sunk costs that cannot be recovered for a long period of time. This is particularly true for the electricity industry, where the technical lifetime of investments in power plants is usually 30 years and more. As a consequence, today's energy-technology decisions determine the characteristics of the energy system for several decades. As pointed out by Roehrl and Riahi (2000), "research, development and demonstration efforts as well as investment decisions in the energy sector over the next two to three decades are critical in determining which long-term technological options in the energy sector may be opened, or which ones may be foreclosed". Due to the long lifetime of investment in power plants and the partly synchronous phasing-out of many investment cycles in the German power sector, there are certain time windows for investments in new plants. These are different in the eastern and the western part of Germany. In the 1990s, replacement and modernization investment in power production was primarily carried out in eastern Germany, whereas little new capacity was constructed in western Germany. In the former, considerable renewal investments will not be necessary during the next 15 to 20 years, whereas in the latter, a significant number of power plants will reach the end of their technical lifetime, at the latest from 2010 onwards. These plants either have to be replaced by new plants or compensated by considerable electricity savings (Enquete-Kommission 2002).

Under the new conditions of liberalized markets, it can be assumed that operators are anxious to reduce existing excess capacity and to extend the lifetime of power plants with low fuel and operating costs by means of maintenance investment that is less capital intensive. However, even under the assumption of such strategies, the Enquete-Kommission (2002) comes to the conclusion that 40 to 60 GW[1] of new power plant capacity needs to be erected between 2010 and 2025, if electricity consumption does not decrease significantly. This investment will mainly take place in the western part of Germany, whereas in the eastern part, capacity replacement will not be required before the end of the 2020s.

With implementation periods of five years and more for planning, approval and construction of new power plants, long-term investment decisions on new generation capacity are likely to coincide with the start of the EU-wide emissions trading system. Given long technical lifetime and considerable investment volume, the electricity sector plays a key role in the sustainable organization of the energy system. In connection with the time-related considerations mentioned above, investment decisions during the coming 15 years – at least for large power plants – will

[1] Which is equivalent to one-third or half of existing electricity generation capacity in the year 2001.

largely determine the level of greenhouse gas emissions in the year 2050. Emissions trading, if well designed, will therefore noticeably contribute to a reduction of this emission level.

The following brief overview of the institutional and techno-economic development of advanced lignite-fired power plants and combined-cycle technology will help to detect possible windows of opportunity for technological change in the German power sector.

3.2 Advanced lignite-fired power plants

Lignite exploitation for power production has a long tradition in the German energy sector. As a domestic fuel, coal (both lignite and hard coal) has always been the most important energy source in Germany, and its utilization in steam turbines remains the dominant path for electricity production. In contrast to hard coal, lignite is not directly subsidized in Germany. However, exemption of coal from the eco-tax and tax on fuel consumption is clearly a preferential treatment of this energy source in comparison with other fuels (for example, natural gas). Lignite-fired power technology has improved considerably during the last century, although radical innovations have not taken place. The actual state of the art in lignite power production is so-called *BoA*[2] technology, which has increased net efficiency to 43%. This improvement has been achieved through economies of scale, the use of advanced supercritical steam cycles and more efficient turbines as well as other incremental improvements. A further 3-5% improvement has been projected for the *BoA-Plus* system through pre-drying coal with waste heat. According to *RWE Rheinbraun*, the major company in the German lignite industry, new lignite-fired power plants, constructed in the present decade, will use the BoA technology. From 2015 onwards, BoA-Plus technology will be commercially available, and after 2020, BoA-Plus plants with higher steam temperatures, resulting in a level of efficiency of >50%, will come into operation (Lambertz 2003).

With regard to emissions trading, the main disadvantage of lignite is that it has the highest direct CO_2 content of all fossil energy sources. In addition, the efficiency of lignite-fired power production is comparably low, thus increasing CO_2 emissions per kWh of electricity produced. Even advanced lignite technologies cannot compete with the efficiency of hard-coal-fired power plants or combined-cycle gas turbines (CCGT). Moreover, specific investment costs and construction times are higher for lignite-fired than for hard-coal fired power plants, and considerably higher than for CCGT plants. On the other hand, lignite-fired power plants have longer lifetimes; but in liberalized markets with short payback periods, this argument is probably of declining importance. Despite these obvious disadvantages in times of liberalized electricity markets and higher public awareness of en-

[2] German abbreviation for "Braunkohlekraftwerk mit optimierter Anlagentechnik" (that is, an advanced pulverized-lignite-fired plant with supercritical steam conditions); in the year 2002, a 950 MW BoA plant commenced operation at the power production site of Niederaußem.

vironmental problems, lignite still plays a dominant role in German power production. In 2002, it accounted for 27.4% (BMWA 2003) of total electricity production, more than any other fossil fuel. This is partly because lignite has low and stable fuel costs and thus guarantees high supply security in the electricity sector.

But additional factors contribute to the importance of lignite in the German energy sector, namely path dependence and lock-in. Parallel to dependence on the technological path (centralized, large-scale steam turbine technology has been the dominant path in power production since the beginning of industrialization), dependence on political and institutional paths can be observed in the lignite industry. The tight network of a few big, vertically-integrated companies in the lignite sector and political decision-makers in the coal regions stabilizes the dominant path and often makes it more profitable for established firms to invest in the dominant technology, rather than in a radical innovation. Moreover, mutual arrangements or traditional relationships between political institutions and companies in the lignite industry create lock-in situations that make it difficult to depart the traditional path. The *Kraftwerkserneuerungsprogramm*[3] (power plant renewal programme), setup in 1994 by RWE Rheinbraun and the government of the federal state of North Rhine Westphalia, is one example of this kind of arrangement. The introduction of emissions trading at a European level influences decisions on investment in new lignite-fired plants, as the example of Rheinbraun shows. It is doubtful, however, whether it will fundamentally change the German power production structure.

3.3 Combined-cycle gas turbines

Among gas-fired power plants, combined-cycle gas turbine technology (CCGT) is at the focus of interest, since it can play an important role in climate protection strategies. Allowing for the conversion of low-carbon natural gas into electricity with a high level of efficiency, CCGT results in low CO_2 emissions per unit of electricity produced. Sartorius (2002) examined the techno-economic development of gas turbines and CCGT technology under the framework conditions of energy and environmental regulation in Germany and the USA. He points out that the techno-economic window for CCGT opened in the early 1980s, when this technology overcame the technological lock-in situation to which it was subjected by existing technology, the steam turbine. Through a recombination with exactly this technology, the advantages of the former could be used without giving up the benefits contributed by the gas turbine, which had already successfully entered the market in the 1960s (at first as an auxiliary or peak-load device, then later becoming competitive in base-load electricity generation).

[3] RWE Rheinbraun plans to gradually replace old lignite power plants by 8-10 new plants with state-of-the-art technology within the coming 30 years. The federal government has supported this project by granting the required permits and assuring favourable conditions. One new power plant has already been built in Niederaußem; at present, further investments are postponed, so long as the final design of the European emissions trading regime has not been decided.

In Germany, however, the economic window of opportunity of CCGT was narrowed down by industrial policy in 1999. In this year, the eco-tax came into force. Despite its name, the new tax does not account for the carbon content of different fossil fuels. Instead, coal is exempted from imposition and the level of the tax on natural gas was increased (gas used for electricity generation being exempted from that increase, however). Nevertheless, the eco-tax and the tax on mineral oil and gas products systematically favours coal compared with other energy sources. Later in 1999, attempts were made to exempt natural gas for power production from the mineral oil tax. There was strong opposition to this attempt on the part of the coal lobby and other "pro-coal" players. In the end, exemption of natural gas from the mineral oil tax was only applicable for CCGT plants with an efficiency of at least 57.5% that start operation within a certain time-frame.[4]

As a result, in the case of CCGT the economic and institutional window of opportunity in the German power sector could be considered to be virtually closed at present. However, under the new conditions of liberalized markets, and with the introduction of emissions trading in Germany, this situation might change in the near future. Combined-cycle technology offers several advantages, such as low investment costs, short construction periods and a high degree of modularity and flexibility that take effect in a liberalized electricity market. Moreover, its comparatively low CO_2 emissions are advantageous under an emissions trading system, so that a window of opportunity for CCGT in Germany might possibly open in the course of the present and the following decade.

The increased use of natural gas for power production in Germany would be a considerable innovation. It might initiate a development away from a base-load-orientated, centralized power production system towards more flexible, decentralized electricity distribution, and could seriously question the protection and subsidization of the German coal industry. This might, in consequence, give rise to more innovation in other energy technologies; for example, renewable energy or small-scale, decentralized power plants.

4 Innovation incentives of different allocation methods

Emissions trading systems are not specifically designed to foster technological change, but they do have innovation effects. By modifying the price of the commodity that creates externalities – that is, fossil fuel – an emissions trading system provides continuous financial incentives for emitters to adopt innovations that reduce their CO_2 emissions. The incentives basically depend on the targets: stronger targets will induce higher prices for emission allowances and, thus, possibly accelerate investment rates and induce more innovation.

The basic hypothesis of this paper is that different design options for an emissions trading scheme create different innovation incentives and thus influence the

[4] For a more detailed description of the conflict on modern CCGT plants in the context of ecological tax reform, see Stadthaus (2001).

level and structure of innovation and technological change in the electricity industry. Central design options that might initiate or delay innovation and technological change are described in the following sections.

4.1 Initial allocation

Two basic methods of allowance allocation to existing plants are at the focus of public discussion: *auctioning* and *free allocation.* Emission allowances might either be auctioned among the emitters of greenhouse gases, who would consequently have to pay for each unit of CO_2 released into the atmosphere, or they can be allocated free of charge, according to a fixed allocation method. Free allocation on the basis of past emissions in a selected baseline period is referred to as *grandfathering*. It has the effect that an equal percentage reduction is required from all emission sources over the same period of time. This percentage is expressed in a fulfilment factor, which adapts the available amount of allowances for different sectors (the cap; top-down approach) to the aggregate past emissions of each individual installation in the baseline period, calculated on the basis of a bottom-up approach. Another option for free initial allocation is the *benchmarking* approach, where emission allowances are distributed on the basis of output in a recent reference period and an overall target, expressed in CO_2 per unit of output. Plants that are more efficient than average state-of-the-art plants would receive more allowances than they need for production, and plants below this standard would have to buy additional allowances. Three alternative benchmarking methods can be differentiated: fuel-specific, average and best available technology (BAT) benchmarking. In case of fuel-specific benchmarking, several benchmarks are calculated for each fuel (lignite, hard coal, natural gas etc.) whereas only one benchmark is calculated for all power generation technologies in the case of average benchmarking. The BAT benchmark is similar to the average benchmark, but is based not on the average technology, but rather on the best available technology.

The method of allowance allocation to existing installations does not basically affect the static efficiency of an emissions trading system. However, Milliman and Prince (1989, 1992) and Jung et al. (1996) argue that auctioned allowances would create greater incentives for technology diffusion and adoption than allowances allocated free of charge, since technology diffusion reduces the price of allowances. The innovator can benefit from this fall in prices, since he will not have to pay as much for the rest of his emissions. Fisher (2000) points out that, on the contrary, in the case of free allocation the fall in prices due to innovation would lower the value of the innovator's allowances, which makes innovation less attractive. Thus, free allowances provide less incentive to innovate than auctioned allowances.

The EU opted in its Directive for an emissions trading scheme for largely free distribution of emission allowances. Auctioning is only permissible for 5% of allowances in the pilot phase (2005 to 2007) and 10% in the second period of European emissions trading, which runs from 2008 to 2012 (2003/87/EC) but will not be applied in Germany before 2012.

While overall investment patterns and dynamic efficiency under such an emissions trading system depend mainly on the stringency of the overall emission cap and, thus, on the resulting price of allowances, the allocation method affects the competitive situation of companies. Some companies that would be net sellers under grandfathering could become net buyers under benchmarking (IEA 2001).

4.2 New entrants

In the case of free allocation of allowances to existing installations, innovation incentives will also be influenced by the way new plants (including the extension of capacity in existing installations) are provided with allowances. Either they will have to buy required emission allowances on the market, or they will also receive their allowances free of charge.

The advantage of the first option is that companies have to incorporate the "cost" of CO_2 emissions into their calculations of profitability and only enter into the market when expected revenue is higher than expected costs, including CO_2 costs. On the other hand, this option puts new entrants at a disadvantage compared to existing operators and may constitute an obstacle to market entry, especially when the price of allowances is high. If the barriers to market entry are high, and this is especially the case when additional costs arise at the beginning of an investment cycle, new companies have less incentive to invest. In this case, new entrants cannot sufficiently "menace" existing operators; the market power of the latter is reinforced. This situation slows down innovation and restructuring processes.

In the case of free distribution of emission allowances to new entrants, the allocation methods discussed in the previous section on initial allocation can be applied. In the case of grandfathering, it has to be assumed that new entrants will receive allowances according to their needs, but reduced by the fulfilment factor.

We have assessed the effects of these different allocation methods on the basis of a levelized cost model[5], which calculates specific electricity generation costs for the main competing base-load technologies in Germany – that is *BoA* and *CCGT* power plants – commencing operation in 2005. The model follows the full-cost principle and thus represents one important criterion for the investment decision of an operator that might decide to invest in a BoA or a CCGT plant. It thus differs from the operational view, where decisions on the output level in existing power plants are based only on variable operating costs. Within our model, allowances allocated free of charge are considered as income that can only be realized when the corresponding investment is carried out whereas allowances needed for operation of the power plant constitute an expenditure. The difference between the expenditure for required allowances and the income represented by freely allocated allowances are the actual costs of a plant's emissions. In the case of benchmarking, these costs can also be negative as a result of "over-allocation". The results of the simulation are shown in

[5] For details on the model design the reader is referred to Schneider (1998).

Table 1. Impact of different methods of allocation on competing electricity generation technologies

Price per allowance	No emissions trading	Auctioning	Grandfathering	Average benchmark	BAT benchmark	Fuel specific benchmark
Change in specific generation cost due to emissions trading						
BoA						
5 €/t CO_2		5.7%	0.5%	2.8%	2.7%	-0.1%
10 €/t CO_2		11.5%	1.0%	5.6%	5.5%	-0.1%
15 €/t CO_2		17.2%	1.5%	8.4%	8.2%	-0.2%
CCGT						
5 €/t CO_2		2.0%	0.2%	-0.7%	0.2%	0.0%
10 €/t CO_2		4.0%	0.4%	-1.5%	0.4%	0.1%
15 €/t CO_2		6.1%	0.5%	-2.2%	0.5%	0.1%
Difference in specific generation cost at a CCGT compared to a BoA plant						
5 €/t CO_2	6.2%	2.5%	5.9%	2.6%	3.6%	6.3%
10 €/t CO_2	6.2%	-0.8%	5.5%	-0.9%	1.1%	6.4%
15 €/t CO_2	6.2%	-3.9%	5.2%	-4.2%	-1.3%	6.6%

Source: Own calculations[6]

In the first place, the results show that auctioning (without redistribution of revenues) causes substantially higher cost increases than free allocation. In the case of grandfathering or fuel specific benchmarking, cost increases are rather limited (less than 2%), even if the price of allowances is relatively high (15 €/t CO_2). Apart from auctioning, only average and BAT benchmarking would cause substantial changes in specific generation costs of new power plants, resulting in an increase in generation costs for BoA plants and a decrease for CCGT plants.

In the present situation without emissions trading, generation costs are about 6% higher in CCGT plants than in BoA plants. If grandfathering were chosen as allocation method, the cost advantage of BoA plants would hardly change. In the case of fuel-specific benchmarking, BoA plants would even gain in competitiveness. Only if auctioning, average or BAT benchmarking were applied, would CCGT plants improve their competitiveness. However, they can only catch up with BoA plants if the expected price of allowances increases to at least 10 €/t CO_2. Surprisingly, the average-benchmark approach leads to almost the same results as the auctioning of allowances. From an innovation perspective, and given

[6] For BoA plants, we assume an efficiency of 44.5%, a capacity of 950 MW and a depreciation period of 30 years. In case of CCGT plants, we assume 57.5% efficiency, 800 MW plant capacity and a depreciation period of 20 years. Operation time is assumed to be 7.000 hours per year for both types of plant.

the political decision to exclude auctioning as an allocation method, only the average and BAT benchmarking approaches would again widen the *window of opportunity* for CCGT plants, whereas fuel-specific benchmarking would narrow it further.

As a result, free allocation of allowances for new entrants is most likely to promote innovation in the electricity industry under the planned European emissions trading system. Average or BAT benchmarking are more likely to stimulate a switch from the use lignite to natural gas in power generation and thus have stronger effects on the German fuel mix than grandfathering or fuel specific benchmarking.

4.3 Plant closure

Allowances allocated free of charge might expire if the plant to which they are allocated is shut down during the commitment period. This might motivate operators to extend the lifetime of their power plants in order to keep their allowances. Alternatively, allowances might retain their value for a certain time after plant closure, at the most until the end of the commitment period. The latter option is often criticized as a premium for plant closure and seems to be politically not acceptable. However, this option gives more incentives to companies to close down CO_2 intensive power plants and replace them by newer, less CO_2 intensive ones. This incentive is reinforced if additional allowances are allocated to new power plants.

A third option might provide a balanced approach between political acceptability and innovation stimulation: Operators can opt to keep the allowances allocated to an old installation if they replace it by a new installation.[7] If they do so, they are not endowed with additional free allowances for the new plant. They might still be interested in this option because the allowances they can retain from the old installation are higher in number than those they would get for a new power plant, because new plants are usually more efficient and thus emit less CO_2. As this option generally results in an over-provision of allowances for the new plant, the innovation incentive is maintained.

5 Conclusions

Today's energy technologies are strongly embedded in nearly all economic and social activities of everyday life. They are part of a complex system of integrated technologies for the production, distribution and use of energy, which interact with the socio-economic system from which they emerge. Investment cycles and sunk costs play a very important role in the energy economy. This system will not change solely through the development of individual alternative technologies or

[7] This option requires adaptation rules where the capacity of the old and new plant are different, or where closure and commencement of operations do not coincide.

the introduction of an emissions trading system. However, the approaching internalization of the environmental cost of greenhouse gas emissions by means of emission allowances is indeed one important instrument of climate change policy, but it should not be the only measure. Considering the main problems of technological change, changes in the price of one factor are not likely to be sufficient to bring about the radical change in power production technologies that is needed from the perspective of sustainability. Instead, a change in the techno-economic paradigm must involve an integrated process of change in science, engineering practice, physical infrastructure, social organization and plant design.

Emissions trading alone will not stimulate innovation in the sense of a radical shift away from the use of fossil fuels. However, changes in the present fuel mix can contribute substantially to the achievement of medium-term mitigation goals. Whether there will be a fuel switch away from carbon-intensive lignite to natural gas largely depends on the design of the emissions trading system. Techno-economic windows can be widened for both: advanced lignite-fired power plants and CCGT technology.

In order to foster innovation in general, it is important that new entrants are treated similarly to existing installations and therefore provided with allowances free of charge. Should auctioning, average or BAT benchmarking be applied for the allocation of allowances to new entrants, it would be more attractive for companies to invest in efficient, gas-fired CCGT power plants. In the case of grandfathering or a fuel-specific benchmarking system, this is less likely to be the case. The treatment of plant closure will also influence innovation incentives. However, with particular regard to this design feature, it is important to draw a balance between political concerns, which might emerge if new installations temporarily receive "double allocation", and difficulties in properly identifying plant closures. Instead of allowances expiring at the end of a year or a commitment period when a plant is shut down, we therefore prefer a hybrid approach that would allow operators to retain their allowances to the end of the commitment period only if they build a new plant.

If these findings are considered in implementation of the European emissions trading directive in Germany, emissions trading might promote a substantial shift within the present fuel mix and towards more climate-friendly technologies. However, to encourage energy technologies that are not based on fossil fuels and that would constitute a radical regime shift, a more integrated policy approach is probably needed, which would make use of the cumulative and self-reinforcing character of technological change.

References

BMWA, Bundesministerium für Wirtschaft und Arbeit (2003) Zahlen und Fakten - Energie Daten 2003, Nationale und internationale Entwicklung, Berlin, available at: http://www.bmwi.de/Navigation/Technologie-und-Energie/energiepolitik.html

David PA (1985) Clio and the Economics of QWERTY, American Economic Review 75: 332-337

Enquete-Kommission (2002) Endbericht der Enquete-Kommission „Nachhaltige Energieversorgung unter den Bedingungen der Globalisierung und der Liberalisierung". Deutscher Bundestag, 14. Wahlperiode, Drucksache 14/9400, available at: http://www.bundestag.de/gremien/ener/schlussbericht/index.htm

European Commission (2003) Directive 2003/87/EC of the European Parliament and of the Council of 13 October 2003 establishing a scheme for greenhouse gas emission allowance trading within the Community and amending Council Directive 96/61/EC

IEA, International Energy Agency (2001) International Emission Trading - From Concept to Reality. Paris

Jung C, Krutilla K, Boyd R (1996) Incentives for Advanced Pollution Abatement Technology at the Industry Level: An Evaluation of Policy Alternatives. Journal of Environmental Economics and Management 30: 95-111

Milliman SR, Prince R (1989) Firm Incentives to Promote Technological Change in Pollution Control. Journal of Environmental Economics and Management 17: 247-265

Milliman, SR, Prince R (1992) Firm Incentives to Promote Technological Change in Pollution Control: Reply. Journal of Environmental Economics and Management 22: 292-296

Lambertz J (2003) Neue Entwicklungslinien in der Braunkohlenkraftwerkstechnik. In: Energiewirtschaftliche Tagesfragen, no. 1/2, Januar/Februar 2003, Essen

Nill J (2002) Wann benötigt Umwelt(innovations)politik politische Zeitfenster? Zur Fruchtbarkeit und Anwendbarkeit von Kindons "policy window"-Konzept. Diskussionspapier des IÖW 54/02

Roehrl RA, Riahi K (2000) Technology Dynamics and Greenhouse Gas Emissions Mitigation - A Cost Assessment. Technological Forecasting and Social Changes 63: 2-3

Sartorius C (2002) Combined-cycle gas turbines - carbon dioxide abatement through efficiency increase. SUSTIME project, 3rd draft, TU Berlin

Schneider L (1998) Stromgestehungskosten von Großkraftwerken - Entwicklungen im Spannungsfeld von Liberalisierung und Ökosteuern. Berlin

Stadthaus M (2001) Der Konflikt um moderne Gaskraftwerke (GuD) im Rahmen der ökologischen Steuerreform. Forschungsstelle für Umweltpolitik, FU Berlin, FFU rep 01-03, Berlin

Zundel S et al. (2003) Entwurf einer ökologischen Innovationspolitik, 2 Zwischenbericht "Innovation, Zeit und Nachhaltigkeit - Zeitstrategien ökologischer Innovationspolitik"

A dynamic game of technology diffusion under emissions trading: an experiment

Ivana Capozza

Ministero dell`Economia e delle Finanze
Unita di Valutazione degli Investmenti Pubblici
Via Nerra 1, 00187 Roma, Italy
ivana.capozza@tesoro.it

Abstract

In this paper we investigate how the interaction between the product and the emission permit markets may affect firms' propensity to adopt less polluting technologies in a non perfectly competitive industry. We develop a model of duopoly, in which firms engage in quantity competition in the output market, behave as price takers in the permit market and can switch to a cleaner production technology at some cost. We set up a dynamic game over an infinite horizon in order to investigate firms' strategic decisions. Technology diffusion is one of the possible stationary equilibria of the game, depending on both the investment cost and the emission cap. We replicate the game in a laboratory experiment. The experimental results suggest that firms' behaviour will eventually lead to innovation diffusion.

Keywords: Tradable permits, technology adoption, oligopoly, laboratory experiments

Acknowledgement: I would like to thank John Hey for his constant and precious guide and Andrea Morone for his helpful suggestions and for the stimulating discussions I had with him. All remaining errors are solely my own responsibility.

1 Introduction

One central concern of environmental policy is how it can stimulate innovation and diffusion of cleaner technologies. Alternative environmental policy instruments can have significantly different effects on the rate and direction of technological change. Most of the related literature has focused on comparing different policy approaches, leading to an almost general agreement on the superiority of market-based instruments (Denicolò 1999; Downing and White 1986; Fisher et al. 1998; Jung et al. 1996; Milliman and Prince 1989). In particular, emissions trading is claimed to be one of the most effective instruments in inducing an "environmental technological change", though this conclusion is still debated.

Only a few studies have dealt with how an emissions trading scheme has to be designed in order to induce innovation (Laffont and Tirole 1996a, 1996b). *What are the variables that affect firms' technology choice under an emissions trading scheme?* In this work, we shall try to answer this question, focusing on the variables that the environmental regulator may control. Moreover, there is little rigorous evidence concerning the ability of tradable permits to encourage innovation and adoption of cleaner technologies, mostly because of the scarcity of available data (Jaffe et al. 2002). This observation has led us to adopt the experimental approach. *Do laboratory subjects decide to innovate as theory predicts?*

As it has been pointed out in several studies (Fershtman and de Zeew 1995; Montero 2002; Requate 1998), considering only the allowance market may lead to biased conclusions. This is because the adoption of a cleaner technology implies both a direct internal cost-reducing effect and strategic effects arising from the interaction between the output and the permit markets. For this reason, we depart from the competitive assumption, explicitly considering the production decision in a Cournot duopoly, where the technology is modelled in terms of emissions per unit of output, rather than in terms of a reduction in marginal abatement costs.

In order to analyse the incentives to innovate in the long run, we set up a dynamic game and we solve it looking for a stationary symmetric equilibrium, which may involve mixed strategies. We find that this game leads to multiple stationary equilibria, which crucially depend on the emission cap (affecting the price of an emission certificate) and on the cost of switching to a cleaner technology. In particular, in one of these possible outcomes both firms adopt the most efficient technology available (*diffusion outcome*). However, it is not possible to predict which outcome will actually occur. An experimental investigation can help us in seeing this. We design and implement an *innovation experiment* that replicates the *innovation game*. Our aim is to see whether subjects tend to converge to a specific equilibrium among those that has been identified in the theoretical analysis.

The rest of this paper is structured as follows. In the next section we develop a static model of duopoly, and we determine how the permit price and the period profits change when firms move to a cleaner technology. In section 3, we model firms' strategic decisions of adopting a cleaner production technology as a dynamic game. Section 4 describes the experiment that was implemented to test the predictions of the theoretical model. The final section concludes the paper, drawing some lessons for environmental policy design.

2 The model

We consider two profit-maximising firms producing a homogeneous good and competing à la Cournot in the output market. Let y_i be firm i's output level, for $i = 1,2$. The inverse demand for output is linear $P(y_1 + y_2) = a - b(y_1 + y_2)$ and firms have identical constant marginal production cost c, with $c < a$.

The production process generates pollution according to a constant emissions-output ratio k_j, for $j = H, L$, such that firm i's emission level is $e_i = k_j y_i$. There are only two possible technologies in this setting, k_H and k_L, with $k_j \geq 0$ for $j = H, L$, and $k_H \leq k_L \leq 2k_H$ [1]. The subscripts H and L indicate the high and low technology efficiency, respectively. Technology is the only issue in which firms may differ from one another. The government implements a market of emission permits implying that each firm must hold one permit to discharge one unit of emissions. The emission cap is fixed at E and firms do not have any initial endowment of licences: both of them are buyers on the allowance market.

We assume discrete time and an infinite horizon. In each period firms can face one of the following situations:

- State 0: they both use the least efficient technology k_L, and an emission reducing technology becomes available for purchase at a (sunk) cost C;
- State 1: only one of them has switched to the most efficient technology k_H, implying that it cannot return to the previous one, neither can it innovate further, whereas the other firm can still make its investment decision; we assume that firm 1 is the innovating firm[2];
- State 2: both firms have adopted the best emissions-output ratio k_H.

In what follows the superscript $s = 0,1,2$ refers to the state, the subscript $i = 1,2$ refers to the firm and the subscript $j = H, L$ refers to the emissions-output ratio. We first analyse the firms' profit maximising decisions in each period and the determination of the equilibrium allowance price, q, for each possible combination of production technologies, disregarding the technology choice.

When both firms use the same technology, they are identical in all respects and the solution of the model is similar to the standard Cournot duopoly case. Let us

[1] This condition is imposed to assure existence and uniqueness of equilibrium. If this condition is not met, the demand for permits of the most efficient firm may be upward sloping.

[2] We use the terms "innovation", "adoption" and "investment" as equivalent, meaning that a firm changes its technology paying a given cost. More rigorously, the term "innovation" is used to indicate the result of R&D, whereas "adoption" is used to denote the switch to a new technology available on the market at some cost (Tirole 1998).

consider firm 1's maximisation problem in state 0 (the analysis of state 2 is analogous):

$$\max_{y_1}[a - b(y_1 + y_2)]y_1 - cy_1 - qk_L y_1 \tag{1}$$

Solving both firms' maximisation problems yields the equilibrium output levels and the demands for permits, implied by the adopted technology[3]. The equilibrium allowance price, q, is derived endogenously by imposing a market clearing condition[4]. In each state, q will be positive only if the permit supply is lower than some strictly positive values, denoted as α, $\hat{\alpha}$ and $\overline{\alpha}$ in state 0, 1 and 2, respectively, with $\overline{\alpha} < \hat{\alpha} < \alpha$ [5]. Hence, there are emission caps for which the adoption of a cleaner technology by one or both firms drives the initially positive price to zero.

Furthermore, there can be cases in which the innovation by one firm drives the other firm out of the market.

Proposition 1. There exists a critical value of the permit supply below which the innovating firm becomes the unique buyer on the permit market and a monopolist on the output market (*drastic* innovation), otherwise both firms stay on the market (*nondrastic* innovation) and make strictly positive profits. We denote this value as $\theta > 0$ [6].

If $E \leq \theta$, the equilibrium allowance price in state 1 is too high for the non-innovating firm, implying that its permit demand drops to zero. We will consider only the cases of non-drastic innovation and strictly positive permit prices. Thus, we will assume that the condition $\theta < E < \overline{\alpha}$ holds throughout this paper.

---

[3] The equilibrium output levels are identical, and equal to $y_1^0(q) = y_2^0(q) = (a - c - qk_L)/3b$, where the term qk_L is an extra bit of the marginal production cost. If only firm 1 innovates, we have $y_1^1(q) = [a - c - q(2k_H - k_L)]/3b$ and $y_2^1(q) = [a - c - q(2k_L - k_H)]/3b$. If everything stays equal, reducing the emissions-output ratio makes both firms' output levels increase.

[4] We can think of an auctioneer who sells the permit at the market clearing price: given a price (previous period price or a price announced by the auctioneer), firms believe they can purchase any amount of permits at that price and announce the quantities they will put on the output market; the auctioneer anticipates the factor demands for permits and announces the market clearing price (Requate 1998).

[5] The thresholds are $\alpha = [2k_L(a-c)]/3b$, $\hat{\alpha} = [(k_L + k_H)(a-c)]/3b$ and $\overline{\alpha} = [2k_H(a-c)]/3b$ in state 0, 1 and 2, respectively. Notice that if the permit price is null, both firms produce an output quantity equal to $(a-c)/3b$. This is the standard result of a Cournot game with identical firms and linear output demand. It is the levels of output each firm would produce if there were no environmental regulation (Business As Usual, BAU). α and $\overline{\alpha}$ are the level of emissions associated to the BAU level of output, when the emissions-output ratios are k_L and k_H respectively.

[6] It is $\theta = [k_H(k_L - k_H)(a-c)]/(2k_L - k_H)b$.

Proposition 2. When both firms adopt the cleanest emissions-output ratio k_H, there exists a critical value $\overline{\beta} > 0$ of the emission cap for which it is $q^2 < q^0$, if $\overline{\beta} < E < \overline{\alpha}$ and it is $q^2 > q^0$, if $0 < E < \overline{\beta}$.

Proposition 3. When only one firm adopts the cleanest technology k_H, it is always $q^1 < q^0$, if $k_L/2 < k_H \leq 2k_L/3$. If $k_H > 2k_L/3$, there exists a critical value $\hat{\beta} > 0$ of the permit supply for which it is $q^1 < q^0$, if $\hat{\beta} < E < \overline{\alpha}$ and it is $q^1 > q^0$, if $\theta < E < \hat{\beta}$, with $\hat{\beta} < \overline{\beta}$[7].

The equilibrium allowance price affects output and profit levels. Let us denote firm i's profits in the sth state with π_i^s, for $i = 1,2$ and $s = 0,1,2$.

Proposition 4. When both firms adopt the same emissions-output ratio it is $\pi_1^0 = \pi_2^0 = \pi^0$ and $\pi_1^2 = \pi_2^2 = \pi^2$ in state 0 and 2, respectively, with $\pi^2 > \pi^0$[8].

Proposition 5. There exists a critical value $\varepsilon > 0$ of E, with $\hat{\beta} < \varepsilon < \overline{\alpha}$[9] such that, in state 1,

- if $\theta < E < \varepsilon$, it is $\pi_2^1 < \pi^0$ and $\pi^2 < \pi_1^1$: innovation by one firm hurts the other firm, whether the firm that adopts k_H is the first or the second one to do so;
- if $\varepsilon < E < \overline{\alpha}$, it is $\pi_2^1 > \pi^0$ and $\pi^2 > \pi_1^1$: innovation by one firm benefits the other, whether the firm that adopts k_H is the first or the second one to do so.

Proposition 6. From propositions 4 and 5, there are two possible rankings of per-period profits:

- if $E < \varepsilon$ then $\pi_1^1 > \pi^2 > \pi^0 > \pi_2^1$ and $(\pi_1^1 - \pi^0) > (\pi^2 - \pi_2^1)$;
- if $E > \varepsilon$ then $\pi^2 > \pi_1^1 > \pi_2^1 > \pi^0$ and $(\pi_1^1 - \pi^0) < (\pi^2 - \pi_2^1)$.

Taking the permit price as fixed, the adoption of a cleaner technology makes increasing a firm's output level more convenient. Whether the firm is the leader or the follower or both firms innovate simultaneously, innovation cuts overall production costs by reducing permit requirement and expenditure on licence purchase per unit of output: given the price, a reduction in the emissions-output ratio is equivalent to a decrease in the marginal cost. Hence, when only one firm adopts the new technology, innovation enables it to increase its market share and profits

---

[7] We have $\overline{\beta} = 2k_H k_L (a-c)/3b(k_H + k_L)$ and $\hat{\beta} = k_L(2k_H - k_L)(a-c)/3bk_H$.

[8] Firms' identical equilibrium profits are $\pi^0 = b(E/2k_L)^2$ and $\pi^2 = b(E/2k_H)^2$, when both firms use k_L and k, respectively. Profits are constrained by the environmental regulation. In state 1, profits are $\pi_1^1 = [(2k_H - k_L)bE + k_L(a-c)(k_L - k_H)]^2/4b(k_H^2 - k_H k_L + k_L^2)^2$ and $\pi_2^1 = [(2k_L - k_H)bE + k_H(a-c)(k_H - k_L)]^2/4b(k_H^2 - k_H k_L + k_L^2)^2$.

[9] It is $\varepsilon = k_H k_L (a-c)/(k_H + k_L)b$.

at the expense of its rival. Despite the higher average abatement by the innovating firm, the allowance price may or may not decrease, depending on the emission cap and on the technology combination (k_L,k_H): the smaller the permit supply is, the bigger the innovation effort must be in order to determine a sufficient decrease in permit requirement and thus a fall in the permit price. Consequently, the non-innovating firm may be advantaged by the other firm's investment: the lower licence price may enable it to increase its production and profits, without investing in a cleaner technology. However, firm 2 can exploit the price drop only if this drop is sufficiently large. This happens for $E > \varepsilon$. On the other hand, if firm 2 follows and adopts the new technology (state 2), its output level is always higher than in state 1, whereas firm 1's production may go up or down. When imitation occurs, the firm that has innovated first may benefit from imitation and may increase its output (and profits) in state 2, if the permit price decreases sufficiently[10]. The condition for this to occur is again $E > \varepsilon$. Finally, if $E < \varepsilon$, the increment in per-period profits a firm can get by adopting first is higher than the increment in profits it can get by being second. The opposite occurs if the permit supply is sufficiently large.

3 The game

In analysing the strategic investment decision, we assume discrete time and infinite horizon. Let us consider the following game, with two players, firm 1 and firm 2, choosing between two possible actions in each period, "innovate" and "not innovate". In every period *t*, each firm may adopt the best available technology k_H if it has not already done so, incurring in a once-for-all cost *C*. If neither firm has innovated at time *t*, each one can still decide to do so in the next period, facing the same decision problem at time $t+1$. Hence, the number of periods the game will last is potentially infinite: the game ends when both firms have changed their technology[11]. In making their choice in each period, firms must consider their profit flow over their infinite life, discounted with a discount factor ρ, $0 < \rho < 1$. Firm *i*'s payoff associated to a given combination of firms' actions is its respective lifetime profit as viewed from the beginning of the period[12].

If both firms decide to adopt technology k_H, they get a profit flow of π^2 for all the remaining periods, net of the investment cost:

[10] We will say that a firm imitates when it is the second to adopt the new technology (follower).

[11] The game ends since in this case players have no strategic decision to take, but they continue to get their per-period profits for the rest of their life.

[12] Both firms know that if they played the game at an arbitrary time *t*, it would be because they will not have innovated before time *t*. Hence they both will have accumulated $\pi^0 \sum_{z=0}^{t-1} \rho^z$ and they will keep this profit for the periods thereafter, whichever decision they take at time *t*. This part of firms' payoff will not affect their choices.

$$\Pi^2 = \pi^2/(1-\rho) - C \tag{2}$$

If one firm has innovated (firm 1), the other one (firm 2) must decide whether or not to follow in the next period. Firm 2's action is chosen solving a single firm optimisation problem. Considering two arbitrary periods, t and $t+1$, firm 2 will adopt the best technology in period t rather than in the next one if the extra-profit it would get is higher than the cost savings due to one period delay[13], i.e. if

$$(1-\rho)C < \pi^2 - \pi_2^1 \tag{3}$$

Proposition 7. Let us denote $(\pi^2 - \pi_2^1)/(1-\rho)$ as $\overline{C}$. If firm 1 innovates, then firm 2 either adopts the new technology next period or never, depending on the investment cost. Two cases are possible:

1. *quick imitation*: if $C < \overline{C}$, firm 2 innovates immediately after the other one has; the lifetime profits of the leader and of the follower are Eqs. (4) and (5), respectively, since state 1 will last only one period and firms pass to state 2

$$\Pi_1^1 = \pi_1^1 - C + \rho\pi^2/(1-\rho) \tag{4}$$

$$\Pi_2^1 = \pi_2^1 + \rho\pi^2/(1-\rho) - \rho C \tag{5}$$

2. *infinite delay*: if $C > \overline{C}$, firm 2 always delays adoption (never adopts); the lifetime profits of the leader and of the follower are Eqs. (6) and (7), respectively, since state 1 will last forever.

$$\Pi_1^1 = \pi_1^1/(1-\rho) - C \tag{6}$$

$$\Pi_2^1 = \pi_2^1/(1-\rho) \tag{7}$$

In each period the payoff Π^0 that each firm gets if no one innovates, depends on what firms will do in the following period(s).

Since we solved the part of the game involving only one firm's decision, we can represent this *innovation game* as a symmetric game with two identical firms-players and two actions ("innovate" and "not innovate") in each period, and with payoffs equal to lifetime profits, net of the investment cost. There are two games, one for the quick imitation case and one for the infinite delay case. We can solve the game looking at each period in which neither player has innovated yet. Our

[13] Considering two arbitrary periods, t and t+1, firm 2 will adopt the best technology in period t rather than in the next one if it gets higher lifetime profits by doing so, i.e. if $\pi_2^1\sum_{z=0}^{t-1}\rho^z + \rho^t[\pi^2/(1-\rho)-C] > \pi_2^1\sum_{z=0}^{t}\rho^z + \rho^{t+1}[\pi^2/(1-\rho)-C]$ which is true when $(1-\rho)C < \pi^2 - \pi_2^1$.

aim is to solve the game for a stationary Nash equilibrium, possibly involving mixed strategies[14].

A mixed strategy equilibrium is a possible solution to this game in the case of both quick imitation and infinite delay[15]. Whether or not this equilibrium actually arises depends crucially on the parameters of the model, since they affect both the per-period and lifetime profits. We will focus on the effects of the emission cap and of the investment cost on the equilibrium, since these are the parameters that the environmental regulator might adjust to speed up the diffusion of cleaner technologies. In order to determine the critical values of these parameters, we will solve the game assuming mixed strategies and then we will see for which values of E and C this assumption is correct.

Let us denote with p_i, $i = 1,2$, firm i's probability of not innovating. Since firms are identical, it is $p_1 = p_2 = p$. We use the expected payoff equating method to find each firm's optimal mixed strategy $(p^*, 1 - p^*)$. Firm i's expected payoff of innovating and of not innovating are Eqs. (8) and (9), respectively[16].

$$V_I = p\Pi_1^1 + (1 - p\Pi^2) \tag{8}$$

$$V_N = [p\pi^0 + (1 - p)\Pi_2^1]/(1 - \rho p) \tag{9}$$

Both V_N and V_I are functions of p, and their exact forms depend on whether there is quick imitation or infinite delay.

Let us define $\hat{C}$ as the minimum value of the investment cost required to have a mixed strategy equilibrium in the quick imitation case, where

---

[14] If such a symmetric stationary equilibrium exists, it is subgame perfect, since all subgames in which both firms can make a choice have the same structure and payoffs.

[15] In the quick imitation case, it is $\Pi^2 > \Pi_2^1$ by propositions 6 and 7. If it is also $\Pi_1^1 > \Pi^0$, "not innovate" is a strictly dominated strategy for both firms, and the unique possible equilibrium is a symmetric Nash equilibrium (innovate, innovate): both firms innovate straight away in the first period and the game ends. If it is $\Pi_1^1 < \Pi^0$, neither firm has a strictly dominated strategy and there are three equilibria: two symmetric Nash equilibria in pure strategies, in which both firms either innovate straight away or never, and one symmetric equilibrium in mixed strategies. In the infinite delay case, it is $\Pi^2 < \Pi_2^1$ by propositions 6 and 7. If it is also $\Pi_1^1 < \Pi^0$, "innovate" is a strictly dominated strategy for both firms, and the unique possible equilibrium is a symmetric Nash equilibrium (not innovate, not innovate). If it is $\Pi_1^1 > \Pi^0$, neither firm has a strictly dominated strategy and there are three equilibria: two asymmetric Nash equilibria in pure strategies, in which a firm innovates straight away and the other never does, and one symmetric equilibrium in mixed strategies.

[16] The expression in the text is determined solving $V_N = p(\pi^0 + \rho V_N) + (1 - p)\Pi_2^1$ for V_N: if one firm does not innovate, which occurs with probability p, the other firm gets π^0 for the current period and the expected value of not innovating for the future periods, discounted with the discount factor ρ. $V_N = V_I$ yields a quadratic expression in p.

$$\hat{C} = \left[\pi_1^1 - \pi^0 - \rho\left(\pi_1^1 - \pi^2\right)\right] / (1-\rho) \tag{10}$$

Proposition 8. In the quick imitation case, the following equilibria arise:

- if $\hat{C} < C < \overline{C}$, there are a symmetric mixed strategy equilibrium $((p^*, 1-p^*), (p^*, 1-p^*))$ and two symmetric pure strategy Nash equilibria (not innovate, not innovate) and (innovate, innovate)[17];
- if $C < \hat{C}$, there is a pure strategy equilibrium (innovate, innovate) in every period.

Proposition 8 implies that each firm innovates for sure if the benefit of waiting for one period (savings in the investment cost obtained by delaying of one period the adoption of k_H) is lower than the forgone profits of being first (extra-profit of being the first to adopt the new technology, net of the profit decrease caused by imitation)[18]. On the other hand, $\hat{C} < C < \overline{C}$ implies that if the cost savings of waiting is higher than the extra-profits of being first but lower than the extra-profits of being second (from Eqs.(3) and (10)), each firm can do better by waiting for the other to invest and investing with one period delay. Therefore, the mixed strategy equilibrium arises when firms are afraid of failing in coordinating themselves on the Pareto dominant outcome (not innovate, not innovate)[19]. There is a discontinuity at $C = \hat{C}$, as it is evident from the example in Fig. 1[20].

Proposition 9. In the quick imitation case, when a mixed strategy equilibrium arises, the probability of not innovating p^* is decreasing in the investment cost C, other things being equal.

Therefore, it is more likely to adopt the best technology when the investment cost gets higher. This result seems counterintuitive. However, when the invest-

---

[17] We leave the case $C = \overline{C}$ undetermined, since $\overline{C}$ is the value of C for which the non-innovating firm should be indifferent between adopting the new technology in the next period or never.

[18] For $C < \hat{C}$ the probability is not defined and we can conclude that there is a pure strategy equilibrium (innovate, innovate): for each firm it is $V_N < V_I$ either if the other firm innovates for sure or if it does not. If one firm adopts k_H with certainty, for the other firm we have $V_N = \Pi_2^1$ and $V_I = \Pi^2$, with $\Pi^2 > \Pi_2^1$. If one firm does not adopt the new technology for sure, for the other firm we have $V_N = \pi^0/(1-\rho)$ and $V_I = \Pi_1^1$, with $\Pi_1^1 > \pi^0/(1-\rho)$ for $C < \hat{C}$.

[19] Indeed, each firm would be better off if they both continue to use the old technology forever (since it is $\pi^0/(1-\rho) > \Pi^2$), but if one adopts k_H the other one has to follow in order to limit its loss. The mixed strategy outcome is worse than the pure strategy (not innovate, not innovate), but it is better than the pure strategy (innovate, innovate).

[20] For $C = \hat{C}$ the solution to $V_N = V_I$ is $p^* = 1$, implying that for this value of C there is a pure strategy equilibrium (not innovate, not innovate) in every period.

ment cost is higher, both firms have a greater incentive to wait and see whether the other invests: a firm must ("threaten" to) innovate with a higher probability in order to leave its rival indifferent between innovating or not.

Changes in E affect the critical cost range: it can be $\hat{C} < C < \overline{C}$ only for $\overline{\varepsilon} < E < \varepsilon$[21].

Proposition 10. In the quick imitation case, the following equilibria are feasible:

- if $\overline{\varepsilon} < E < \varepsilon$, there are a symmetric mixed strategy equilibrium $((p^*, 1-p^*), (p^*, 1-p^*))$ and two pure strategy Nash equilibria (not innovate, not innovate) and (innovate, innovate), provided that $\hat{C} < C < \overline{C}$[22];
- if $E < \overline{\varepsilon}$ or $E > \varepsilon$, there is a pure strategy equilibrium (innovate, innovate) in every period.

For E outside of the range $]\overline{\varepsilon}, \varepsilon[$, the relative magnitude of per-period profits is such that the net benefit of being first is always greater than the benefit of being second[23].

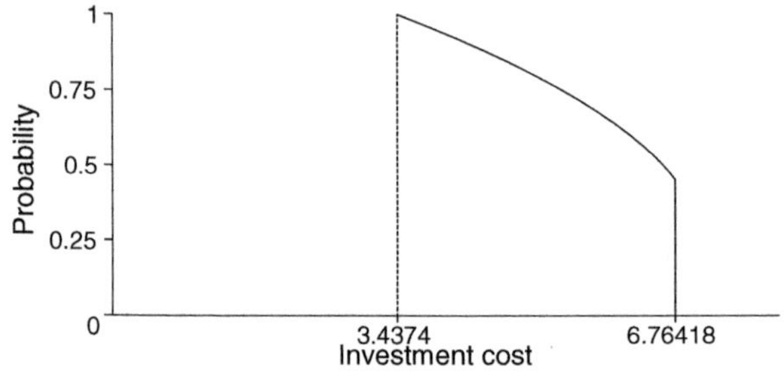

Fig. 1. Quick imitation: optimal probability of not innovating as a function of the investment cost, for fixed E and technology parameters. In this example, $\hat{C} = 3.44$ and $\overline{C} = 6.76$

---

21 $\overline{\varepsilon} = \left(k_L k_H (k_H - k_L)(a-c)\left((\rho-1)k_L^2 - k_H^2\right)\right) / \left(k_L^2\left((k_L - k_H)^2 + 2k_H^2\right)\rho + (k_L + k_H)(k_H - k_L)^3\right)$.
$\overline{\varepsilon}$ is higher or lower than θ depending on the discount factor and on the technology parameters.

22 Recalling proposition 6, if $\overline{\varepsilon} < E < \varepsilon$, the decrease in per-period profit caused by being followed partially compensates the extra-profit of innovating first, so that a firm can get a higher increment of the per-period profit by adopting k_H with one period delay: there exist values of the investment cost $C < \overline{C}$ such that the cost savings of waiting is higher than the extra-profit of being first, but lower than the extra-profit of being second.

23 Indeed, from proposition 6, if $E < \overline{\varepsilon}$ (the emission cap is sufficiently strict), the extra-profit of innovating first net of the decrease in per-period profit caused by being followed is higher than the extra-profit of being second. If $E > \varepsilon$ (the permit supply is sufficiently large), the extra-profit of being second is higher than the extra-profit of being first; however, imitation leads to an increase in the per-period profit of the first adopter, so that being first is still better.

Proposition 11. In the quick imitation case, when a mixed strategy equilibrium arises, the probability of not innovating p^* is increasing in the emission cap E, other things being equal.

When the permit supply increases, the benefit of adopting the new technology increases slightly more than the expected value of not innovating; hence, the incentive to adopt increases accordingly, even though the best outcome for both firms is still associated to both staying with the old technology. A firm must thus ("promise") not to innovate with a higher probability in order to leave its rival indifferent between innovating or not.

We will now consider the infinite delay case. Let us denote $\widetilde{C}$ the maximum value of the investment cost for which a mixed strategy equilibrium arises in the infinite delay case, where

$$\widetilde{C} = (\pi_1^1 - \pi^0)/(1-\rho) \tag{11}$$

Proposition 12. In the infinite delay case, the following equilibria may arise:

- if $\overline{C} < C < \widetilde{C}$, there are a symmetric mixed strategy equilibrium $((p^*,1-p^*),(p^*,1-p^*))$ and two asymmetric pure strategy Nash equilibria (innovate, not innovate) and (not innovate, innovate);
- if $C > \widetilde{C}$, there is a pure strategy equilibrium (not innovate, not innovate)[24].

Proposition 12 implies that each firm does not innovate if the benefits of waiting for one period are higher than the forgone profits of being first. Therefore, the condition $\overline{C} < C < \widetilde{C}$ implies that if the cost savings of waiting are lower than the extra-profits of being first, provided that the investment cost is sufficiently high to avoid imitation (recall proposition 7), each firm can do better by trying to be the first and only one to adopt the new technology, i.e. by *preempting* its rival. Each firm can find it convenient to play a mixed strategy in order to confuse its rival and to be able to reach the innovator's payoff. There is a discontinuity at $C = \widetilde{C}$, as it is evident from the example in Fig. 2.

[24] Indeed, for some values of $C > \widetilde{C}$ the probability is not defined and we can easily conclude that there is a pure strategy equilibrium (not innovate, not innovate). For some other values of $C > \widetilde{C}$, there are two solutions to the equation $V_N = V_I$ which implies that there is no mixed strategy equilibrium. However, for $C > \widetilde{C}$ it is always $V_N(0) > V_I(0)$ and $V_N(1) > V_I(1)$: there is a unique pure strategy equilibrium (not innovate, not innovate). If one firm adopts k_H with certainty, for the other firm we have $V_N = \Pi_2^1$ and $V_I = \Pi^2$, with $\Pi_2^1 > \Pi^2$. If one firm does not adopt the new technology for sure, for the other firm we have $V_N = \pi^0/(1-\rho)$ and $V_I = \Pi_1^1$, with $\Pi_1^1 < \pi^0/(1-\rho)$ for $C > \widetilde{C}$.

Proposition 13. In the infinite delay case, when a mixed strategy equilibrium arises, the probability of not innovating p^* is increasing in the investment cost *C*, other things being equal.

If the investment cost increases, the advantage of being the first to adopt the new technology decreases; hence, both the incentive to invest for preemption and the probability of innovating decrease accordingly.

Proposition 14. In the infinite delay case, the following equilibria are feasible:

- if $E < \varepsilon$, there are a symmetric mixed strategy equilibrium $((p^*,1-p^*),(p^*,1-p^*))$ and two asymmetric pure strategy Nash equilibria (innovate, not innovate), provided that it is also $\overline{C} < C < \widetilde{C}$ [25];
- if $E > \varepsilon$, there is a pure strategy equilibrium (not innovate, not innovate).

As before, changes in the emission cap affect the critical cost range. It can be $\overline{C} < C < \widetilde{C}$ only for $\theta < E < \varepsilon$. For $E > \varepsilon$, the relative magnitude of per-period profits is such that the benefit of being first is always lower than the benefit of being second.

Therefore, the mixed strategy equilibrium arises when the switching cost and the emission cap are such that the payoff profiles imply a *preemption game*: each firm gets the highest payoff if it is the only one to innovate, and the lowest payoff if both do so.

Proposition 15. In the infinite delay case, when a mixed strategy equilibrium arises, the probability of not innovating p^* is decreasing in the emission cap *E*, other things being equal.

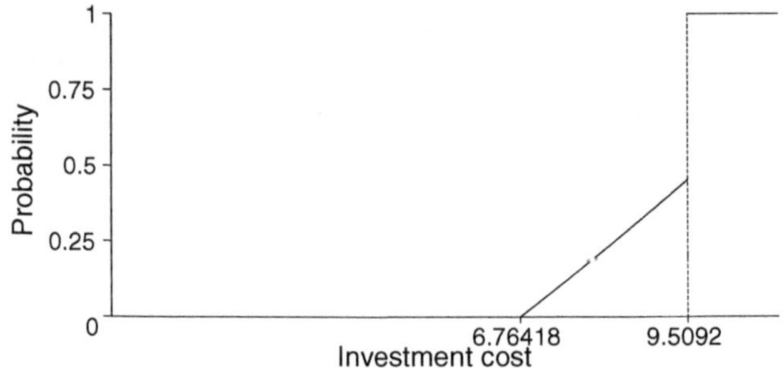

Fig. 2. Infinite delay: optimal probability of not innovating as a function of the investment cost, for fixed *E* and technology parameters. In this example, $\overline{C} = 6.76$ and $\widetilde{C} = 9.51$

[25] Recalling proposition 6, if $E < \varepsilon$, there exist values of the investment cost $C < \widetilde{C}$ such that the cost savings of waiting for one period are lower than the forgone extra-profit of being first, provided that the investment cost is sufficiently high to avoid imitation: a mixed strategy equilibrium is feasible.

This may appear counterintuitive. However, an increase in the permit supply, provided that it is not excessive (such that the condition $E < \varepsilon$ is met) makes the innovating firm's profit higher: the incentive to adopt first becomes stronger and then the probability of innovating goes up.

4 The experiment

We designed and implemented a computerized experiment that replicated the innovation game described in the previous section[26]. The decision problem involves two subjects, representing the two duopolistic firms. Subjects play a dynamic game that ends after a random number of periods. In the initial round of the game, the two players are both in a state denoted as "state A", which is associated to a symmetric combination of payoffs. The players have to decide whether to remain in "state A" or to switch to a state denoted as "state B". Once a player has decided to change state, he/she cannot return to the previous one. The decision of moving from A to B corresponds to the decision of innovating[27].

Subjects are paid the payoff corresponding to the combination of states they are in when the game finishes. When deciding to change state, players incur a once-for-all cost, which is deducted from their payoff in the round in which a subject decides to move from A to B[28]. The number of rounds is randomly determined and subjects do not know which round is going to be the final one. Each player's decision problem ends when he/she has already moved from A to B, even though the game continues until the randomly determined number of rounds is over.

The decision problem faced in the experiment and that faced in the theory are strategically equivalent. In the experiment, there is a probability λ that the game ends at each round, and a probability $(1-\lambda)$ that the game will go on to the next period. Let Π_t denote the payoff that the subject would get if the game finished in period t. Then, considering a stream of payoffs $\Pi_1, \Pi_2, \ldots, \Pi_n$, each player ex-

---

[26] The experiment was computerized using the Z-Tree software developed at the University of Zurich by Urs Fischbacher and was run at the Laboratory of the Centre for Experimental Economics (EXEC) at the University of York.

[27] Hence, "state A" corresponds to the old technology and "state B" corresponds to the new technology. We leave the setting as abstract as possible, in order to avoid any influence on subjects.

[28] Only if this round happens to be the final one, the subject that has moved will get a lower payoff (i.e. it will pay the once-for-all-cost), otherwise the subject will get the "full" payoff. In the instructions, all the payoffs are referred to as being potential, unless they are the payoffs corresponding to the state the subjects are in when the game ends (actual payoffs). It is not explicitly stated that there is a switching cost; however, this is evident from the payoff structure.

pected payoff is[29] $\lambda\sum_{t=1}^{\infty}(1-\lambda)^{t-1}\Pi_t$. Therefore, a discount factor ρ in the theory is equivalent to a probability $(1-\lambda)$ that at each round the game continues to the next one. The only difference between the theoretical and the experimental decision problem is the scaling factor λ, which does not affect decision. In the experiment, the payoff functions and the probability $\lambda = 0.1$ are common knowledge. At the end of any round each subject is told his/her rival's decision.

We implemented 4 treatments, differing in terms of the investment cost and/or the emission cap, as summarised in Table 1.

Table 1. Treatments

	High $E = 22$	Low $E = 18.5$
Quick imitation $C = 1000$	T1	T2
Infinite delay $C = 1500$	T3	T4

Each treatment requires 9 subjects: 8 subjects, divided into 4 pairs, play the game 5 times, each of which with a different opponent. By making the subjects playing several times we aim at collecting a sufficient number of observations, whereas the "absolute stranger" matching should control for correlation[30]. For the scope of the data analysis, we make the assumption that the 5 plays of the game in each treatment are independent.

The ninth subject is the "Round Determinator" (Allsopp 2002), who is in charge of determining the number of rounds each of the 5 games lasts, according to a random mechanism described in detail in the instructions[31]. This number is not revealed to the other participants in the experiment.

The "Round Determinator" is elected by other subjects, who can check that he/she has performed the job described[32].

The experiment consisted of 4 sessions, one for each treatment. Participants in the experiment were all undergraduate students, except for two postgraduate students. There were no trial periods. At the end of each session, subjects were paid in cash.

[29] As viewed from period 1 the probability of the game stopping in period 1 (and getting Π_1) is λ, the probability of it stopping in period 2 (and getting Π_2) is $\lambda(1-\lambda)$, the probability of it stopping in period t is $\lambda(1-\lambda)^{t-1}$.

[30] The "absolute stranger" matching implies that at any game each subject plays with a new partner, who is not known, and two subjects would never play against each other more than once. However, the 5 games will always be correlated, because of the learning effect.

[31] The instructions of the experiment and the payoff tables of the 4 treatments are available from the author.

[32] This mechanism has the advantages of being transparent and of increasing the "trust degree" towards the mechanism itself (because the number of rounds is determined by someone other than the experimenter). The risk associated to subjects' unfamiliarity with probabilities remains.

In all treatments, subjects tended to quickly follow the player that had already moved: most players followed immediately or waited one round to switch to state B. The same behaviour was observed in the infinite delay treatments, despite the negative expected benefits from switching. Possible explanations might be that there was a herd behaviour component or that subjects competed between each other: the players that had not moved were unwilling to let their rivals earn more[33]. Ten players over all the treatments always chose to move to state B when both subjects were in state A. We regard these subjects as playing a pure strategy "move to B (innovate)" in any round and denote them as "type B subjects"[34]. On the other hand, there are no subjects that always chose to remain in state A, when both players in the pair were in A. Hence, we conclude that no subjects played a pure strategy "not move to B (not innovate)".

We estimate the probability of not moving to B over the whole sample. The sample size of each treatment is given by the total number of rounds in which both subjects were in state A, i.e. both could make a decision. From each treatment, we estimate the probability that at least one player decides to remain in state A when both players are in A. Let us denote these probabilities as $\hat{p}_\tau$, $\tau = 1,\dots,4$. This is the sample proportion of the number of times at least one player's choice is A when both are in A. Given the sample sizes of our treatments, $n_1 = 40$, $n_2 = 54$, $n_3 = 54$, $n_4 = 40$, we can approximate the sample distribution of each proportion as

$$\hat{p}_\tau \sim N\left(p_\tau^*, p_\tau^*(1-p_\tau^*)\Big/ n_\tau\right) \tag{12}$$

where p_τ^*, $\tau = 1,\dots,4$, are the theoretical probabilities. The null hypotheses we want to test are $H_0 : \hat{p}_\tau = p_\tau^*$, $\tau = 1,\dots,4$. The rejection rule for a two-tailed hypothesis test will lead us to reject H_0 when $|z_\tau| > c_\gamma$, where z_τ is the usual z-score, c_γ is the $(1-\gamma)\%$ critical value, and γ is the significance level[35].

Analogously, we estimate the probability of not moving to B over the "mixed strategy sample", i.e. excluding the "type B subjects" from the sample size of each treatment. The players that are not type B are assumed to have played a mixed strategy. The total number of rounds in which both subjects (of this type) were in state A represents our sample sizes: $n_1 = 25$, $n_2 = 44$, $n_3 = 49$, $n_4 = 20$. For each treatment, let us denote the probability that at least one (mixed strategy) player decides to remain in state A when both (mixed strategy) players are in A as $\widehat{p}_\tau$, $\tau = 1,\dots,4$. As before, $\widehat{p}_\tau$ is the sample proportion of the number of times at

---

[33] Indeed, the risk of being caught by the end of the game was low and for a moderately risk loving person, this competitive (or envy) component might affect behaviour.

[34] To be more precise, these subjects could also have adopted a mixed strategy, but in those plays it happened that they picked the B decision. We do not have enough observations to be reasonably sure that these players adopted a pure strategy.

[35] The z-score is $z_\tau = (\hat{p}_\tau - p_\tau^*)\Big/\sqrt{p_\tau^*(1-p_\tau^*)/n_\tau}$

least one (mixed strategy) player's choice is A when both are in A. The null hypotheses are $H_0 : \hat{p}_\tau = p^*_\tau$, $\tau = 1,...,4$. The results are summarised in Tables 2 and 3, where "*" and "**" indicate that we reject the null hypothesis that the two probabilities are not statistically different at 5% and 1% significance level, respectively. If the cell reports no "star", then for these two significance levels we cannot reject H_0.

Table 2. Estimates of the probability of not moving to B over the whole sample

Treatment	n	$\hat{p}_\tau$	p^*	z	Result
1	40	0.25	0.84072	-10.2095	**
2	54	0.555556	0.66667	- 1.73211	
3	54	0.611111	0.24572	6.236798	**
4	40	0.25	0.40506	- 1.99767	*

Table 3. Estimates of the probability of not moving to B over the "mixed strategy sample"

Treatment	n	$\hat{p}_\tau$	p^*	z	Result
1	25	0.4	0.84072	-6.0218	**
2	44	0.681818	0.66667	0.213152	
3	49	0.673469	0.24572	6.954958	**
4	20	0.5	0.40506	0.86494	

Comparing the estimated probabilities $\hat{p}_\tau$ and $\hat{p}_\tau$, $\tau = 1,...,4$, we get the same results: we cannot reject the null hypothesis only in treatments 2 and 4. As for the comparison between quick imitation and infinite delay treatments, for the same permit supply, the probability of not innovating should be lower in treatments 3 and 4, but the estimated probability follows this trend only for treatments 2 and 4, corresponding to the low permit supply. As regards the comparison between high and low permit supply treatments, for the same cost, the probability of not innovating should be increasing in E in the quick imitation treatments and decreasing in E in the infinite delay ones, but the estimated probability does not fit this trend in any treatment.

The most frequent outcome was (State B, State B) in all the treatments. Players appeared to be "attracted" by changing state for reasons that are out of the strategic structure.

Both players in state B (in any period) is the Pareto inferior pure strategy equilibrium in the quick imitation case; hence, we conclude by saying that subjects failed in coordinating on the Pareto superior equilibrium in each period. On the other hand, both players in state B is not a pure strategy equilibrium in the infinite delay case, since in this case pure strategy equilibria are asymmetric. However, (B,B) is the most likely outcome when subjects play a symmetric mixed strategy[36].

[36] In this case the probability of (B,B) arising in any period is $(1-p^*)(1-p^*)$, equal to 0.569 and 0.354 in treatment 3 and 4, respectively.

5 Concluding remarks

In a non-competitive setting, firms' investment decisions have a strategic component. We modelled this strategic interaction as a two-firm dynamic game, and we solved it looking for symmetric stationary equilibria. We have seen that the stationary equilibria to this game crucially depend on both the cost of switching to the cleanest technology available and the emission cap. Given the directions in which the investment cost and the emission cap affect the feasible stationary equilibria, an environmental regulator aiming at speeding up the diffusion of an environmental friendly technology has to adjust these variables in an appropriate manner.

It is intuitive that, given the emission cap, subsidising the duopolists in order to lower the investment cost will induce both firms to adopt the superior technology. However, we can expect this diffusion outcome only if the cost of switching is sufficiently low (such that the cost savings from delaying adoption are lower than the net benefits from being the leader). For some higher values of the investment cost, diffusion is only one of the possible outcomes. Given the permits supply, the regulator should try to push the investment cost as close as possible to the critical value that splits the quick imitation and the infinite delay cases. Below this critical value the savings from delaying imitation are lower than the gains from following; above this critical value the opposite occurs. The closer the switching cost is to this value, the higher the probability that any one firm innovates in a period. If the investment cost is above this critical value, the regulator can make joint adoption more likely by subsidising firms. Alternatively, given the cost, it can increase the permit supply! In this case the objective is to make innovation as profitable as possible, increasing both firms preemption incentive. On the contrary, if the investment cost is below this critical value, the regulator may expect a quick imitation and can push diffusion by taxing the investment! Alternatively, given the cost, it can decrease the permit supply. In this case the objective is to make firms afraid of being preempted, so that each firm is not induced to adopt because it will be better off by doing so, but it will be tempted to invest in order to avoid to be worse off if its rival moves first. In other words, the regulator's objective is to make it more difficult for firms to coordinate on the non adoption outcome (without communication).

When the cost and emission cap values are such that diffusion is only one of the possible stationary equilibria, the diffusion is not the dominant outcome. Game theory does not help us in predicting which outcome will actually occur. In particular, not necessarily firms will manage to coordinate on the Pareto superior one.

Our pilot experiment provided interesting initial results as well as revealing some points in the experimental design and parameter set that could be worthy to revise. The mostly observed equilibrium was not the Pareto dominant one. This result is consistent with other experimental investigations, which support the conclusion that a Pareto dominant equilibrium does not necessarily represent a focal one (Ochs 1995). The experiment suggests that when quick imitation is expected, firms may fail to coordinate on not innovating, since for each of them the fear of the other innovating first apparently prevails. In the infinite delay case, when a preemptive equilibrium is the dominant one, the preempted firm does not apparently leave the other firm maintain this advantage position, and tend to imitate, even though for a firm (maximising its expected payoff) it would be better not to do so.

In interpreting these initial results we must pay special attention; in particular, we should consider that subjects might not have completely understood the round determination mechanism or they might not have believed it, so that this mechanism failed in inducing discounting; moreover, the probability of stopping the play at each round could have been too low and the difference between payoffs corresponding to different outcomes is not striking; finally, it is difficult to implement a mixed strategy.

If the experimental results proved to be robust to changes in parameter set and further experimental sessions, we could conclude that (innovate, innovate) is the most likely outcome of this innovation game when the permit supply and the investment cost are such that a stationary symmetric mixed strategy equilibrium may arise in any period. This is an encouraging result from the environmental regulator's point of view.

References

Allsopp L (2002) Search and externalities. A pilot experiment. Working Paper, Adelaide University

Denicolo V (1999) Pollution-reducing innovations under taxes or permits. Oxford Economic Papers 51: 184-199

Downing P B, White L J (1986) Innovation in pollution control. Journal of Environmental Economics and Management 13: 18-29

Fershtman C, deZeeuw A (1995) Tradeable emission permits in oligopoly. Tilburg Center for Economic Research Discussion Paper

Fischer C, Parry IWH, Pizer WA (1999) Instrument choice for environmental protection when technological innovation is endogenous. Resources for the Future Discussion Paper 99-104

Jaffe AB, Newell RG, Stavins RN (2002) Environmental policy and technological change. Environmental and Resource Economics 22: 41-70

Jung C, Krutilla K, Boyd R (1996) Incentives for advanced pollution abatement technology at the industry level: An evaluation of policy alternatives. Journal of Environmental Economics and Management 30: 95-111

Laffont JJ, Tirole J (1996a) Pollution permits and compliance strategies. Journal of Public Economics 62: 85-125

Laffont JJ, Tirole J (1996b) Pollution permits and environmental innovation. Journal of Public Economics 62: 127-140

Milliman SR, Prince R (1989) Firm incentives to promote technological change in pollution control. Journal of Environmental Economics and Management 17: 247-265

Montero JP (2002) Permits, standards, and technology innovation. Journal of Environmental Economics and Management 44: 23-44

Ochs J (1995) Coordination problems. In: Kagel JH, Roth AE (eds) The handbook of experimental economics. Princeton University Press, Princeton New Jersey, pp. 195-252

Requate T (1998) Incentives to innovate under emission taxes and tradeable permits. European Journal of Political Economy 14: 139-165

Tirole J (1988) The theory of industrial organization. MIT Press, Cambridge Massachusetts

Sustainability entrepreneurship in the context of emissions trading

Anne Gerlach

University of Lueneburg
Centre for Sustainability Management (CSM)
Scharnhorststr. 1, 21335 Lueneburg, Germany
anne.gerlach@gmx.de

Abstract

Fundamental innovations are needed in order to achieve sustainable development. The actors who implement sustainability innovations are called sustainability entrepreneurs. In the first part of this paper, a conceptual framework for sustainability entrepreneurship is developed. It is suggested to take a view of power and politics in order to identify crucial actors and barriers within sustainability innovation processes. In the second part, this conceptual framework is applied to the context of emissions trading. The findings suggest that organisations that want to gain advantage from innovative emissions reductions have to foster sustainability entrepreneurship.

Keywords: Sustainability entrepreneurship, sustainability innovations, emissions trading, intrapreneurship

1 Introduction

For a successful implementation of sustainable development, radical social change is required. Incremental adjustments of current behaviour are not sufficient. Thus, without fundamental innovations, implementation of sustainable development cannot be achieved. This paper focuses on business as a major actor in both sustainable development and innovation processes. From the business point of view, successful sustainability innovations provide competitive advantages, i.e. economic success and at the same time contribute to ecological and/or social objectives. The paper is based on the assumption that emissions trading systems (ETS) have a dual-innovative nature. Firstly, their implementation is an innovation process itself, and secondly they are instrumental in fostering sustainability innovations. Innovations are carried out by actors. In the context of corporate sustainability, these actors are called *sustainability entrepreneurs*. The course of innovation processes and the different roles of actors within these processes are described by the promoter model (Witte 1973). According to this model, innovations are impeded by *barriers*. Promoters are actors who take certain *roles* within the innovation process to overcome these barriers. This paper aims to identify crucial barriers, roles of actors, and cooperations between actors in the context of sustainability innovations, and of emissions trading (ET) in particular. In the *first part* of this paper, a conceptual framework for sustainability intrapreneurship (i.e. sustainability entrepreneurship within existing organisations) is created by integrating the concepts of sustainability innovation, sustainability entrepreneurship and the promoter model. The results advise taking a look at power and politics in order to evaluate the processes of sustainability innovations and to identify crucial actors and their power bases. In the *second part*, the conceptual framework is applied to ET. Existing ETS at the national and organisational level as well as the European ETS are investigated as examples of innovation processes. Concerning concrete innovation processes of emissions reductions, the question is raised whether ETS can help to foster sustainability intra- and interpreneurship. The findings of the paper implicate that sustainability intrapreneurs play a crucial role for sustainability innovations and thus should be encouraged by organisations that want to gain advantage from innovative emissions reductions.

2 Conceptual framework

In this part of the paper a conceptual framework for the investigation of actors and barriers of sustainability innovations in the context of emissions trading is developed. The basic concepts of sustainability entrepreneurship and sustainability innovation are presented. Then, the promoter model is applied to sustainability innovations. Thus a framework is created to analyse barriers and actors of sustainability innovation processes.

2.1 Sustainability entrepreneurship

The discussion on sustainability entrepreneurship started with some contributions on *ecopreneurship* and *social entrepreneurship*. Many of these contributions refer to the concept of sustainability. However, they are mostly restricted to either the ecological or the social dimension.[1] Sustainability entrepreneurs can be referred to as actors who initiate and implement sustainability innovations and who pursue social or ecological objectives in addition to seeking economic success (Gerlach 2003).[2] Accordingly, the innovation process as well as the results of the activities of sustainability entrepreneurs can be called sustainability entrepreneurship. Three types of sustainability entrepreneurs can be distinguished: *founders and owners* of sustainability start-ups; *intrapreneurs* who foster sustainability innovations within existing business organisations and *interpreneurs* who draw on cooperations and networking to foster innovations on an inter-organisational level. This contribution focuses on sustainability intrapreneurs and interpreneurs who initiate and implement sustainability innovations within existing organisations.

2.2 Sustainability innovation

In this section, firstly the term innovation is discussed briefly before looking particularly at sustainability innovations. *Innovations* have been investigated from many perspectives. Accordingly, various definitions exist. Two main research dimensions can be distinguished. The first dimension is the *object-oriented* dimension concentrating on the result of innovation, for instance a new idea, process or product (Zaltman et al. 1984, p. 8). The second dimension is *process-oriented* and focuses stages of identification and implementation. Within this dimension, again two views can be distinguished. From the *view of technological development* (macroeconomic view) and *diffusion research* the following stages are distinguished: (1) invention, (2) innovation (primary economic exploitation), (3) adoption (by a certain actor), (4) diffusion and (5) imitation (Mensch 1971; Rogers and Shoemaker 1971). This paper adopts a *view of business organisations* which considers *"[...] only the process whereby an existing innovation becomes a part of an adopter's cognitive state and behavioral repertoire."* (Zaltman et al. 1984, p. 8; see also Knight 1967, p. 478). A crucial point of discussion arising from this definition is the degree of objectiveness of an innovation. Is it a sufficient criterion if the innovation object is new from the adopter's view (Zaltman et al. 1984, p. 10) or does it have to be objectively new considering other organisations (Knight 1967, p. 478 and Becker and Whisler 1967, p. 463). In this paper the view of

[1] For an overview of existing approaches of ecopreneurship and social entrepreneurship see Gerlach (2003) and Hockerts (2003).

[2] Hockerts (2003) takes a similar approach. He defines sustainability entrepreneurship as "[...] the identification of a sustainability innovation and its implementation either through the foundation of a start-up or the radical reorientation of an existing organization's business model so as to achieve the underlying ecological or social objectives" (Hockerts 2003, p. 50).

Knight (1967) is adopted, who views the innovation process *"as a special case of the process of change in an organization. The two differ only in the novelty of the outcome"* (Knight 1967, p. 479).

Innovations are considered to play a crucial role for implementing sustainable development (SD) (Vollenbroek 2002; Deutscher Bundestag 1998). The discussion on SD has been largely influenced by the definition of the Brundtland-report "Our Common Future" which defines SD as a *"[...] development that meets the needs of the present without compromising the ability of future generations to meet their own needs."* (WCED 1987, p. 8). A crucial point in the discussion on implementing the normative concept of SD is the idea that sustainability encompasses three dimensions – an economic, an ecological and a social one – and that these dimensions influence each other and have to be treated in an integrated fashion.[3] From the business perspective, the idea of three dimensions has been adopted by the "triple-bottom-line" concept (Elkington 1997). Dyllick and Hockerts (2002) apply the Brundtland definition of SD to firms and their stakeholders. They define corporate sustainability as *"[...] meeting the needs of a firm's direct and indirect stakeholders [...], without compromising its ability to meet the needs of future stakeholders as well. Towards this goal, firms have to maintain and grow their economic, social and environmental capital base [...]"* (Dyllick and Hockerts 2002, p. 131-132). Based on this concept, Hockerts (2003) defines sustainability innovations as *"[...] any process of social change which increases the proceeds derived from current natural, social, and economic capital, while at the same time protecting and enhancing the underlying capital stock"* (Hockerts 2003, p. 45). This definition is adopted for this paper.

Considered from a *macroeconomic point of view,* a sustainability innovation comprises the stages from invention to diffusion of the innovation object. Diffusion plays a crucial role for sustainability innovations, as the positive impact is enhanced or the negative impact is decreased to the degree to which the innovation is spread (Klemmer et al. 1999, p. 39).

2.3 The promoter model

After having discussed the basic terms sustainability entrepreneurship and sustainability innovation, this section follows the questions *how to specify (1) important innovation barriers as well as (2) the roles that sustainability intra- and interpreneurs take within the process.* The basis for answering these questions is formed by the promoter model developed by Witte (1973). A fundamental assumption of this model is that innovations are hampered by barriers, and promoters are actors who cooperate in order to overcome these barriers. Four promoter roles are distinguished in literature: power promoters, expert promoters, process promoters, and

[3] Particularly concerning the German context see Deutscher Bundestag (1998).

relationship promoters.[4] All four promoter roles are characterised by a high degree of activity concerning the innovation process (Witte 1973, p. 29; Folkerts 2001, p. 33). However, the roles differ according to their power base and their contribution to the innovation process. Table 1 gives an overview of the promoter roles and the barriers they overcome, their source of power and their contribution to the innovation process[5].

The barrier of willingness as well as the barrier of capacity are first of all *internal barriers*. The willingness and capability to achieve an innovation are matters of cognitive and motivational processes. Contrarily, whether or not the actor is allowed to implement the innovation is a matter of his relations to intra- and inter-organisational stakeholders. Thus the barrier of allowance can be called an *external barrier*.

Concerning the external control of organisational behaviour by interest groups, Pfeffer and Salancik (1978, p. 18)[6] state that *"behind every constraint there is an interest group that has managed to have that constraint imposed"* and that *"Since many constraints derive from the actions of others, one important function of management is influencing these others [...]"* (Pfeffer and Salancik 1978, p. 18). The purpose of utilising influence is to gain important resources such as financial support, motivation (e.g. incentives), acceptance, legitimacy, knowledge, and information. Transferred to innovative behaviour, this means that the actors (or promoters) who pursue the innovation are dependent on interest groups who control resources that are important for the innovation.

Table 1. Promoter roles and their criteria

	Barrier	Power base	Contribution for the innovation process
Power promoter	Barrier of willingness and acceptance	Hierarchical power; authority	– defining innovation objectives – providing important resources – assuring innovation support – tolerance towards mistakes – creating incentives for innovation – protection against opponents
Expert promoter	Barrier of capacity and knowledge	Expert knowledge	– generating ideas – analysing new solutions – developing alternative solutions

[4] Concerning power promoters and expert promoters see Witte (1973), concerning process promoters see Hauschildt and Chakrabarti (1988), and concerning relationship promoters see Gemünden and Walter (1995).

[5] For a detailed description of the contribution of power promoters, expert promoters and process promoters see Folkerts (2001, p. 34-37) and Witte (1973, p. 17-19). For a detailed description of the contribution of relationship promoters see Gemünden and Walter (1995).

[6] Although the approach of Pfeffer and Salancik is not focused on innovative behaviour, it might be useful to integrate their "resource dependence view" into a framework of sustainability intra- and interpreneurship.

			– Searching and processing information and argumentation for the discussion with proponents and opponents
Process promoter	Barrier of allowance and intra-organisational cooperation	Communicative skills; knowledge of the organisation	– mediate between power promoter and expert promoter – communication with organisational members affected by the innovation – controlling innovation process – developing action plan
Relationship promoter	Barrier of allowance and inter-organisational cooperation	Existing network, networking skills	– networking – searching the dialogue – bridging intercultural, interdisciplinary, interhuman distance – conflict regulation – creating inter-organisational teams

2.4 Roles and barriers of sustainability innovations

Promoters are actors who cooperate to put innovation plans into action. Sustainable entrepreneurs have been defined as innovative actors who foster the implementation of sustainability innovations. This implicates that the roles of sustainable entrepreneurs for implementing sustainability innovations are comparable to the roles of promoters for the innovation process. Thus, applying the promoter model to sustainability innovations might be helpful in order to specify the roles of sustainability entrepreneurs.

Table 2. Overlappings between promoters and sustainability intra- and interpreneurs

Sustainability entrepreneurship level / Promoter level	Power promoter	Expert promoter	Process promoter	Relationship promoter
Intrapreneur	X	X	X	
Interpreneur				X

Two levels can be distinguished, the sustainability entrepreneurship level and the promoter level. Table 2 shows these levels and their overlappings.

This paper suggests that sustainability intrapreneurs assume the tasks that are described by the profiles of power promoters, expert promoters or process promoters and that the contribution of sustainability interpreneurs can be characterised by the profile of relationship promoters. The *power promoter* employs his hierarchical potential to support ecological, economic and sociological innovation objectives. The *expert promoter* uses his specialist knowledge about ecological, economic and sociological topics in order to solve the complex problems within the scope of sustainability innovations.[7] *Process promoters* and *relationship promoters* undertake the communication and coordination of relationships with internal (e.g. managers and employees) and external (e.g. customers, suppliers, legislators, NGOs, associations) stakeholders. For organisations engaged in sustainability management, it is not sufficient to merely react to external pressure by the time stakeholders point out ecological, economic or social problems. These problems have to be addressed before they arise. (Schaltegger et al. 2003). To encourage such a proactive strategy, organisations seek the discussion with their stakeholders, set up relationships with important interest groups and communicate with them.

The promoter model not only differentiates between promoter roles but also describes different *innovation barriers* which shall also be applied to sustainability innovations. *Barriers of willingness* in the context of sustainability innovations are characterised by a lack of motivation and a lack of incentives to address ecological, economic or social innovation objectives. *Barriers of capacity* occur when the required expertise concerning ecological, economic, or social topics is not available. The *allowance barrier* plays a crucial role for sustainability innovations. Critical interest groups such as residents or environmental organisations can have the power to create major constraints to organisational behaviour. The influence of internal and external stakeholders can be based on laws or rules as well as on lacking acceptance. The example of offshore wind turbines illustrates the interrelations between innovation barriers. To realise this innovation, apart from the technological knowledge concerning the installation or the capacity of the power supply network, extensive specialist knowledge of ecological impacts on the animal and plant world is required. Expert opinions on these topics are partially demanded by law. Thus biological knowledge is required to overcome both knowledge barriers and allowance barriers. To meet the requirements of technological and biological knowledge probably more than one expert promoter is required. In addition it might be necessary to involve external consultants. These would act as relationship promoters. This example shows that a knowledge barrier can easily become an allowance barrier. The acceptance of inhabitants is another potential allowance barrier. Thus a *barrier system* (Müller Phillips Sohn 1976, p. 167) is formed.

2.5 Fostering sustainability intrapreneurs by management with power

The preceding discussion results in two theses which are relevant for further idea development. The first thesis is that organisations which want to gain advantage

[7] Witte (1973) investigates technological innovations and describes the expert promoter as technological specialist. In contrast, sustainability innovations call for specialist knowledge in ecological, economic, *and* social questions. Thus it seems most probable that more than two or more expert promoters appear within the process of a sustainability innovation.

by sustainability innovations have to foster sustainability intra- and interpreneurs. The second thesis is that this is done by providing required resources including power bases.

Pfeffer defines power "as the potential ability to influence behavior, to change the course of events, to overcome resistance [...] Politics and influence are the processes, the actions, the behaviors through which this potential power is utilized and realized" (Pfeffer 1992, p. 30). From this view, implementing sustainability innovations is a matter of (1) deciding on personal innovation goals, (2) diagnosing the "political landscape" including important actors and their power bases, (3) developing additional bases of influence to gain more control over the situation, (4) evaluating appropriate strategies and tactics, and (5) choosing a course of action (Pfeffer 1992, p. 293). Thus, encouraging sustainability intrapreneurship can be seen as a matter of influencing the motivation and interests of actors. Further, it is a matter of increasing action control and decreasing dependence on resources controlled by others (Neuberger 2002).

3 Sustainability intra- and interpreneurship in the context of emissions trading

In this part of the paper, the conceptual framework of sustainability intra- and interpreneurship that has been developed in the first part, is applied to emissions trading (ET). The Kyoto protocol signed in 1997 triggered off a radical increase in the discussion of ET. Economic, law, and political articles broadly discuss emissions trading systems (ETS) as an instrument of environmental policy that uses market mechanisms *"to ensure that emissions reductions required to achieve a predetermined environmental outcome take place where the cost of reduction is the lowest"* (Commission of the European Communities 2000, p. 8). Market instruments, such as ET, are distinguished from instruments of technical regulation on the one hand and environmental agreements on the other hand.[8]

The literature on ET reviewed for this paper can be classified according to their main perspective. Four categories can be distinguished. The *first category* comprises contributions on the design and implementation of ETS and related problems (e.g. Frenz 2003a; Frenz 2003b; Commission of the European Communities 2000; Europäisches Parlament und Rat der Europäischen Union 2003). The *second category* is formed by contributions on designing and implementing ET on a national level (e.g. Klemmer et al. 2002; Schafhausen 2002b; Matthes et al. 2002; Bode 2002; Derwent 2001; Sorensen 2001; Butzengeiger et al. 2001; Lübbe-Wolff 2001; 2001). The contributions of the *third category* discuss ET from the perspective of industrial sectors (e.g. Stronzik et al. 2002; Pocklington 2002; Hillebrand 2003). These articles focus on the inclusion of, the impact on, or the position of industrial sectors concerning ETS. The contributions of the *fourth category* take a

[8] For a categorisation of instruments of environmental policy see Rennings et al. (1997, pp. 80).

business perspective (e.g. Schleich et al. 2003; Ott and Santarius 2002; Santarius and Ott 2002; Trautwein 2002; Flatnitzer et al. 2001; Wieler 2001; Grohmann 2001; Lewis 2001). Their main issues are what opinion business organisations have of ET, how they are involved in and prepared for the implementation of ETS and which strategies there are to reduce their greenhouse gas emissions, including the strategy to implement an ETS on organisational level (Grohmann 2001; Lewis 2001; BP 2003a). Many contributions, regardless of their category, evaluate advantages and disadvantages of ET as a political instrument compared with environmental agreements and technological regulation, or analyse the possibility of combining it with other instruments. Another much discussed issue is the question of design options for ETS and their outcome.

The following *criteria* are a common framework to analyse instruments of environmental policy: certainty of ecological outcome, economic efficiency, dynamic efficiency, social acceptance, aspects of competition policy, and administrative feasibility (Trautwein 2002; Endres 1994).[9] The criterion of *dynamic efficiency* is crucial for investigating sustainability innovations in the context of ET. It is also called *innovation efficiency* (Rennings et al. 1997) and describes the degree to which an instrument helps to encourage innovative activities.

3.1 Emissions trading as an incentive for innovations

Many authors state that ET functions as an *incentive for innovations*. Some speak of the creation of pressure to innovate (Lübbe-Wolff 2001; Stronzik et al. 2002), others emphasise that market instruments in general are characterised by innovation friendliness and that innovative behaviour is rewarded by ET (Stronzik et al. 2002). According to Gagelmann and Hansjürgens (2002), the prospect that future emission reductions might be turned into revenue by selling emission allowances can lead to radical changes of production processes and to technological innovations. The European Commission, too, highlights that *"[...] as emissions trading will induce competition between companies to find cost-effective ways to reduce their emissions, an additional boost will be given to environmentally friendly technologies"* (Commission of the European Communities 2000, p. 8).

As it is an economic instrument, the degree to which ET functions as an incentive for emissions reduction innovations is determined by the price of allowances (e.g. Klemmer et al. 2002; BP 2003a). The underlying model of decision making is purely rational. The author agrees with the idea that economic incentives are an important situational factor to influence the motivation of actors to implement sustainability innovations. But it is not assumed to be sufficient. Rather, as pointed out in the first part of the paper, sustainability innovations are accomplished by

---

[9] For another framework consisting of the criteria effectiveness (goal conformity), system conformity, economic efficiency (including cost efficiency and dynamic efficiency) and institutional controllability see Rennings et al. (1997, p. 20) and Stronzik et al. (2002, p. 196).

means of politics. This implies that for decisions concerning processes of emissions reductions, a political model of decision making should also be considered.

Based on the framework of sustainability intrapreneurship, in the following, sustainability innovations in the context of ET shall be investigated from the standpoint of politics. The aim of looking at innovation processes concerning ET in this paper cannot be to precisely analyse the important actors, their power bases, emerging barriers and strategies and tactics to overcome the constraints of certain emissions reduction projects, as this would require one to make an assessment of detailed project data. Rather, the aim is to develop ideas about which variables are relevant for such an investigation of empirical data of sustainability innovations in general and emissions reduction projects in particular.

Prior to this, it has to be specified whether ETS are to be addressed as sustainability innovations themselves or if they meet the relevant criteria at all. As it was shown, it is widely agreed that ET provides an incentive for innovations as well as relative certainty of environmental outcome. Still, some authors note that it does not of itself reduce emissions, (Commission of the European Communities 2000; Lewis 2001) and that to achieve emission reductions technological innovations and process innovations are required (Grohmann 2001). ETS do not directly contribute towards an *enhancement of the ecological capital stock,* and therefore their implementation shall not be treated as a *sustainability* innovation here. However, there are many European countries, which have little or no experience with ETS so far. Thus, implementing ET can be referred to as an innovation process itself.[10] At the same time ETS are instrumental in fostering sustainability innovations. Consequently, ET can be said to have a *dual innovative nature*. In the following sections the framework of sustainability intrapreneurship will be employed to analyse ET from these two perspectives.

3.2 Implementing emissions trading as innovation process

ETS have been implemented at the European (Europäisches Parlament und Rat der Europäischen Union 2003), the national (e.g. in the UK, Derwent 2001, and in Denmark, Sorensen 2001), and the organisational level (e.g. in the BP Group, BP 2003a). In this section, examples of ETS at these three levels will be investigated.

As ETS have been identified as instrumental in encouraging sustainability innovations, and power promoters are thought to provide incentives for innovation by authority, it seems rather obvious that *power promoters* exist in all of the examples. In existing national ETS as well as in the European ETS, the government takes the role of a power promoter whose main power base is authority and who has to overcome *barriers of acceptance*[11]. Such barriers are mentioned regarding the implementation of the UK scheme: The opportunity of adopting specific tar-

[10] For a different opinion see Klemmer et al. (2002, p. 5) who state that emissions trading belongs to the standard instruments of environmental policy.

[11] For an overview of barriers of acceptance concerning the implementation of environmental allowance trading in general see Gawel (1998).

gets on a per unit of output basis *"helped the scheme's designers to overcome political resistance [...]"* (Derwent 2001, p. 48).[12] The European Commission emphasises the importance of *stakeholder consultation processes* (Commission of the European Communities 2000, 7). This indicates the existence of *process promoters*, and *relationship promoters* in particular. Also the Danish scheme was implemented *"after hearings with the industry and potential participants"* (Sorensen 2001, p. 54). For the implementation of the ETS in the UK, the "emissions trading group" was formed (Derwent 2001). In Germany, a similar group has been established (Schafhausen 2002a). The necessity of *expert knowledge* is addressed by emphasising the possibility to profit from early practical experience (e.g. Commission of the European Communities 2000; BMU 2003). No one would doubt that on this level the innovative behaviour of implementing an ETS is a matter of politics. For instance, negotiation processes are evident, especially concerning design options of ETS and reduction targets (Pocklington 2002; Derwent 2001; BMU 2003). Especially the article of Pocklington (2002), by considering the various position papers of unions and organisations addressing the European proposal for a directive of ET, gives an idea of how coalitions of actors advocate their interests. Watkins et al. call processes of politics "the influence game" and point out that influencing government and policy-making is a matter of managing relations, communication, and forming coalitions (Watkins et al. 2001).

If one compares the processes of implementing ETS on national or European level with the scheme implemented by the BP Group as an example for ET on organisational level, some similarities and differences can be identified. Again, the power promoter role seems to be the most obvious. CEO Sir John Browne commits himself personally to the emissions reduction goals of his company (BP 2003b; Grohmann 2001; Lewis 2001). This implicates that in addition to authority another power base is employed – the power of identification. An interesting question in this context is, if and why this power base plays a more important role for innovative behaviour on an organisational level than on a national or European level. It could be assumed that the power of identification helps to enhance the commitment of members to organisational goals. However, to the same degree to which actors are committed to *organisational* goals, they are restricted to pursue their *personal* goals and strategies of power,[13] which might be required in order to gain resources and overcome innovation barriers. Therefore, it is suggested that a balance should be aimed at between *restricting* the use of politics – e.g. by fostering commitment to organisational goals and rules – on the one hand and *allowing* the use of power in order to gain critical resources for creating emissions reductions on the other hand. Similar to the expert groups on national level, the internal network "Green Operations" has been implemented in the BP Group in order to support *"business unit efforts to identify and implement the most effective and economic greenhouse gas reductions, by sharing best practice and undertaking research and development"* (Lewis 2001, p. 79). As the "Green Operations" con-

[12] As the resistance was overcome by choosing certain design options, it can be argued that the barriers of acceptance were closely linked to barriers of knowledge.

[13] This is the case, at least, when personal goals do not match with the organisational goals.

sists of a network of experts, it can be seen as an indication for the existence of process promoters as well as expert promoters. In addition, similar to the national and European level, the expert knowledge gained by implementing an ETS at an early stage is seen as an important advantage (BP 2003a).[14]

3.3 Does emissions trading foster sustainability intrapreneurship?

The discussion in the preceding section has shown that there are several examples for actors taking important roles for encouraging innovative behaviour concerning emissions reductions and providing relevant resources. Does this mean that ET is indirectly able to foster sustainability intrapreneurs? And more generally, what other factors encourage their occurrence? To answer these questions, it is necessary to look at the project level of innovations concerning emissions reductions.[15] Important variables have to be identified in order to learn more about how to encourage actors who initiate and implement emissions reductions. As a first step of an investigation on project level, it has to be specified, if the investigated project meets the criteria of a sustainability innovation. Two criteria seem to be crucial. The first one is the degree of novelty.[16] The second criterion is the degree to which the project contributes towards achieving ecological, social, and economic benefits (e.g. significant emissions reductions). A second step is the investigation of the innovation process according to the promoter model. Who are the actors – can they be classified as sustainability intrapreneurs? What are important barriers constraining the sustainability innovations? What patterns of power bases (and other resources), and which influencing strategies can be identified? But most importantly, having identified sustainability intrapreneurs, what motivates them to make use of their power bases and to gain the critical resources for the innovation? This seems to be a crucial question in order to find out if sustainability intrapreneurship can be fostered by (e.g. economic) incentives or by other means or resources.

4 Conclusions

In the first part of this paper, a framework of sustainability intrapreneurship has been presented. It has been stated that sustainability intrapreneurhsip has to be fostered in order to accomplish sustainability innovations. In the second part of the paper, the implementation of ETS has been investigated according to the previously developed framework and the question has been raised if ETS help to foster sustainability intrapreneurship.

[14] Concerning *expert promoters* on an organisational level in general, the investigation of Santarius and Ott (2002) shows that in many German business organisations, expert promoters are required to overcome barriers of knowledge.

[15] For instance, completed CDM or JI projects could serve as investigation objects.

[16] To operationalise this criterion it might be useful to think of sustainability innovations and corporate sustainability as two poles of a continuum.

The findings indicate that ETS can be seen as part of the institutional framework of innovative emissions reduction projects. In order to investigate the potential of ETS to encourage sustainability intrapreneurship, the requirement to analyse concrete examples of emissions reductions projects concerning a motivational view has been pointed out.

References

Becker SW, Whisler TL (1967) The innovative organization: a selective view of current theory and research. Journal of Business 40: 462-469

BMU (2003) Arbeiten der Arbeitsgruppe "Emissionshandel zur Bekämpfung des Treibhauseffektes" (AGE) (Stand: Januar 2003) Bundesministerium für Umwelt, Naturschutz und Reaktorsicherheit. (http://www.bmu.de/files/treibhauseffekt_arbeiten.pdf on 05.10.2003)

Bode S (2002) Emissionshandel und Selbstverpflichtung. Zur Kompatibilität eines Emissionsrechtehandels mit der Selbstverpflichtungserklärung der Industrie. Energiewirtschaftliche Tagesfragen 52: 283-287

BP (2003a) Emissions Trading (www.bp.com/environ_social/environment/clim_change/emissions.asp on 20.05.03)

BP (2003b) Environmental and Social Review 2002.

Butzengeiger S, Betz R, Bode S (2001) Making GHG Emissions Trading work - crucial Issues in designing national and international Emissions trading Systems. Hamburg: Hamburgisches Welt-Wirtschafts-Archiv

Commission of the European Communities (2000) Green Paper on greenhouse gas emissions trading within the European Union, COM (2000) 87. Brussels

Derwent H (2001) Greenhouse Gas Trading in the UK. In: BMU (ed) Emissions Trading and Joint Implementation as a Chance for the Central and Eastern European Countries. BMU, Berlin, 47-51

Deutscher Bundestag (1998) Abschlussbericht der Enquete-Kommission "Schutz des Menschen und der Umwelt - Ziele und Rahmenbedingungen einer nachhaltig zukunftsverträglichen Entwicklung" des 13. Deutschen Bundestages: Konzept Nachhaltigkeit. Vom Leitbild zur Umsetzung. Dt. Bundestag, Referat Öffentlichkeitsarbeit, Bonn

Dyllick T, Hockerts K (2002) Beyond the business case for corporate sustainability. Business Strategy and the Environment 11: 130-141

Elkington J (1997) Cannibals with Forks: the Triple Bottom Line of 21st Century Business. Capstone, Oxford

Endres A (1994) Umweltökonomie. Eine Einführung. Wissenschaftliche Buchgesellschaft, Darmstadt

Europäisches Parlament und Rat der Europäischen Union (2003): Richtlinie 2003/87/EG Des Europäischen Parlaments und des Rates vom 13. Oktober 2003 über ein System für den Handel mit Treibhausgasemissionszertifikaten in der Gemeinschaft und zur Änderung der Richtlinie 96/61/EG des Rates, Amtsblatt der Europäischen Union, 25.10.2003, L275/32-46.

Flatnitzer KH, Herrmann M, Mohnhaupt M, Scholtissek S (2001) Energieversorger müssen weiterhin mit "Kyoto" rechnen. Energiewirtschaftliche Tagesfragen 51: 338-341

Folkerts L (2001) Promotoren in Innovationsprozessen. Deutscher Universitäts-Verlag, Wiesbaden

Frenz W (2003a) Zertifikathandel, freiwillige Anstrengungen der Wirtschaft und Stand der Technik - Teil 1. Energiewirtschaftliche Tagesfragen 53: 524-528

Frenz W (2003b) Zertifikathandel, freiwillige Anstrengungen der Wirtschaft und Stand der Technik - Teil 2. Energiewirtschaftliche Tagesfragen 53: 594-597

Gawel E (1998) Akzeptanzprobleme von Zertifikaten: In: Bonus H (ed) Umweltzertifikate. Der steinige Weg zur Marktwirtschaft. Analytica, Berlin, 113-134

Gemünden HG, Walter A (1995) Der Beziehungspromotor. Schlüsselperson für interorganisationale Innovationsprozesse. Zeitschrift für Betriebswirtschaft 65: 971-986

Gerlach A (2003) Sustainable entrepreneurship and innovation. Conference Proceedings of Corporate Social Responsibility and Environmental Management Conference 2003 in Leeds.

Grohmann WR (2001) Klimaschutz durch Emissionshandel - Sichtweisen und Erfahrungen der BP: In: Rengeling HW (ed) Klimaschutz durch Emissionshandel. Achte Osnabrücker Gespräche zum deutschen und europäischen Umweltrecht am 26./27. April 2001. Heymanns, Köln, Berlin, Bonn, München, 179-185

Hauschildt J, Chakrabarti AK (1988) Arbeitsteilung im Innovationsmanagement. Forschungsergebnisse, Kriterien und Modelle. Zeitschrift Führung und Organisation 57: 378-388

Hillebrand B (2003) Neuer Schub für Bio-Energien. DLG-Mitteilungen 118: 78-79

Hockerts K (2003) Sustainability Innovations: Ecological and Social Entrepreneurship and the Management of Antagonistic Assets. University St. Gallen, Dissertation

Klemmer P, Hillebrand B, Bleuel M (2002) Klimaschutz und Emissionshandel - Probleme und Perspektiven. Essen: Rheinisch-Westfälisches Institut für Wirtschaftsforschung

Klemmer P, Lehr U, Löbbe K (1999) Umweltinnovationen: Anreize und Hemmnisse. Analytica, Berlin

Knight KE (1967) A descriptive model of the intra-firm innovation process. Journal of Business 40: 478-496

Lewis R (2001) BP: Practical Experiences with GHG Emissions Trading: In: BMU (ed) Emissions Trading and Joint Implementation as a Chance for the Central and Eastern European Countries. BMU, Berlin, 77-79

Lübbe-Wolff G (2001) Der britische Emissionshandel - Vorbild für Deutschland? Energiewirtschaftliche Tagesfragen 51: 342-345

Matthes FC, Cames M, Deuber O, Koch M, Harnisch J, Kohlhaas M, Schumacher K, Ziesing HJ (2002) Analyse und Bewertung eines europäischen Emissionshandelssystems für Deutschland. Berlin: Öko-Institut e.V., DIW Berlin, ECOFYS

Mensch G (1971) Zur Dynamik des technischen Fortschritts. Zeitschrift für Betriebswirtschaft 41: 295-314

Müller Phillips Sohn H (1976) Determinanten der Innovationsfähigkeit. Versuch einer empirischen Überprüfung. Universität Stuttgart (Technische Hochschule), Dissertation

Neuberger O (2002) Führen und führen lassen: Ansätze, Ergebnisse und Kritik der Führungsforschung. Lucius & Lucius, Stuttgart

Ott HE, Santarius T (2002) Emissionshandel in Deutschland - wie denken Unternehmen darüber? Energiewirtschaftliche Tagesfragen 52: 280-282

Pfeffer J (1992) Managing with Power. Politics and Influence in Organizations. Harvard Business School Press, Boston, Massachusetts

Pfeffer J, Salancik GR (1978) The External Control Of Organizations: A Resource Dependence Perspective. Harper & Row, New York

Pocklington D (2002) European Emissions Trading - the Business Perspective. European environmental law review 11: 209-218

Rennings K, Brockmann KL, Koschel H, Bergmann H, Kühn I (1997) Nachhaltigkeit, Ordnungspolitik und freiwillige Selbstverpflichtung. Physica, Heidelberg

Rogers EM, Shoemaker FF (1971) Communication of Innovations. A Cross-Cultural Approach. Second Edition, Free Press, New York

Santarius T, Ott HE (2002) Meinungen in der deutschen Industrie zur Einführung eines Emissionshandels. Wuppertal: Wuppertal Institut für Klima, Umwelt, Energie

Schafhausen F (2002a) Bericht über die Beratungen der Arbeitsgruppe "Emissionshandel zur Bekämpfung des Treibhauseffektes". Ergebnisse der Phase I (Januar bis Dezember 2001). Berlin: Arbeitsgruppe Emissionshandel zur Bekämpfung des Treibhauseffektes

Schafhausen F (2002b) Der Emissionshandel als klimaschutzpolitisches Instrument. Zwischen Ideologie und praktischem Einsatz. Energiewirtschaftliche Tagesfragen 52: 563-568

Schaltegger S, Burritt R, Petersen H (2003) An Introduction to Corporate Environmental Management - Striving for Sustainability. Greenleaf, Sheffield

Schleich J, Ehrhart K-M, Hoppe C, Seifert S (2003) Üben für den Ernstfall: der Emissionsrechtehandel als Planspiel. Energiewirtschaftliche Tagesfragen 53: 104-108

Sorensen MP (2001) Emissions Trading in Danmark: In: BMU (ed) Emissions Trading and Joint Implementation as a Chance for the Central and Eastern European Countries. BMU, Berlin, 53-56

Stronzik M, Bühler G, Lambrecht U (2002) Ansatzpunkte für einen Emissionshandel im Verkehrssektor. Zeitschrift für Energiewirtschaft 26: 193-208

Trautwein S (2002): Chancen und Probleme des betriebsinternen CO_2-Zertifikatehandels - am Beispiel des Otto Versand, Hamburg. Lüneburg: Centre for Sustainability Management

Vahrenholt F (2001) Verkehr und Haushalte einbeziehen. Interview mit Fritz Vahrenholt. Umweltmagazin: 13

Vollenbroek FA (2002) Sustainable development and the challenge of innovation. Journal of Cleaner Production 10: 215-223

Watkins M, Edwards M, Thakrar U (2001) Winning the influence game. What every Business Leader should know about Government. John Wiley, New York

WCED (1987) Our Common Future. Oxford University Press,

Wieler B (2001) Nur heiße Luft? Umweltmagazin: 35-36

Witte E (1973) Organisation für Innovationsentscheidungen. Otto Schwartz, Göttingen

Zaltman G, Duncan R, Holbeck J (1984) Innovations and organizations. Robert E. Krieger, Malabar, Florida

Part B

Investment and management strategies under emissions trading

Optimal strategies for emissions trading in a Putty-Clay Vintage Model

Peter Letmathe[I], Sandra Wagner[II]

[I, II] University of Siegen
School of Economic Disciplines
Value Chain Management in Small and Medium-sized Enterprises
Hoelderlinstraße 3, 57068 Siegen, Germany
peter.letmathe@uni-siegen.de
sandra.wagner@uni-siegen.de

Abstract

In December 2002, the European Union decided to set up a carbon dioxide trading system to meet the goals of the Kyoto protocol. The trading system applies to energy producing companies and to firms with high energy usage. Firms will obtain the allowance to emit a given amount of carbon dioxide which will be reduced step by step on a yearly basis. The firms will have to buy additional allowances if the given amount does not satisfy their actual needs and they can sell allowances if they have excess allowances. In order to improve their market position, companies may invest in technical progress leading to fewer emissions.

The article examines which strategy companies should choose to adjust optimally to the trading system. The theoretical background builds Solow's hypothesis that technical progress is embodied in capital goods which leads to different production functions in each period (Vintage Production Functions). In combination with the Putty-Clay Model, it is shown that technical progress rates, timing of investment and prices of allowances play an important role in defining the optimal strategy of firms in order to cope with emissions trading.

Keywords: Emissions trading, theory of entrepreneurship, production planning, Vintage Production Functions, Putty-Clay Model, investment planning

1 Introduction

The European Commission's directive (Directive establishing a scheme for greenhouse gas emission allowance trading within the Community, in respect to the Kyoto Protocol's project mechanisms: COM 2003) on greenhouse gas emissions trading determines the beginning of an internal European emissions trading scheme in 2005 starting in a relatively small number of industry sectors that contribute significantly to total emissions, e.g. the energy sector (i.e. energy production and energy consumption) and other energy intensive industries like the metal industry or the paper industry. The scheme is limited to carbon dioxide in the first three-year period, but may be extended to other greenhouse gases later. According to the European Commission, the emissions trading scheme "will contribute towards the cost-effective fulfilment of the EU's commitments under the Kyoto Protocol." (COM 2003, p. 4).

The flexible market mechanism of buying and selling emission allowances gives an incentive to firms to produce fewer emissions by investing in and using more efficient technologies. Hence, short-term production planning and long-term, dynamic planning models – taking into account technical progress – have to be adjusted to the directive's requirements. Short-term production planning has to be extended by integrating the costs of emissions trading within given capacities, i.e. machines and installations as well as other factors of production. Long-term production planning includes the consideration of technical progress within the concepts of Engineering Production Functions and Vintage Production Functions which result in the Putty-Clay-Model (Johansen 1972; Bosworth 1976; Chenery 1949; Kistner 1993) within which the costs of emissions trading under the condition of variable capacities can be included.

2 Emissions trading from the firm's perspective

The European Commission's directive on emissions trading applies to between 4.000 and 5.000 energy-intensive installations which account for approximately half of the European Union's total carbon dioxide emissions. For this reason the impact of emissions trading on the European economy is highly relevant and affects companies from many different industry sectors as well as in all sizes (Öko-Institut, DIW Berlin and ECOFYS 2003).

The system's market mechanisms of buying and selling, of demand and supply of emission allowances establishes a market price per ton (unit) of carbon dioxide – and perhaps later as well for other greenhouse gas emissions – which has a high but yet partially undetermined impact on entrepreneurial decision making. The internalized costs for one unit of carbon dioxide, representing the price (in terms of opportunity costs) of the equivalent emission allowance, have to be figured in and may change future decisions. According to initial estimations, one ton of emitted carbon dioxide will cost approximately between 15 and 20 euro with the outcome that the variable costs of power generation currently amounting to between two and three cents would increase by one cent, or at least by one third, per kWh (Wietschel et al. 2002, p. 125). Furthermore, there are other economic consequences like transaction costs or penalty costs that have to be considered. Transaction costs occur in the form of information costs, trading costs and others. Information costs include, for example, the costs of gathering information concerning the decision whether to invest in new technology or not. Trading costs appear, for example, in terms of occupied time or personnel capacities for establishing, conducting and checking the emissions trading system within the company. Penalty costs emerge if a company does not own the emission allowances equivalent to its amount of emission. For every ton of emission that is not covered by an allowance, the company has to pay a penalty of 40 euro in the period from 2005 to 2007 and 100 euro thereafter (COM 2003).

With regard to the economic relevance of emissions trading, different alternatives for entrepreneurial action can be derived. Against the background of the basic entrepreneurial functions according to Schneider these alternatives can be analyzed and classified (Schneider 1997). Adopting income risks when the entrepreneur starts to hire employees is the first function. Schneider defines this function as the constitutive function of an enterprise. Moreover, the entrepreneur fulfils the function of maintaining the institution (enterprise) which is directed inwards as well as outwards. The internal enforcement of changes defines the inwardly-directed function, and the search for and exploitation of arbitrage opportunities are subject to the outwardly-directed function of maintaining an enterprise. For the analysis of possible actions within the scope of emissions trading, the functions of maintaining an enterprise play an important role as they define the main options for coping with regulatory or market changes. However, market changes originating from emissions trading may also offer opportunities for entrepreneurs to start new companies and to assume income risks when engaging employees.

The starting point of the following considerations is the fact that emissions trading has the same effect as a change of factor prices. The emission of carbon dioxide becomes economically relevant because each emission unit generates costs equal to the value of the equivalent allowance. Alternatives of adaptation to this factor price change can be found in both the short-term and the long-term perspective. The short-term options contain the adoption of market signals by optimizing the utilization of given technologies in terms of time, intensity, performed tasks etc. Basic approaches in this field are measures of factor substitution, paying special attention to energy saving in factor employment, as well as measures of process utilization and optimization in order to reduce the amount of carbon dioxide emission per product unit. Furthermore, next to pure internal measures a company can take outward steps by passing on additional costs to the costumer or other institutions. Another option may be the passing of energy intensive tasks to external suppliers. But this type of outsourcing may also have a negative effect on the procurement of future allowances through the national allocation plan.

The long term perspective also contains inwardly- and outwardly-directed options. Inwardly-directed long-term options help to cope with the corporate consequences of emissions trading and include the creation and application of new knowledge, the investment in technical progress and new technologies. The internal adjustment may have external effects since the long-term adoption of market signals could strengthen the company's competitive advantage on existing markets as well as on new or emerging markets. Here, investment in new technologies, e.g. power generation with renewable resources, plays an important role. Companies either apply emission-avoiding or emission-reducing technologies (Fichtner 1999) with the result that they can sell surplus allowances on the market at a profit, or that they have to buy additional allowances in order to meet their actual amount of emitted carbon dioxide (IEA 2001). The EU Directive itself represents this decision as follows: „The Community scheme gives installations' operators flexibility either to invest in abatement technology or to acquire EU allowances on the market to match their emissions, whichever is the cheaper" (COM 2003, p. 4).

With regard to the entrepreneurial functions defined by Schneider the different options can be summarized and arranged as follows:

Table 1. Short-term and long-term options of firms within the scope of emissions trading

Functions of an entrepreneur	Inwardly-directed maintenance of an enterprise (internal enforcement of changes)	Outwardly-directed maintenance of an enterprise (search for arbitrage potentiality)
Short-term options	• Optimized utilization of given technologies (factor substitution, utilization and optimization of processes)	• Passing on additional costs to costumers, etc. • Outsourcing
Long-term options	• New application knowledge • Investment in technical progress and new technologies	• Competitive advantage on existing markets • Introduction of new products as well in new or emerging markets

Both, short-term as well as long-term options can not be chosen without having an impact on the other option. There are interdependencies which should be identified and considered. If a company, for example, decides to optimize its utilization of existing technology, decision makers may neglect opportunities of knowledge creation and decide to adopt technical progress. On the other hand, investment in new technologies may lead to overcapacities and could prevent the company from using the given capacities in an efficient way.

3 Short-term production planning

Short-term production planning (Letmathe and Balakrishnan 2005) deals with the adoption of market signals by optimizing the utilization of existing technologies. In the model, we assume that the firm is producing a single product, which is typical in the power generating industry. However, it would not cause any difficulty to extend the model to multiple products. Within the scope of profit maximization the company identifies the product quantities x_τ in subperiod τ (τ = 1,...,D) for which the difference of revenues and costs – including the costs of emissions trading – reaches its maximum. At first, the factor costs of the input factor quantities r_i (i = 1,...,I) assessed with the corresponding factor prices q_i as well as the fixed costs K_F and the costs of emissions trading are subtracted from the revenues represented by the revenues functions $E(x_\tau)$. The costs of emissions trading are calculated by the total amount e_m of emissions m (m = 1,...,M) times the selling price r_m per unit of emission m. The emission price should be interpreted as the selling price of an emission type already reduced by possibly existing transaction costs. This formulation leads to the conclusion that the firm can sell each emission unit, which is not actually used for their own purposes, at the price r_m. If the firm has to buy additional allowances $\tilde{e}_m$, it can purchase them for the price $\tau_m + \tilde{\tau}_m$. In this context, $\tilde{\tau}_m$ reflects the transaction costs of emissions trading per unit of emission type m. If the firm's emissions of a given type exceed the given allowances plus the amount $\tilde{e}_m$ bought in the market, it will have to pay penalty costs. The costs per unit exceeding the legal amount are calculated through $\tau_m + \tilde{\tilde{\tau}}_m$ and are multiplied by the amount of emissions $\tilde{\tilde{e}}_m$ which were not allocated through the national allocation plan or bought in the market. $\tau_m + \tilde{\tilde{\tau}}_m$ stands for the penalty costs per uncovered unit of emission type m. Figure 2 presents an example of a calculation of emissions trading costs. In the example, the allocation plan assigns allowances of 1.000 tons compared to an actual emission of 1.500 tons of carbon dioxide to the firm considered here. In addition to the allocated allowances, the firm buys extra allowances of 300 tons. The selling price of one ton in the market is 15 euro and the transaction costs of buying one ton are assumed to amount to two euro. In this case, the firm has to cover emissions trading costs of 28.100 euro as illustrated in Table 2. Note that 15.000 euro – the value of carbon dioxide assigned to the firm through the national allocation plan – do not reflect actual payments, but represent opportunity costs because these 1.000 tons could have been sold if the respective emissions had not been generated in the firm's own production.

Table 2. Example calculation of emissions trading costs

Opportunity costs for not selling allowances assigned through the national allocation plan	$\tau_m \cdot e_m^A = 15 \cdot 1.000 = 15.000$
Costs of buying additional allowances	$(\tau_m + \tilde{\tau}_m) \cdot \tilde{e}_m = (15+2) \cdot 300 = 5.100$
Penalty costs (2005–2007)	$(\tau_m + \tilde{\tilde{\tau}}_m) \cdot \tilde{\tilde{e}}_m = (15+25) \cdot 200 = 40 \cdot 200 = 8.000$
Total costs of emissions trading	**28.100 euro**

Using the given parameters and the decision variables as explained above, the function of profit maximization for one period consisting of the sub-periods $\tau = 1,\ldots D$ may be expressed as follows:

$$G=\sum_{\tau=1}^{D} E(x_\tau)-\sum_{i=1}^{I} q_i \cdot r_i - \sum_{m=1}^{M}\left(\tau_m \cdot e_m + \tilde{\tau}_m \cdot \tilde{e}_m + \tilde{\tilde{\tau}}_m \cdot \tilde{\tilde{e}}_m\right) - K_F \Rightarrow \max \qquad (1)$$

In order to identify the mix of processes which maximizes the firm's profit, the following constraints have to be adhered to:

The amount of emissions e_m of type m results from the quantities of different input factors r_i, from the utilization of operating procedures y_p ($p = 1,\ldots,P$) necessary to produce the product quantities x_τ, and from the output quantity x itself multiplied by the factor-related, process-related and product-related environmental impacts b_{mi}, c_{mp} and d_m respectively:

$$e_m = \sum_{i=1}^{I} b_{mi} \cdot r_i + \sum_{p=1}^{P} c_{mp} \cdot y_p + d_m \cdot x \quad \text{with m} = 1,\ldots,\text{M} \qquad (2)$$

The quantities $y_{p\tau}$ of the employed operating procedures in the periods $\tau = 1,\ldots,D$ can be calculated by multiplying the output quantities for each period x_τ by the respective coefficients of processes a_p^{TP} and shares z_τ of the operating procedure p in period τ. The coefficient of a process a_p^{TP} indicates how often an operating procedure p is carried out in order to produce one product unit.

$$y_p = \sum_{\tau=1}^{D} y_{p\tau} = \sum_{\tau=1}^{D} a_p^{TP} \cdot x_\tau \cdot z_\tau \quad \text{with p} = 1,\ldots,\text{P} \qquad (3)$$

Furthermore, the amounts of input factors that are used in the periods $\tau = 1,\ldots,D$ result from multiplying production coefficients a_{ip}^{FT} by the quantities $y_{p\tau}$ of employed operating procedures p in period τ. The coefficient a_{ip}^{FT} describes the required amount of production factor i per operating procedure p.

$$r_i = \sum_{\tau=1}^{D} r_{i\tau} = \sum_{p=1}^{P}\sum_{\tau=1}^{D} a_{ip}^{FT} \cdot y_{p\tau} \quad \text{with i} = 1,\ldots,\text{I} \qquad (4)$$

The shares of employed operating procedures $z_{p\tau}$ causing different amounts of emissions in period τ have to be 1:

$$\sum_{p=1}^{P} z_{p\tau} = 1 \quad \text{with } \tau = 1,\ldots,\text{D} \qquad (5)$$

Beyond this, the emissions trading scheme requires that the quantity of carbon dioxide e_m (for example) emitted by the firm is equal to or lower than the assigned quantity of allowances e_m^A. If this is not fulfilled, the company can either buy extra allowances and / or pay a penalty for exceeding the assigned quantity of emission m. However, the following constraint ensures that the sum of the assigned quantity of allowances e_m^A plus the purchased allowances $\widetilde{e}_m$ plus the penalized emissions $\widetilde{\widetilde{e}}_m$ is smaller or equal than the actual emission quantity e_m of emission type m:

$$e_m - \widetilde{e}_m - \widetilde{\widetilde{e}}_m \leq e_m^A \quad \text{with } m = 1,\ldots,M \tag{6}$$

In addition to this, some input factors i can only be utilized to an upper limit in period τ, e.g. a machine may have a given capacity that can not be exceeded within one period:

$$r_{i\tau} \leq \bar{r}_{i\tau} \quad \text{with } i = 1,\ldots,I \tag{7}$$

The output quantities for the products are given or forecasted for each period respectively. The total quantity of the product results from the aggregation of different product quantities in the relevant periods:

$$x_\tau = \bar{x}_\tau \quad \tau = 1,\ldots,D \quad \text{and } x = \sum_{\tau=1}^{D} x_\tau \tag{8}$$

Furthermore, the non-negative constraints for utilizing operating procedures have to be considered:

$$z_{p\tau} \geq 0 \quad \text{with } p = 1,\ldots,P \text{ and } \tau = 1,\ldots,D \tag{9}$$

The profit maximizing solution of this model provides information about which operating procedure should be used to what extent in order to achieve the given product quantities. Sensitivity analysis can reveal the consequences of changes of emissions trading costs, of different output quantities, of the availability of additional operating procedures and of other alterations. As a result, the model enables companies to cope with the issue of emissions trading within given capacities cost-efficiently. In many respects, it is easy to extend the model, for example by introducing multiple products or by transforming it into a multiple period model. It can also provide some valuable insights if strategic and investment decisions are considered, as will be discussed in the next chapter.

4 Long-term production planning with the Vintage Production Functions and the Putty-Clay Model

In the long-term perspective, certain influential factors like legal requirements and technical progress that change the firm's framework play an important role in choosing the appropriate technology. Long-term options in terms of investment in technical progress and new technologies can be mapped out in long-term, dynamic production functions. Those production functions which are appropriate and relevant, in reference to decision making regarding emissions trading, are Engineering

Production Functions (Chenery 1949; Smith 1961), Vintage Production Functions (Solow 1960) and the Putty-Clay Model (Johansen 1972; Bosworth 1976). Steven dealt with the integration of environmental aspects into these approaches (Steven 1994).

The Engineering Production Function aims at the description of long-term production alternatives. Basic principle is the mapping of input-output-relations in the production function by including functional correlations based on physical, chemical or biological principles. The results show the different possibilities of installations' configuration or substitution of input materials due to specific, technical circumstances of input factors at a certain point in time.

Vintage Production Functions include the temporal aspects because they do not only represent operating procedures (activities or processes) in a certain period, but moreover they can be used to estimate future emissions and costs related to new operating procedures embodied in new techniques. The availability of these improved operating procedures is only given if a firm invests in techniques which reflect the newest level of technical progress in the investment period.

Initial point of the Vintage Production Function is the fact that, with different status of technical progress at hand, the same quantities and values of input factors can produce varying output quantities. This includes the perception that certain possibilities for technical progress can only be realized if new installations are used – this kind of technical progress is named embodied technical progress (Fischer 1980).

Therefore, each investment decision is related to the question of how to configure the whole plant to ensure future cost-efficiency and compatibility to existing technologies. The relevant decision criteria are the current price ratios between input factors and the status quo of embodied technical progress. This can lead to different results concerning optimal investment in machines etc. if price ratios change unexpectedly. As stated before, emissions trading affects factor prices with the consequence that the unit price of carbon dioxide influences long-term decision making as well. The optimal ratio of input factors will change because a company can switch to other operating procedures, change capacity utilization or even invest in more energy-efficient technology. If we assume a Cobb-Douglas-Production-Function, this would change the ratios of optimal factor quantities r_i. We further assume that technical progress can be expressed by the term $e^{\beta \cdot \tau}$ with e^{β} representing the technical progress rate per period and τ the period. With the standard constant α_0 and the elasticities of production factors α_i, a Vintage Production Function can be expressed as follows:

$$x_{j\tau} = \alpha_o \cdot e^{\beta \cdot \tau} \cdot r_1^{\alpha_1} \cdot \prod_{i=2}^{I} r_i^{\alpha_i} \qquad \sum_{i=1}^{I} \alpha_i \leq 1,\ \alpha_i \geq 0 \quad (i = 1, \ldots, I) \tag{10}$$

The relevance of technical progress can be illuminated by a simple example. We assume that a firm employs a given operating procedure which can be used to produce one product item ($x = 1$). The process time of this operating procedure is five time units ($r_1 = 5$) and the input quantity of a fossil resource is eleven tons

($r_3 = 11$). Each ton of the fossil resource generates two tons of carbon dioxide. This leads to the following process:

$$y_1 = (r_1^1, r_2^1, r_3^1, x^1, e^1) = (5, 0, 11, 1, 22)$$

If we assume for example that technical progress is embodied in the employed machine and leads to a lower input quantity of fossil resources (which means that eco-efficiency increases) and $\beta = 0.019062$ then we have a progress rate of 1.0192245 per period or 1.10000 in five periods. The investment in a new technology of the same type after five years would therefore lead to the following process:

$$y_2 = (r_1^2, r_2^2, r_3^2, x^2, e^2) = (0, 5, 10, 1, 20)$$

As can be seen above, technical progress leads to a saving of the fossil resource of one unit and a reduction of carbon dioxide emissions of two tons per output unit. Hence, Solow also uses the term factor-increasing technical progress.

For each vintage and type of technology, or set of technologies displaying embodied technical change, there exist different production functions. If technologies of different vintages and types can be separated, their outputs can be added:

$$x_t = \sum_{\tau=t-T}^{t-1} \sum_{j=1}^{J} x_{j\tau} \tag{11}$$

Going back to the example, this means that we can combine both operation procedures. If we have a capacity constraint of five time units for each technology, we can achieve an output of two product units by fully exploiting both processes which would result in the following operating procedure:

$$\tilde{y} = y_1 + y_2 = (r_1, r_2, r_3, x, e) = (r_1^1 + r_1^2, r_2^1 + r_2^2, r_3^1 + r_3^2, x^1 + x^2, e^1 + e^2) = (5, 5, 21, 2, 42)$$

Combining both operation procedures – adopting the assumptions mentioned above – results in an input quantity of 21 tons and an emission quantity of carbon dioxide of 42 tons. The aggregation of the input factor, output and emission quantities gives, therefore, an idea of feasible combinations of inputs and outputs and their impacts e.g. on cost-efficiency or emission-reduction in each period.

The Putty-Clay Model distinguishes between two production functions: the ex ante and the ex post production function. The ex ante production function (putty component) represents the potentialities of technological options depending on the status quo of technical progress in a certain period. Thus, the ex ante production function represents possible operating procedures before facilities are installed. The variety of possibilities can be compared to putty which is deformable and modifiable. These substitution potentialities can be mapped in an engineering production function. After the company has invested in new production technology, the set of operating procedures is given. By choosing one configuration, the firm determines the utilization of specific procedures and resources which are by then fixed and unchangeable to a high degree – like fired clay. Hence, this set of operating procedures determines the ex post production function (clay component) in the Putty-Clay Model.

Further on, technology-oriented investment decisions are made at different points in time, between which technical progress advances. These decisions lead to operating procedures from investments in different vintages. As shown, the task of short-term production planning is to determine the optimal utilization of these operating procedures available in a given period. Long-term planning has the task of identifying the optimal mix of technologies which would allow a company to simultaneously achieve cost and environmental goals in future periods.

The processes which can be realized and their combinations in the ex post production function represent all operating procedures and, therefore, form the set of technologies to be interpreted as knowledge about accomplishable production. The possibilities that occur within this set can be explored by using the activity analysis (Koopmans 1951; Feldmann 2002). This instrument defines an activity as a combination of input factors and products. Integrating emission reduction within the emissions trading scheme makes it necessary to add different types of emissions as unwished by-products as a component to these activities.

Since it is impossible to forecast all important future parameters correctly, e.g. the prices of emission allowances in each period, production flexibility is one important goal. Flexibility is given if a firm is able to adjust cost-efficiently to parameter changes at each point in time. This could be achieved with a set of operating procedures provided by totally different types of technologies, e.g. operating procedures provided by nuclear power plants, coal power plants and power generating facilities using renewable resources. On the other hand, advantages of specialization may be realized with a very narrow set of operating procedures. Such a specialization strategy could be successful if environmental policy unilaterally subsidizes a single type of technology reflecting this narrow set of procedures. However, the second strategy is less flexible and much more risky since decisions of environmental policy makers are difficult to forecast.

5 Integrated investment and production planning

The previous considerations have shown the high degree to which production planning and investment decisions are connected. Using the function of profit maximization described in chapter 4 and adding the investment costs by multiplying the number of units g_{jt} of technology j purchased in period t with costs for each unit of C_{jt}, the present value PV of a set of different investments in each period can be calculated:[1]

$$PV = \left[\sum_{t=1}^{T} \frac{\left(E(x_t) - \sum_{i=1}^{I} q_{it} \cdot r_{it} - \sum_{m=1}^{M} \left(\tau_{mt} \cdot e_{mt} + \tilde{\tau}_{mt} \cdot \tilde{e}_{mt} + \tilde{\tilde{\tau}}_{mt} \cdot \tilde{\tilde{e}}_{mt} \right) - C_{jt} \cdot g_{jt} - K_{Ft} \right)}{(1-i)^{-t}} \right] \cdot \Rightarrow \max! \qquad (12)$$

[1] i is the firm's internal interest rate relevant for investment decisions.

This method integrates emissions trading costs in the same way as in the short-term production planning model. Technical progress is embodied in a broader set of operating procedures available after investments in new technologies have been made. Without considering the constraints of this objective function – many of them are similar to the constraints in the short-term production planning model – it becomes clear that this optimization model – if formulated realistically – is much more difficult to solve since it includes non-linearities and integer variables. This makes it necessary to exclude some investment options:

- If a company follows a given strategy, e.g. the strategy of flexibility with a broad set of operation procedures or the strategy of specialization with a narrow set of operation procedures, several technological options, which do not fit into the strategy, can be excluded. For example, if a power generating company wants to position itself as an environmental pioneer, it may only consider technologies using renewable resources. Since a given strategy reduces the number of investment alternatives, it also helps to decrease complexity of long-term planning.
- Some technological options can be evaluated with simple cost comparisons. If we consider – for example – an old technology represented through the following operating procedure

  $y_{old} = (r_1, r_2, r_3, x, e) = (5, 0, 15, 1, 30)$

  and a new technology represented through the following operating procedure

  $y_{new} = (r_1, r_2, r_3, x, e) = (0, 5, 10, 1, 20)$

  we can easily calculate the cost savings per period if we invest in the new technology and close down the old facility. The production of 1.000 product units would lead to a consumption of 15.000 or 10.000 tons of fossil resources (e.g. black coal) and to an emission quantity of 30.000 or 20.000 tons of carbon dioxide per period. If the price of one ton of the fossil resource considered here is 80 euro and the cost of one ton of carbon dioxide is 15 euro, an early investment in the new technology would lead to cost savings of 550.000 euro per period.

With the present value method, it can be calculated if cost savings justify investing in the new technology. In order to compare only two investment alternatives, the objective function above combined with the constraints of the model in chapter 4 can be used. The model has to be calculated twice. The first model would be applied while still having the old technology in place and investing in the new technology after the end of the useful life of the old technology. In the second model, it would be assumed that the old technology would be replaced by the new one immediately. In this case, the only extension of the model in chapter 4 would be to ensure that the constraints are satisfied in each period of the planning horizon.

Although the optimization problem seems to be very complex, there are opportunities to reduce complexity. However, since investment decisions in new technologies have serious consequences, a higher methodical effort seems to be appropriate.

6 Conclusions

In this paper we presented different approaches for how to include consequences of emissions trading in short-term production planning as well as in long-term planning. Whereas it is simple to plan with given technologies represented by a given set of operating procedures, long-term planning which includes strategy and investment decisions is far more complex. Since the functioning and outcomes of emissions trading are still uncertain, it is difficult to choose the right technologies in the present. This may force power generating companies to postpone their investment decisions and, therefore, to not make use of the current technological progress.

References

Boehringer C (2000) Industry-level Emission Trading Between Power Producers in the EU. Applied Economics 34: 523-533

Bosworth DL (1976) Production Functions. Lexington, Massachusetts

Chenery HB (1949) Engineering Production Functions. The Quarterly Journal of Economics 63: 507-531

COM (Commission of the European Communities) (2003) Proposal for a Directive of the European Parliament and the Council amending the Directive establishing a scheme for greenhouse gas emission allowance trading within the Community, in respect of the Kyoto Protocol's project mechanisms, COM (2003) 403. Brussels

Feldmann M (2002) Handelbare Umweltzertifikate in der Linearen Aktivitätsanalyse. Zeitschrift für Betriebswirtschaft 72: 673-693

Fichtner W (1999) Strategische Optionen der Energieversorger zur CO_2-Minderung. Berlin

Fischer KH (1980) Empirische Anwendungen der Produktionstheorie. Zeitschrift für Betriebswirtschaft 50: 314-335

IEA, International Energy Agency (2001) International Emission Trading - From Concept to Reality. Paris

Johansen L (1972) Production Functions. Amsterdam

Kistner KP (1993) Produktions- und Kostentheorie. 2. Auflage Physica, Heidelberg

Koopmans TC (ed) (1951) Activity Analysis of Production and Allocation. Yale University Press, New Haven, London

Letmathe P, Balakrishnan N (2005) Environmental considerations on the optimal product mix. European Journal of Operational Research 167: 398-412

Öko-Institut, DIW Berlin, ECOFYS (2003) Auswirkungen des europäischen Emissionshandelssystems auf die deutsche Industrie. Berlin Köln

Schneider D (1997) Betriebswirtschaftslehre, Bd. 3: Theorie der Unternehmung. München Wien

Smith VL (1961) Investment and Production. Cambridge, Massachusetts

Solow RM (1960) Investment and Technical Progress. In: Arrow KJ, Karlin S, Suppes P (eds) Mathematical Models in the Social Sciences. Stanford, pp. 89-104

Steven M (1994) Produktion und Umweltschutz. Gabler, Wiesbaden

Wietschel M, Fichtner W, Enzensberger N, Rentz O (2002) Zur Relevanz eines europaweiten CO_2-Zertifikatehandels für strategische Unternehmensentscheidungen. In: Fichtner W, Geldermann J (eds) Einsatz von OR-Verfahren zur techno-ökonomischen Analyse von Produktionssystemen. Peter Lang, Frankfurt am Main

Strategic production management of companies participating in the European greenhouse gas emission allowance trading scheme

Wolf Fichtner

University of Technology Cottbus
Chair of Energy Economics
Institute of Energy Technology
Walther-Pauer-Straße 5, 03046 Cottbus, Germany
Wolf.Fichtner@tu-cottbus.de

Abstract

The European greenhouse gas emission allowance trading scheme will result in a new input factor for CO_2 emission intensive companies. Therefore, it is the first objective of this paper to characterise this new input factor. The second objective of this paper is to present different analysis tools for certificate trading issues. First a European energy model for the quantification of the impacts that emissions trading may have on electricity prices, technology choices, allowance prices and inter-regional power exchanges is developed. Furthermore, a model for the economic assessment of CO_2 emission reduction technologies and strategies on a detailed company level is presented. Accounting for the fact that emission reduction measures in industrial production companies are rather limited, inter-company energy supply concepts are integrated into the energy planning process of these production companies. Finally, it is shown how project-based flexibility mechanisms (JI/CDM) can be considered in investment strategies of companies affected by emissions trading.

Keywords: Production management, optimisation models, European electricity market, investment and long-term production planning, cooperation approaches

1 Introduction

The greenhouse gas emission allowance trading scheme agreed upon by the European Community will affect energy-intensive companies (e.g. in the energy sector all companies having combustion installations with a rated thermal input exceeding 20 MW (COM 2003)) all over Europe. In order to remain competitive, the strategic (production) management of these companies has to consider the new framework conditions. The first objective of this paper will therefore be to characterise the new intangible input factor *emission allowances*. The paper will then focus on how quantitative procedures can contribute to the development of an appropriate strategy for companies affected by the European emissions trading scheme. To realise this, a package of different energy models called PERSEUS (Programme Package for Emission Reduction Strategies in Energy Use and Supply) has been developed and applied at the Institute for Industrial Production (IIP) in order to provide analysis tools for certificate trading issues.

2 Characterisation of the new production factor emission allowances

Due to the fact that the European member states have to ensure that energy-intensive companies affected by the European emissions trading scheme on the one hand hold greenhouse gas emission permits for their installations, and on the other hand the emission allowances, actually two new basic factors of production will emerge. But as an applicant for a greenhouse gas emissions permit only has to deliver information on his installation and the measures planned to monitor and report emissions the more critical and scarce input factor seems to be the CO_2 emission allowances. In addition to this intangible input factor the atmosphere - the place for the accumulation of trace gases - is, of course, still needed to run CO_2 emitting production processes.

A CO_2 emission allowance is the right to emit the equivalent of one ton of carbon dioxide during a specified period – due to the rather small quantity of one ton this input factor can be characterised as divisible (see Table 1). Allowances lose their productive effect after the emission of the appropriate quantity of CO_2, therefore allowances are consumable factors of production (Corsten 1994, p. 9). For the three-year period beginning in 2005 allowances will be allocated free of charge. By the end of April of each year, the operator of each installation has to surrender a number of allowances equal to the total emissions from that installation during the preceding calendar year. Member states have to ensure that operators who do not hold sufficient allowances to cover their emissions shall be held liable and forced to pay excess emissions penalties. A way to substitute allowances would be a technology switch, for example, to renewable energies to produce electricity. Due to their homogeneity and their transferability, the purchase and sale of allowances can be varied at short notice, they are marketable as well as sellable and a continuous production quality (Haak 1982, p. 144) can be guaranteed. Furthermore, CO_2 emission allowances are not an integral part of the product to be manufactured (Busse von Colbe and Laßmann 1975, pp. 66f.). Due to the fact that allowances are no intermediate products produced within the companies, they are received from outside the company and can therefore be characterized as primary input factors (Bohr 1979, p. 1483). Allowances can be used for completely different production processes, for example for electricity production in a hard coal power plant as well as for the production of cement clinker in rotary kilns. Therefore, CO_2 emission allowances can be characterised as flexible. Another reason for this characterisation is the fact that it is intended to enlarge the European CO_2 emissions trading scheme into a European greenhouse gas emission allowance trading scheme by enabling a conversion of greenhouse gases into CO_2 equivalents by calculating their Global Warming Potential (GWP) (see e.g. (IPCC 2001)).

Table 1. Characterisation of the new input factor *CO_2 emission allowances*

criterion	characteristics		
scheduling feasibility	directing activity	basic factor of production	additional factor
category	personnel	non-personnel	
divisibility	divisible	indivisible	
consumption	potential factor of production	consumable factor of production	
replacement	substitutional	limitational / fixed	
changeability	long-term	short-term	
marketability	marketable	non-marketable	
resell-ability	alienable	inalienable	
repeatability	possible	impossible	
integral part of the product to be manufactured	direct factor	indirect factor	
materiality	substantial	intangible	
stage of production	primary	secondary	
elasticity	elastic	inelastic	

3 A model for investment and production planning within electric utilities considering the framework of the European CO_2 emission allowance trading

Recent liberalisation and re-regulation efforts in the European electricity sector have created new market structures and a new competitive environment. Due to this liberalisation process the entire European electricity market nowadays is relevant for the investment and production planning of an electric utility. As a result of the greenhouse gas emission allowance trading scheme agreed upon by the European Community, the production of CO_2 emissions will also have to be integrated into the investment and production planning process of electric utilities.

It is thus the general objective of the energy system model PERSEUS-CERT (–CERtificate Trading) to provide an analysis tool for the quantification of the economic and technological impacts that a CO_2 trading system may have on electricity prices, technology choices, certificate (allowance) prices and interregional power exchanges. PERSEUS-CERT is an energy and material flow model applying a multi-periodic linear programming approach. The target function requires a minimisation of all decision-relevant costs within the entire energy supply system. This basically comprises fuel supply and transport costs, transmission fees, fixed and variable costs of the physical assets (operation, maintenance, load variation costs etc.) and investment costs for new plants. The relevant techno-economic characteristics of the real supply system have been considered by implementing further equations covering technical, ecological and political restrictions. The most important technical restrictions are:

- Physical energy and material balances: match of demand and supply, taking into account storage options and time structures of electricity and heat demand (load curves).
- Capacity restrictions: transmission capacities, availability of installed capacities, (de)commissioning restrictions, technical lifetime of physical assets.
- Plant operation: maximum/minimum hours of full load operation, fuel options, cogeneration options, load variation restrictions.

The PERSEUS-CERT model consists of 42 regions covering all Western, Middle and Eastern European countries, in some case even splitting one country into several regions. All power generators within one region compete on a free market for the regional electricity demand that has to be satisfied. However, at the same time there is also direct competition between the different regions given that neighbouring regions are connected by interconnection lines (represented as an interregional high voltage grid, see Figure 1).

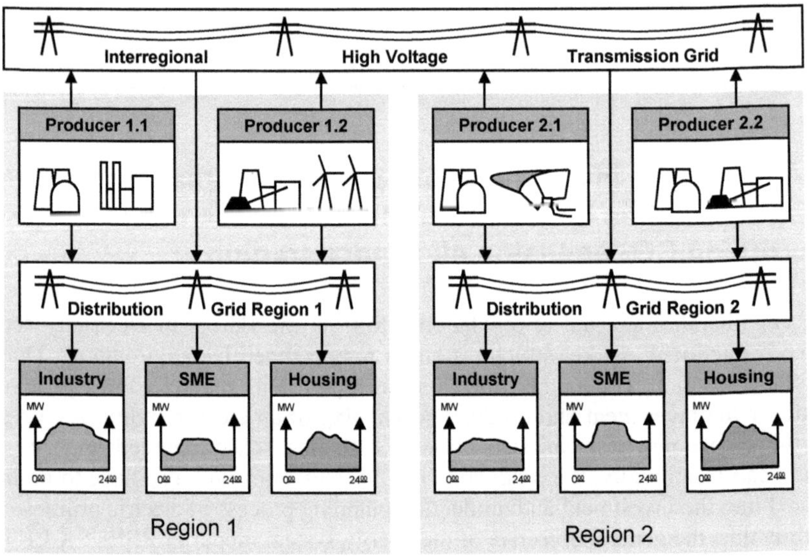

Fig. 1. Regional model structure

The consideration of transmission capacities and losses as well as transmission fees ensures a realistic representation of the real power exchange characteristics within the model. The interregional electricity market model – as it has been described until this point – has already been successfully used for different analyses and studies (see e.g. Enzensberger 2003).

However, the objective of analysing the impact of an emissions trading scheme on the physical electricity market requires the integration of a second market layer, the certificate market. Given the obviously strong interdependencies between both markets, the linkage has to be adequately reflected by the chosen modelling approach.

Every regional supply system is made up of a different set of physical power and heat generating facilities offering different operation modes and options (availability, load variation possibilities, costs) and especially the appropriate emission factors. Moreover, the total amount of emission allowances, i.e. the respective emission reduction targets, also varies between the different countries and regional energy supply sectors. So, any power exchange directly impacts emission balances and subsequently available emission allowances. Furthermore, the direct links between power or heat generation and CO_2 emissions result in complex (price) interdependencies between the electricity and the certificate market.

The certificate market has been integrated into this linear programming model by a set of additional equations. As the basic purpose of this model is the simulation of different market scenarios in order to support planning and consulting activities for industry representatives and policy makers, it contains a large set of design options for certificate trading schemes that can be (de)activated according to the user's preferences and/or expectations.

Within the model it is assumed that market prices of homogeneous goods in an open and fully competitive market are set by the marginal costs. We, therefore, use the marginal costs of the mathematical model to indicate both the resulting electricity prices as well as certificate prices. Electricity prices can be analysed for each time slot of a day (a year is represented by eight characteristic days – four seasons with working day / weekend, each represented with a simplified load curve of 6 (working day) respectively 3 time slots (weekend)). Certificate prices are given on regional and on a sector level.

The model described above has been implemented as a PC version that can be run on most commercial PCs. However, due to its high complexity and the resulting large problem size, it requires state-of-the-art hardware components. The model is equipped with an MS Access based data management system that permits easy data handling and a fully automated link to the mathematical module. The model itself is programmed in GAMS (Brooke et al. 1988). Formatted and structured results become available in MS Excel spreadsheets. In order to solve the problem, commercial solvers like CPLEX can be applied. The total model accounts to around 3 million variables. Calculation time ranges from 1 to 20 hours, basically depending on the chosen time horizon.

In the following, some exemplary results of the PERSEUS-CERT model are highlighted. These results are based on two scenarios: a reference scenario without any emission reduction obligation and an emissions trading scenario based on the

EU burden sharing targets (EU15-BS) without the possibility of using project-based flexibility mechanisms (Joint Implementation (JI) and Clean Development Mechanism (CDM)). The emissions trading scenario assumes an international trade that is limited to the present 15 EU member states. Scenarios that consider a possible participation of the accession countries are not discussed in the paper.

Considering the future emissions trading framework, these externally fixed emission reduction targets require an optimal allocation of the available emission allowances within the system. This implies the creation of market prices for this new commodity as well as important adaptations of the existing energy supply system to the new planning problem. With regard to the first issue, model results indicate certificate prices for the trading period 2008-2012 of approximately 15.5 €/t CO_2. Due to increasing primary energy prices the certificate prices rise to 21.2 €/t CO_2 in 2013-2017 and 24.5 €/t CO_2 in 2018-2022 (see Table 2). These price calculations are based on the assumption that the CO_2 emission caps as agreed upon in the EU burden sharing targets for the period 2008-2012 will continue to be valid for the following periods.

Table 2. Certificate prices in the EU15-BS scenario

	Trading Periods			
	2005 - 2007	**2008 - 2012**	**2013 - 2017**	**2018 - 2022**
Price (€/t CO_2)	2.7	15.5	21.2	24.5

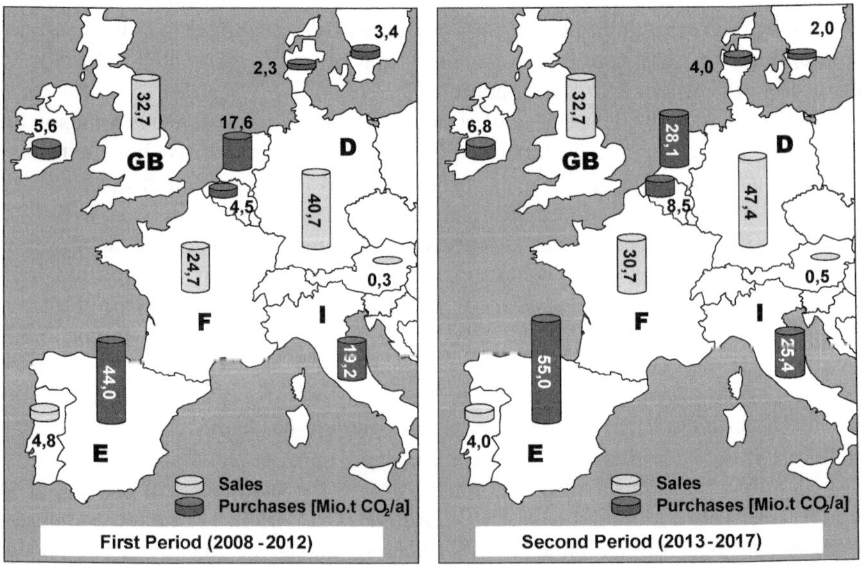

Fig. 2. Sales and purchases of CO_2 emission certificates

Within the European market, the different national electricity sectors show very different trade characteristics regarding their net sales volumes. Electric utilities from Germany, Great Britain and France become the major net sellers of emission allowances whereas companies from Spain, Italy and the Netherlands evolve to be the principal net purchasers in the market. The respective position of each national sector depends on the past emission trends, the national EU burden sharing emission reduction target and the specific emission reduction options in each national supply system. Beside these national characteristics the international electricity exchanges also have a crucial impact on the certificate market and vice versa.

As seen in Figure 2, German electric utilities become significant net sellers of emission allowances in the market (about 40.7 Mio t CO_2/a in the first commitment period). These net sales, however, are to a large extent based on a significant reduction of the internal power generation. Figure 3 indicates net interregional electricity exchanges between Germany and its neighbouring countries in the year 2010.

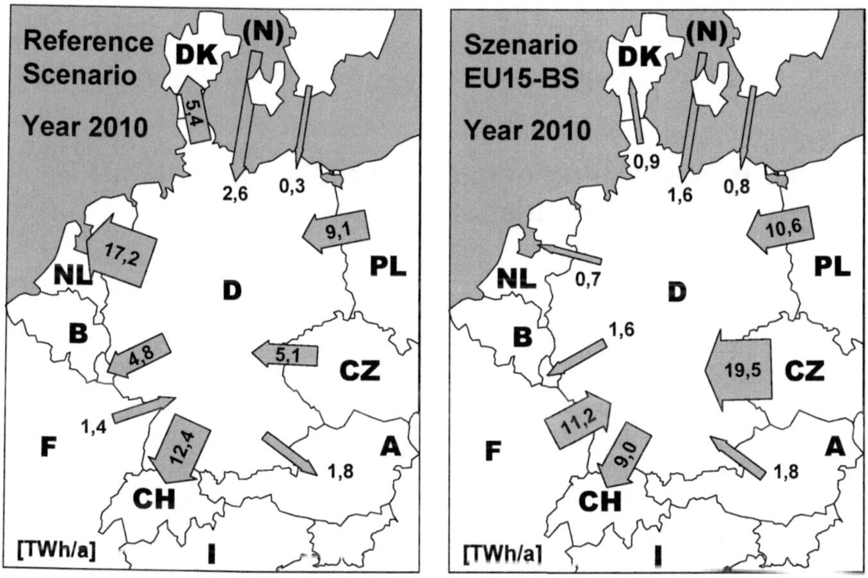

Fig. 3. Interregional electricity exchanges between Germany and its neighbouring countries in the year 2010

Along with this reduction of the internal net generation, the technical supply systems of the electric utilities are also affected. However, it has to be mentioned here that the results with regard to company-specific investment and production plans can only be briefly outlined on a national level due to the confidential nature of the data. Figure 4 compares the development of the electricity generation by fuel in Germany between the reference and the emissions trading scenario. The model results indicate a sharp change in the future electricity supply structure due to the externally fixed emission reduction targets. After the year 2010, nuclear power generation in Germany declines due to the politically decided nuclear phase-out. In the reference scenario, this production reduction is compensated by the construction of new coal and lignite-fired steam power plants. In the case of an additional emission reduction obligation, however, this capacity loss is basically compensated by new gas-fired combined-cycle power plants as well as significantly increased net electricity imports.

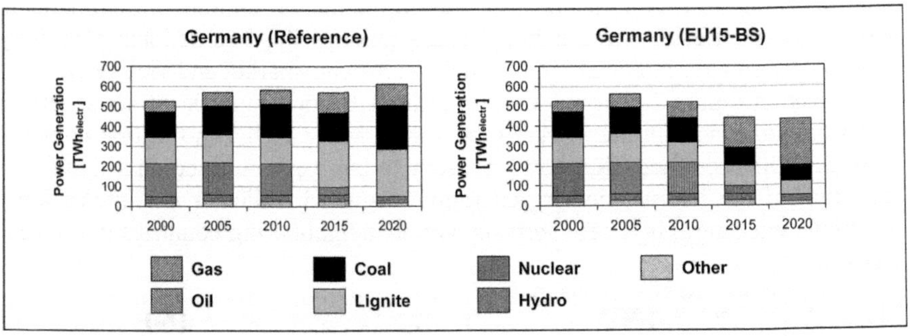

Fig. 4. Development of electricity generation by fuel in Germany in the different scenarios

4 Long term planning of energy supply concepts in energy-intensive production companies

Although the European energy supply sector is responsible for the largest portion of total CO_2 emissions in Europe, industry sectors like iron & steel, cement, paper & pulp, glass, chemicals and refineries also emit significant amounts of CO_2, on the one hand by auto-generation of the necessary energies, on the other hand by process emissions. However, emission reduction measures in these companies are rather limited, due to the fact that their energy supply system is usually made up of only a few plants. Accounting for that fact, inter-company energy supply concepts and their integration into the energy planning process of production companies will be described in the following, an option which considerably enlarges the decision areas of these companies. In addition to investments in boilers, turbines etc. realised independently by the companies, various forms of energy supply concepts (including different alternatives of energy transportation) realised in co-operations with possible partners – such as the development of new power plants – can be taken into account.

For the economic assessment of CO_2 emission reduction technologies and strategies on a detailed company level, investment planning alone is not sufficient, as future payments connected to a new energy supply concept strongly depend on the utilisation of this energetic unit. In turn, this utilisation depends on the existence of other energetic units available in the company. To calculate the emission reduction costs within energy intensive production companies an optimisation model has been developed in which existing technologies and future investment options are represented as discrete variables. Furthermore, this model to analyse inter-company energy supply concepts needs to be able to represent the production of different qualities of steam with regard to temperature and pressure, since each company tends to have their own standard. In addition, it must be possible to identify which units of a company should not be operated at full-load in order to be able to compensate for a breakdown of a unit in one of the other companies of the network. In order to assess the effects of inter-company energy concepts, due to the combination of different load profiles for example, it is indispensable to represent the electricity and heat demand profiles of the companies as related to their machines and appliances.

For this task, the multi-period mixed-integer linear optimisation approach PERSEUS-ICC (Program Package for Emission Reduction Strategies in Energy Use and Supply – Inter-Company Concepts) has been developed (see for example Frank 2003). Considering energy conversion technologies, which already exist in the analysed system, as well as future investment options, it is the objective of this model to provide information on how the exogenously given energy demand needed to maintain production in the firms involved could be met in an economically optimal way.

In order to assess economic effects of inter-firm energy supply concepts, the model represents the energy production in the firms on a very detailed level - taking into account the widely varying forms of energy conversion processes. Individual technologies or purchase contracts as well as demand side measures are represented by technical, economic and ecological parameters. In addition to existing energy conversion units, opportunities to fulfil the energy demand are presented as future investment options. The model uses a discrete-time approximation in which the planning horizon is broken down into periods and each period is divided into time intervals representing the load curves of typical days. This mixed-integer linear optimisation model has been complemented by a process simulation tool to determine the necessary data.

To elaborate on recommendations for the future energy system structure, a system optimisation is performed. The decision variables of this multi-period mixed-integer linear optimisation model are, on the one hand, process activities in energy conversion units and energy flows at typical time intervals – for example between 6:00 and 8:00 a.m. on a summer working day - and, on the other hand, new capacities for energy conversion and energy transport. The target function of the model is the minimisation of all decision-relevant costs within the whole supply system over the time horizon using the net present value method.

In the following, some exemplary results of the PERSEUS-ICC model are highlighted for the case-study *Karlsruher Rheinhafen.* In this case-study, supported by the German Federal Ministry of Education and Research (BMBF), the PERSEUS-ICC model has been applied to evaluate the economic and ecological effects of inter-company energy supply concepts between five energy-intensive production companies (in addition, the municipal utility of Karlsruhe is integrated in this case study). All companies are located in the area near the Rhine harbour of Karlsruhe with a maximum distance of about 4 km between them. The model has been applied to the energy systems of the firms involved. All existing boilers and turbines have been modelled on the basis of existing data, possible inter-firm investment options have been adapted to the specific circumstances of the case at hand (e.g. different natural gas fired Combined Cycle Power Plants (CCPP) for electricity and heat production, diverting unused energy flows from one company into another company to meet the demand for low temperature heat etc.) by using the process simulation tool Aspen Plus (Aspen 1994). On the demand side, the future development of energy demand in the different companies has been estimated to stay nearly constant on the level of the year 2000. In addition, an adequate level of aggregation of the daily load curves had to be found. Therefore, data on the energy demand of the companies for each hour of the year 2000 has been analysed and aggregated. Finally, some energy-economic framework assumptions had to be made including projections of primary energy carrier prices and legal framework conditions for cogeneration.

Computational results show that if inter-company energy supply concepts are excluded (reference case), the companies will continue to meet their electricity and heat demand partly by auto-production and partly by purchase from utilities. Due to the pay-back-times required there will only be smaller investments, e.g. for peak load boilers. Necessary CO_2 emission allowances will be bought on the market.

If investments in inter-company energy supply concepts can also be chosen, the model results show that under the assumed framework conditions (concerning prices for natural gas and the German combined heat and power law, for example) the installation of the combined-cycle power plant, which would produce heat and power for all partners involved, is the minimum cost option. Compared to the reference case, the discounted decision-relevant costs of energy supply needed to maintain production in the companies can be reduced by about 25 percent – even without considering possible revenues due to the sale of surplus emission certificates. Furthermore, the model results illustrate that the combined heat and power plant is operated at almost base-load to produce electricity and heat. Therefore, the production plan for this inter-firm energy supply concept has only small variations over the time of day, leading to a high utilisation of the new energetic unit. The remaining peak demands are covered by existing boilers and turbines.

Compared to the autonomous development without an inter-firm cooperation to link energy flows, the installation of the combined-cycle power plant brings about a reduction of CO_2 emissions amounting to approx. 30%. These surplus emission allowances could be sold under the European CO_2 emissions trading scheme, making the installation of said combined-cycle power plant even more attractive. Besides this reduction of CO_2, the inter-company energy supply concept would also result in considerable improvements in other ecological problem areas, such as the protection of natural resources or the reduction of pollutants like SO_2.

Finally, one major problem of inter-company strategies should be mentioned here. The model identifies possible economic and environmental benefits of inter-company energy supply concepts by determining the global optimum for the whole network instead of the optimum for each company. For this reason, the decisions recommended are not necessarily the same as the decisions from the point of view of individual players. For example, the model will identify the installation of a new combined heat and power plant as the optimal solution, even if this leads to higher costs for one partner company compared to its reference case. In order to realise the overall benefits – economic savings as well as surplus CO_2 emission allowances -, the companies placed at a disadvantage would have to be compensated by the other partners. Therefore, it is the subject of further research to tackle this problem of assignment using game theoretic approaches, which have been successfully used for similar problems in other areas (see e.g. Nouweland et al. 1996).

5 A model to analyse the efficiency of international cooperation in mitigating climate change

To integrate international cooperation approaches into investment strategies of companies affected by the European emissions trading scheme, there are two possible alternatives for coping with this inherently specific problem, which consists of the question of how to model the energy systems of different companies in different countries:

- on the one hand, by developing a separate model for each company and linking these models by using a decomposition algorithm (see e.g. Nurminski and Balabanov 1983) or,
- on the other hand, by developing a single model representing the energy systems of all companies involved.

Due to the fact that there are only a few interdependencies between the models of the different companies, it seems to be more advantageous to develop different models for each company and to link these models by using a decomposition algorithm. Then, in an iterative process, the marginal costs for CO_2 reduction, which constitute the demand price for JI / CDM projects, and the marginal costs - including transaction and risk coverage costs - for CO_2 reduction in the JI / CDM host company, which constitute the supply price, have to be brought to an equilibrium.

An analysis of JI / CDM potentials for German investors in the Russian Federation, India, the Republic of South Africa, and Indonesia as host countries shows that, beside European allowance prices, the following basic conditions have a significant impact on the number and extent of economical cooperation projects:

- the acknowledgement of sinks as JI / CDM reduction options,
- the total potential of CO_2 reduction options world-wide,
- the energy-economic framework in the country of the investor,
- the level of the reduction target of the investor country as well as
- the level of transaction costs.

For high reduction targets in 2020 the mitigation costs for German companies can be cut by roughly 70%, with about 40% of the total emission reduction resulting from JI / CDM projects. Half of the remaining expenditures should be invested in Germany, the other half in JI / CDM projects. If measures for the sequestration of CO_2 (so-called sinks) are accredited as JI / CDM options without restrictions, the costs are even lower (about 10% of the original mitigation costs). Apart from sinks, which are unequivocally the most cost-effective reduction measures, a considerable potential for economical emission reduction is represented by projects in the energy supply sector (power plant projects). This applies especially to the expansion of hydropower plants and the increased use of natural gas fired combined-cycle plants.

6 Summary

The greenhouse gas emission allowance trading scheme agreed upon by the European Community will result in a new input factor for CO_2 emission intensive companies. In order to integrate these new framework conditions into the company specific planning problems, the energy system analysis models of the PERSEUS model family have been developed and used.

In a first case-study, it has been shown how this type of energy model can be amended in order to provide an analysis tool for certificate trading issues, especially for the investment and production planning of electric utilities in the context of a European electricity market. The model results indicate that German electric utilities will become significant net sellers of emission allowances in the market. However, these net sales would partly be based on a significant reduction of power generation in Germany.

Although it is the European energy sector that is responsible for the largest share of the total CO_2 emissions in Europe, industry sectors will also be affected by the European greenhouse gas emission allowance trading scheme. For these production companies a model has been elaborated, which integrates inter-company energy supply concepts into the energy planning approach, enlarging the decision areas of these companies considerably. This model has been used to identify economic and ecological implications of inter-company energy supply concepts in a network of five industrial companies located in the area near the Rhine harbour of Karlsruhe. The model results show that, under the chosen framework assumptions, the installation of a custom-designed combined-cycle power plant would be the most promising solution for meeting the energy demands needed to maintain production in the companies involved. This strategy would result in considerable cost savings and CO_2 emission reductions, which in return could lead to additional benefits through the sale of emission allowances in the market.

Finally, a model has been developed, with the help of which JI / CDM projects can be considered. The analyses show that cooperation approaches potentially result in significant decreases in the total expenditures for emission reduction in German companies affected by the European greenhouse gas emission allowance trading scheme.

References

Aspen (1994) Aspen Plus™: Getting Started, Cambridge, Massachusetts

Bohr K (1979) Produktionsfaktorsysteme In: Kern, W. (ed): Handwörterbuch der Produktionswirtschaft, Stuttgart: Poeschel Verlag, 1979, pp. 1481-1493

Brooke A, Kendrick D, Meeraus A (1988) GAMS - A User's Guide, Redwood City/CA: Scientific Press

Busse von Colbe W, Laßmann G (1975) Betriebswirtschaftstheorie, Band 1: Grundlagen, Produktions- und Kostentheorie, Berlin et al.: Springer-Verlag, 1975

COM (2003) Commission of the European Communities: Common position adopted by the Council on 18 March 2003 with a view to the adoption of Directive of the European Parliament and of the Council establishing a scheme for greenhouse gas emission allowance trading within the Community and amending Council Directive 96/61/EC, Brüssel

Corsten H (1994) Produktionswirtschaft, 4. Auflage, München et al: Oldenbourg Verlag

Enzensberger N (2003) Entwicklung und Anwendung eines Strom- und Zertifikatmarktmodells für den europäischen Energiesektor, VDI-Verlag

Frank M (2003) Entwicklung und Anwendung einer integrierten Methode zur Analyse von betriebsübergreifenden Energieversorgungskonzepten, Dissertation, Karlsruhe: Universität Karlsruhe, Fak. für Wirtschaftswissenschaften

Haak W (1982) Produktion in Banken: Möglichkeiten eines Transfers industriebetrieblich-produktionswirtschaftlicher Erkenntnisse auf den Produktionsbereich von Bankbetrieben, Frankfurt a.M. et al.: Peter Lang Verlag

IPCC (2001) Intergovernmental Panel on Climate Change: Climate Change 2001: The Scientific Basis, Cambridge: Cambridge University Press

Nouweland A, Borm P, Brouwers W, Bruinderink R, Tijs S (1996) A Game Theoretic Approach to Problems in Telecommunication, Management Science, 1996(2): pp. 294-303

Nurminski E, Balabanov T (1983) Decomposition of a Large-Scale Energy Model, Large Scale System, 1983(4): pp. 295-308

Decision making in the emissions-market under uncertainty

Gorden Spangardt[I], Michael Lucht[II], Christian Wolf[III], Christian Horn[IV]

[I, II, III, IV] Fraunhofer Institute for Environmental Safety and Energy Technology UMSICHT
Osterfelder Straße 3, 46047 Oberhausen, Germany
gorden_spangardt@gmx.de

Abstract

In this paper a stochastic optimization model for decision making in the emissions market under uncertain boundary conditions is presented. This model aims at finding a strategy for profit optimal emissions trading/emissions reduction. The uncertainties in the emissions market are modelled via a scenario approach considering the price risk in the emissions market as well as the project risk of a potential emissions reduction project.

Keywords: Emissions trading, stochastic optimization, decision making, conditional value at risk, risk, uncertainty

1 Background

The presented research results were obtained within the scope of the "*jupiter emissions trading training*[1]" where emissions trading strategies are to be evaluated by Fraunhofer UMSICHT on behalf of several industrial clients. The project consists of a realistic emissions trading simulation, which took eight months of time in order to map the years 2005-2012 on one month each. This simulation is framed by technical, economical and legal courses and consulting (Spangardt and Lucht 2003), where the handling of make-or-buy decisions under the uncertain boundary conditions of the emissions market is the most important aspect.

The European emissions trading system affects large plants in several industrial branches. At the beginning of every year a certain number of emissions certificates will be allocated to every plant by the government. On the 30th of April of every year the companies that operate the plants have to "pay" for the last year's emissions of their plants with the equivalent amount in certificates. There are two different ways to comply with this requirement when the allocated amount of certificates does not cover the amount of emitted carbon dioxide: Buying emissions certificates from other companies or reducing emissions by a technical emissions reduction project. Both options include certain risks.

Deciding to buy lacking certificates at the emissions market means having to take a price risk, deciding to invest in a technical emissions reduction project on the other hand means having to take the project risk, as the project costs may exceed the calculated investment budget.

As the decision must aim at maximization of the expected profits and minimization of the risk exposure, a multi-criterial stochastic optimization approach has been developed in order to support decision makers in an objective and transparent way.

2 Modelling the decision process

The characteristic approach to make-or-buy decisions in the emissions market can be essentially regarded as a sequential process:

1. The decision maker comes to a certain decision, e.g. buying certificates and/or developing an emissions reduction project.
2. A random event occurs, e.g. the certificate price rises or falls.
3. The decision maker comes to a follow up decision.
4. Again a random event occurs ... go to (3).

All decisions have to be made based only on the information about the realization of stochastic parameters in the past, disregarding their subsequent realization in the future. This principle is well-known as non-anticipativity.

---

[1] For further Information see www.jupiter-nrw.de.

Concerning the stochastic price progression this decision process can be modelled by a scenario tree, which covers the relevant time horizon from 2005 to 2012 using a yearly time discretisation. Figure 1 shows a binomial scenario tree of the stochastic price progression.

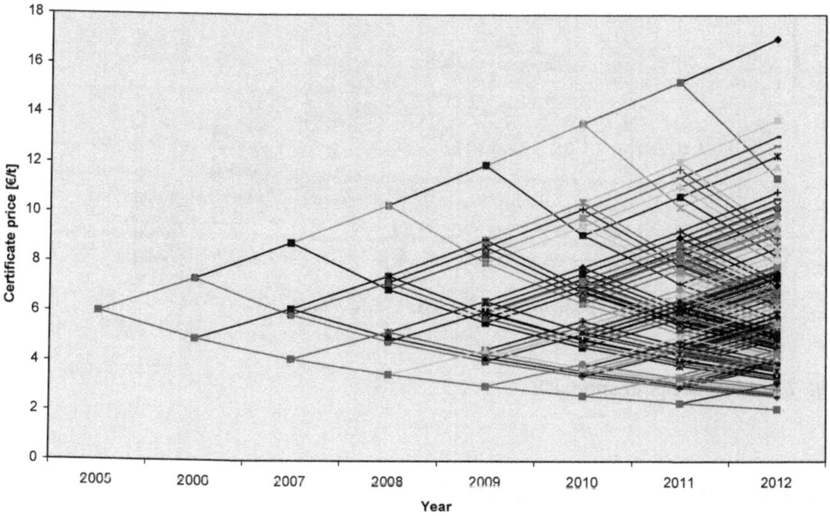

Fig. 1. Scenario tree of the stochastic price progression

In the tree shown one certain scenario is described by a path from the root node to one of the leaf nodes. Each node of the scenario tree represents a point of decision, where certificates can be sold or purchased or an emissions reduction project can be set up.

The project risk cannot be incorporated in this scenario tree as the project outcome can not be assigned to a certain time step. Therefore the scenario tree is cloned several times representing the price progression in different project outcome scenarios. The project cost is a measure for the - stochastic - project outcome; in figure 2 the developed approach is demonstrated.

The described decision process and its structural representation by the decision trees is the basis for the optimization model, which is mainly composed of submodels defined by the price scenario trees and formalized by multistage stochastic optimization models (Birge and Louveaux 1997; Kall and Wallace 1994) coupled by some non-anticipativity constraints. The non-anticipativity of trading-decisions has to be modelled explicitly (see part 4, equation 8), when the project outcome is not known at the time of decision making.

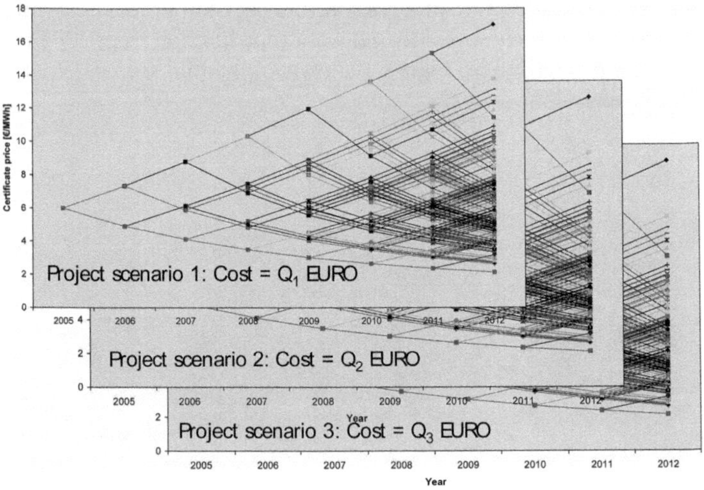

Fig. 2. Project scenario approach

3 Modelling the stochastic variables

The certificate price progression is modelled as mean-reversion process described by equation (1) (Clewlow and Strickland 2000), where yearly discretization is applied again.

$$x_t = x_{t-1} + \alpha(\bar{x} - x_{t-1}) + \sigma\varepsilon \quad (1)$$
$$\varepsilon \in N(0,1)$$
$$x = \ln P$$

with P: Certificate Price, α: Mean reversion rate, $\bar{x}$: Long term mean, σ: Standard deviation of the price returns, ε: Standard normal variable

Mean Reversion price models are usually applied on commodity markets. The main feature is that the price shows the trend to return to a long-term mean value. Strong mean reversion behaviour can be observed in many energy markets for example. If necessary, this model could be extended by assigning time dependency to the long term mean so that price trends can be modelled in this exemplary study the long term mean is assumed to be constant.

Usually the model parameters α, $\bar{x}$ and σ have to be estimated analyzing historical market data with regression methods. As the emissions market does not yet exist for the following exemplary calculations, the model parameters are assumed as follows:

$\alpha = 0.1$; $\bar{x} = 1.79 = \ln 6$; $\sigma = 0.2$

The chosen values induce the shape of the resulting price tree to appear realistic with respect to the actual limited market knowledge.

The scenario tree is built up successively beginning with the root node. The price steps are calculated using (1) with $\varepsilon = 1$ for upside steps and $\varepsilon = -1$ for downside steps. The node probabilities in every stage are equal and the sum of all node probabilities in one stage is 1. The resulting scenario tree has already been shown in Figure 1.

The project risk is usually estimated by analyzing the project outcome - respectively the project cost - of similar projects; an approach that needs a large project database. In this example the project risk is estimated using general experience in the field of plant construction. It is assumed that in the planning stage of a project the cost estimate accuracy is –15% to +30% with a confidence level of 80% (Jung 1993).

The mode[2] is assumed to be the estimated cost, so that the cost is more likely underestimated than overrated. Modelling this skew in the project cost distribution the project cost C_P is assigned to be lognormal distributed.

$$\ln C_P \sim N(\mu, \sigma^2)$$

For the following example the project cost estimation is set to 8.000.000 €. As the time horizon of the model is limited to 8 years (2005-2012), the investment costs are handled as variable cost of 1.000.000 € annually. Otherwise the earnings for a project which is started late in the time horizon cannot be determined adequately.

Considering the assumed cost estimate accuracy and the mode, it can be shown that $\sigma = 0.1624$ and $\mu = 13.84$. For the exemplary calculations 5 project cost samples are taken from the project cost distribution to represent 5 project scenarios. The probabilities for the five scenarios have to be normalized so that the sum of the project scenario probabilities equals one. Figure 3 and Table 1 show the 5 project scenarios and its normalized probabilities.

[2] The most likely cost.

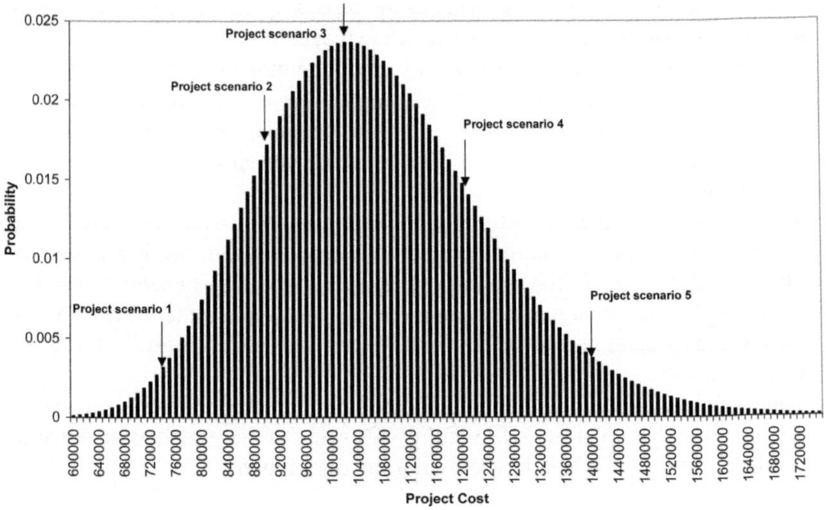

Fig. 3. Probability distribution of the project cost

Table 1. Project scenarios and normalized probabilities

Project Scenario No.	Project cost (€)	Probability
1	750.000	0.06
2	900.000	0.27
3	1.000.000	0.37
4	1.200.000	0.23
5	1.400.000	0.06

4 Optimization model

The optimization model for decision making in the emissions-market is formulated as a multi-criterial mixed-integer linear multi-stage stochastic optimization model. The basis for the model are the trees as described in part 2 of this paper. Every constraint has to be modelled for every node in the price tree for every project scenario. In the following the objective and the constraints of the model are described in detail. For clarity in the following equations model variables are printed in bold letters.

Objective

Objective of the optimization model is the simultaneous maximization of the expected profit and the minimization of risk. This can be expressed in the objective function by a weighted sum of the expected profit and a risk measure. As in this model the cash-flow is calculated for each node in the price tree for each project scenario. The expected profit can be calculated as the weighted sum of the cash-flows in the leaf nodes of the scenario trees.

The Conditional Value at Risk (CVaR) is chosen as risk measure as it is convex and can be modelled with only linear restrictions (Uryasev 2000). In short the CVaR is the mean loss of the (1-ß)-worst part of a loss distribution (see also (9)). Moreover, the CVaR-calculation in optimization models allows the estimation of the more well-known risk measure Value at Risk[3] (VaR). Since CVaR has to be minimized, the CVaR is added into the objective with a negative sign.

$$Max\ (1-\alpha)\frac{1}{N_L}\sum_{n\in L}\sum_{s\in S} p_s \mathbf{CF_{n,s}} - \alpha \mathbf{CVaR} \tag{2}$$

with α: risk aversion, N_L: Number of leaf nodes, L: Set of all leaf-nodes of the price-tree, S: Set of all project scenarios, p_s: Probability of the project scenario s, $CF_{n,s}$: Cash-flow in node n and project scenario s $CF_{n,s} \in IR$, CVaR: Conditional Value at Risk $CVaR \in IR$.

Defining the individual risk aversion by setting the risk aversion parameter α is rather arbitrary as the optimization solution could be quite sensitive to slight variations in the weighting parameter. A better way – especially in times of powerful IT-resources – is to calculate the optimization several times with different weighting parameters to generate the so called "efficiency frontier". In part 5 of this paper the generation of the efficiency frontier is demonstrated.

Certificate balance

The number of certificates in one node of the price tree for one project scenario equals the number of certificates in the parent node plus the allocated certificates minus the certificates to "pay" the own emissions plus the certificates traded in the emissions market.

$$\mathbf{c_{n,s}} = \mathbf{c_{parent(n),s}} + a_n - \mathbf{e_{n,s}} + \mathbf{m_{n,s}} \quad \forall n = 1,\dots,N; s = 1,\dots,S \tag{3}$$

with $c_{n,s}$: Number of certificates in node n and project scenario s $c_{n,s} \in IR^{\geq 0}$ ($c_{n,s}$ can be interpreted as the amount of banked certificates after the "payment" of the emissions), a_n: Number of allocated certificates in node n (Remark: The allocation is not project scenario dependent), $e_{n,s}$: Emissions in node n and project scenario s $e_{n,s} \in IR^{\geq 0}$, $m_{n,s}$: Number of sold or purchased certificates from the emissions market in node n and project scenario s, positive values represent a purchase $m_{n,s} \in IR$.

---

[3] The VaR is the maximum loss that can occur within a certain confidence level ß.

Cash balance

The cash-flow in one node of the price tree for one project scenario equals the cash-flow in the parent node minus the expenses in the emissions market minus the cost for the emissions reduction project if it is already set up, plus the return from money put into a risk free investment.

$$\mathbf{CF_{n,s}} = \mathbf{CF_{parent(n),s}} - \mathbf{m_{n,s}} \cdot P_n - \mathbf{E_n} \cdot Q + \mathbf{I_{parent(n),s}} \cdot \mathrm{r} \quad \forall n = 1,\ldots,N; s = 1,\ldots,S \tag{4}$$

with P_n: Price for one certificate in node n (Remark: The price is not project scenario dependent), E_n: Decision variable $E_n \in \{0;1\}$, 0 means the emissions-reduction project is not set up, 1 means the project is set up (Remark: the decisions variable must not be project scenario dependent because the project outcome is not known before the project is set up), Q: Yearly net-cost of the emissions-reduction project, $I_{n,s}$: Risk free investment sum $I_{n,s} \in IR^{\geq 0}$, r: Interest rate.

Budget constraint

It is assumed that there is a certain start budget B for the activities in the emissions market. Furthermore, it is assumed that all money that is not invested in the emissions reduction project or in emissions certificates is put into a risk free investment.

$$\mathbf{I_{n,s}} = B + \mathbf{CF_{n,s}} \tag{5}$$

As $I_{n,s}$ is a positive variable (5) also implies a lower bound for the cash-flow, namely the negative start budget.

Emissions calculation

For the emissions in one node of the price tree for one project scenario, there are two possibilities. Either the emissions reduction project is already set up, or it is not. If the project is set up, the emissions are the forecasted emissions minus the emissions reductions by the project. Otherwise, the emissions equal the forecasted emissions.

$$\mathbf{e_{n,s}} = e_{base} - \mathbf{E_n} \cdot R \quad \forall n = 1,\ldots,N; s = 1,\ldots,S \tag{6}$$

with e_{base}: Forecasted baseline emissions without the emissions-reduction project, R: Yearly emissions-reduction of the project.

Project constraint

For the decision-variables of the emissions reduction project another constraint has to be implemented that ensures that when a project is set up, it exists in all subsequent nodes.

$$\mathbf{E_n} \geq \mathbf{E_{parent(n)}} \quad \forall n = 1,...,N \tag{7}$$

This means: If E_n=1 then E_{n+1} must be 1, too. If E_n=0 then E_{n+1} can be either 0 or 1.

Non-anticipativity of the traded certificates

One non-anticipativity restriction has to be formulated explicitly. It has to be required that the number of certificates in the same nodes in the price tree for different project scenarios has to be equal as long as the project is not set up. When the project is not set up the project outcome is unknown and, therefore, there is no reason why the market activities should be different.

$$\begin{aligned} &\mathbf{c_{n,s}} = \mathbf{c_{n,s-1}} + \mathbf{slack_{n,s}} \\ &\mathbf{slack_{n,s}} \leq c_{\max} \cdot \mathbf{E_n} \\ &\mathbf{slack_{n,s}} \geq -c_{\max} \cdot \mathbf{E_n} \\ &\forall n = 1,...,N; s = 2,....,S \end{aligned} \tag{8}$$

CVaR-calculation

The following equations are needed to model the CVaR for linear stochastic models with a discrete representation of the stochastic variables via scenarios (Uryasev 2000; Rockafellar 2002).

$$\begin{aligned} &\mathbf{z_{n,s}} \geq (g - \mathbf{CF_{n,s}}) - \xi \quad \forall n \in L; s = 1,...,S \\ &\mathbf{z_{n,s}} \geq 0 \quad \forall n \in L; s = 1,...,S \\ &\mathbf{CVaR} = \xi + \frac{1}{(1-\beta)N_L} \sum_{n \in L} \sum_{s \in S} \mathbf{z_{n,s}} p_s \end{aligned} \tag{9}$$

with g: Target-win, β: Confidence level, L: Set of all leaf-nodes of the price-tree, $z_{n,s}$, ξ: Auxiliary variables.

As already mentioned the CVaR represents the mean loss of the (1-ß)-worst part of a loss distribution. Now let ß be 0.8. In the scenario representation the CVaR can then be interpreted as the mean loss in the 20% worst scenarios of the loss distribution. In this model the loss distribution is calculated by subtracting the leaf-cash-flow from a certain target win. In the following the target win is assumed to be 0 €

5 Exemplary results

In the presented example a plant causing annual emissions of 1.000.000 t CO_2/a is considered. The gratis certificate allocation for this plant amounts to 950.000 t CO_2eq/a. The emissions reduction project is assumed to reduce 220.000 t CO_2/a, so that the specific CO_2 reduction cost varies, depending on the project scenario from 3.41 €/t to 6.36 €/t. The start budget is assumed to be 8.000.000 €, the risk free interest rate is fixed at 3%. The confidence level for the CVaR Calculation β is set to 0.8. The stochastic variables certificate price and project costs are approximated as described in part 3. The optimization is calculated 20 times with different values for the weighting parameter α to generate the efficiency frontier (see figure 4).

The solutions in the upper right part of the graph are fundamentally different from the other solutions. In this risky but most profitable solution, the emissions reduction project starts in 2005. In all other solutions the project starts later, depending on the price developments.

It is obvious that the efficiency frontier is very advantageous for decision making when both the profit and the risk have to be regarded. Choosing a proper solution according to the individual risk aversion can be done either intuitively or by applying methods from decision theory (Laux 1995), due to the shape of the efficiency frontier.

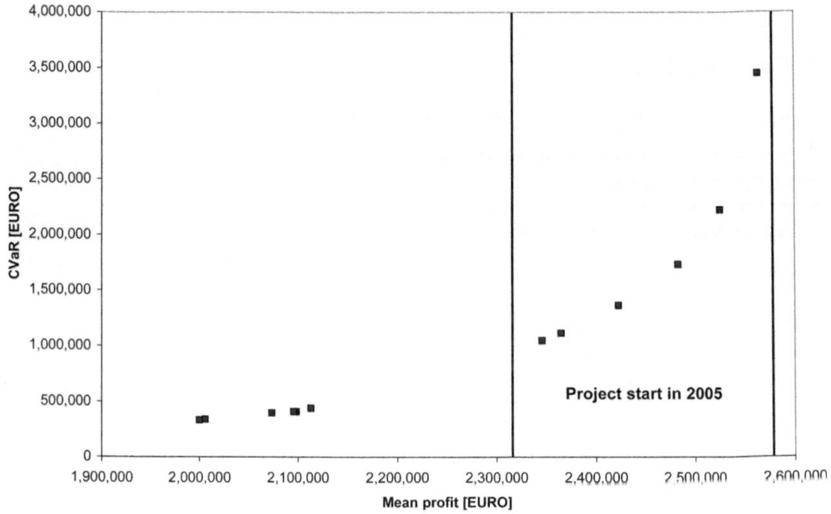

Fig. 4. Efficiency frontier

Next, a deeper view into the solution will be provided. Figures 5-8 show the results for the project decisions and the certificate portfolio for the different nodes in the decision trees. Here the most and least risky solutions (top right and bottom left solutions in the efficiency frontier) are displayed. Remember that the project decisions are project scenario independent, whereas the certificate portfolio can be scenario independent for nodes in which the project outcome is already known. In the following the certificate portfolio is exemplarily shown for the project scenario 3. For the project decisions figure a positive project decision is identified by a larger square in the decision tree, the size of the squares in the certificate portfolio figure denotes the amount of certificates in the portfolio.

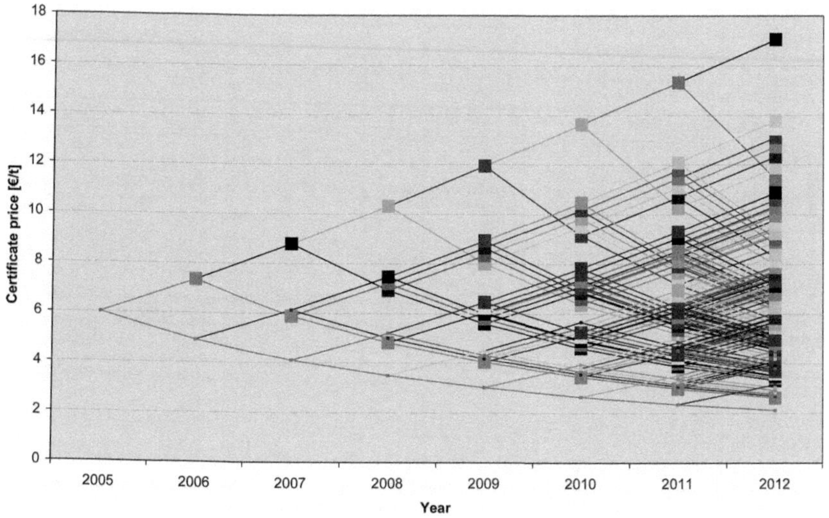

Fig. 5. Project decisions in the low risk solution

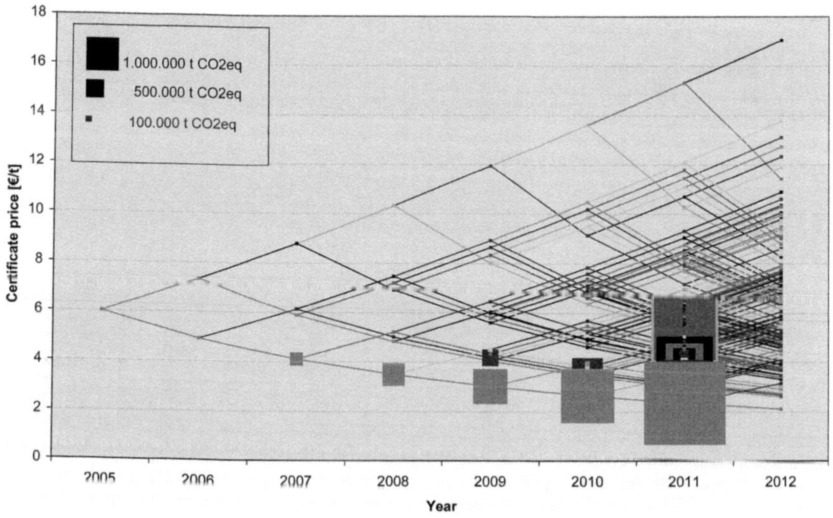

Fig. 6. Certificate portfolio in the low risk solution (project scenario 3)

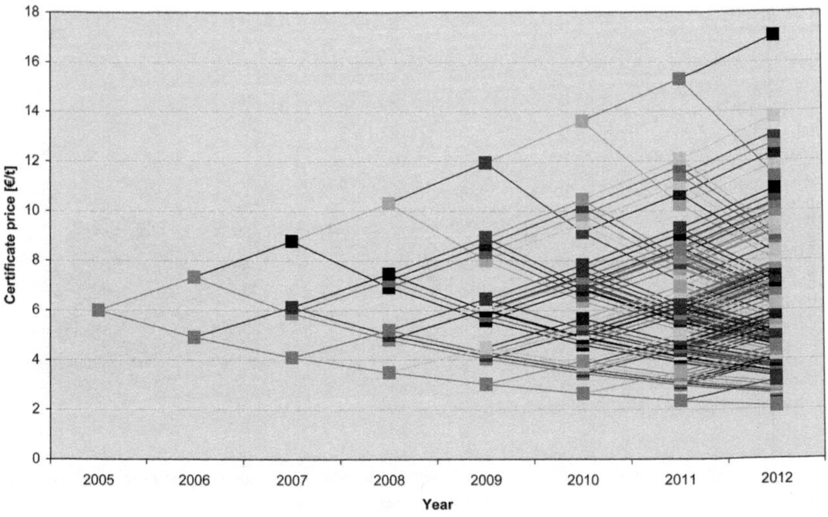

Fig. 7. Project decisions in the high risk solution

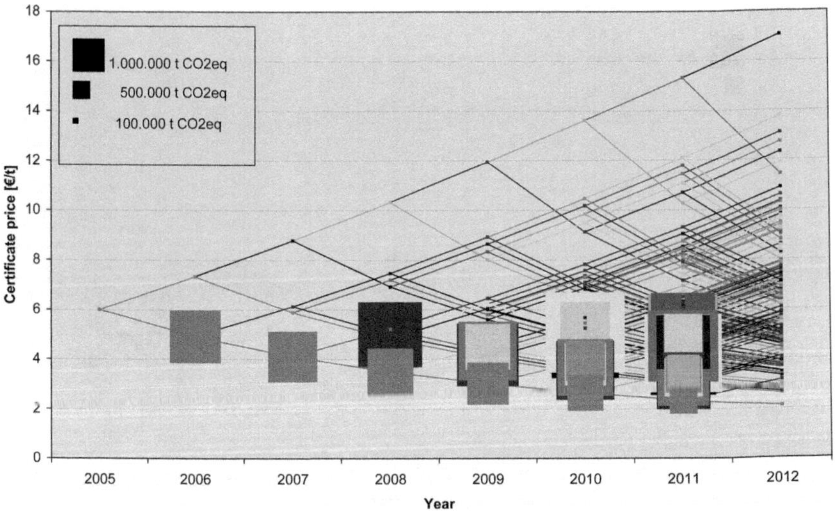

Fig. 8. Certificate portfolio in the high risk solution (project scenario 3)

When analyzing the optimization results it has to be kept in mind that the only practically relevant information of a stochastic optimization solution is the first stage solution (here: Year 2005), because the first stage variables are the only deterministic variables in the model. The further stages are only required for evaluation of the first stage decisions. Follow up decisions in 2006 or 2007 can not be extracted from this solution. They have to be determined by continuous optimization of the decision model with actual model data and a rolling optimization time horizon.

In this case the first stage solution shows that a risk-averse decision maker would decide not to set up the project immediately. If the decision maker is up to taking more risk, he would set up the project immediately.

At last a more detailed presentation of the cash-flow development for the different scenarios is given. For this purpose the cash-flows for project scenario 3 in the low risk solution are shown in Figure 9.

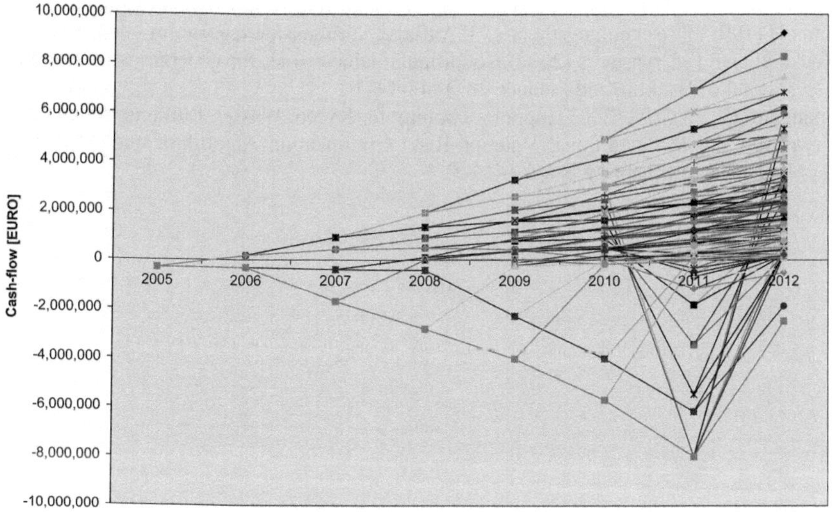

Fig. 9. Cash-flows in the low-risk-solution (project scenario 3)

6 Summary and outlook

The presented stochastic decision model supports make-or-buy decisions in the emissions market, while taking into account price risks of the emissions market and project risks of technical emissions reduction. The model provides several different solutions with different profit and risk levels forming an efficiency frontier.

The developed approach represents a basic method for decision making in the emissions market that can be extended for specific project types by a deeper analysis of the project risk, especially when projects aim at profits outside the emissions market, e.g. fuel saving. In these cases these profits and their underlying risk should be incorporated.

References

Birge JR, Louveaux F (1997) Introduction to Stochastic Programming. Springer Verlag, New York

Clewlow L, Strickland C (1998) Energy Derivatives: Pricing and Risk Management. Lacima Group, London

Jung J (1993) Angewandte Kosten- und Wirtschaftlichkeitsrechnung bei der Anlagenprojektierung. Begleitmaterial zur Vorlesung. Fachbereich Chemietechnik, Universität Dortmund

Kall P, Wallace SW (1995) Stochastic Programming. John Wiley & Sons, Chichester

Laux H (1995) Entscheidungstheorie. 3. Auflage, Springer-Verlag, Berlin

Rockafellar RT, Uryasev S (2002) Conditional value-at-risk for general loss distributions. Journal of Banking and Finance 26: 1443-1471

Spangardt G, Lucht M (2003) jupiter - Training für Kyoto. Wasser, Luft und Boden 08

Uryasev S (2000) Conditional Value-at-Risk: Optimization Algorithms and Applications. Financial Engeneering News 14: 1-5

The impact of climate policy on heat and power capacity investment decisions

Harri Laurikka

Helsinki University of Technology
Laboratory of Energy Economics and Power Plant Engineering
P.O. Box 4100, 02015 TKK, Finland
harri.laurikka@tkk.fi; harri.laurikka@greenstream.net

Abstract

Climate policy has become a major source of uncertainty in energy investments. This paper explores how different instruments of climate policy, such as emissions trading and taxes, affect heat and power capacity investment decisions. I start here with an analysis on the role of climate policy instruments in an investment decision process. Secondly, I examine how climate policy instruments affect the key components of a quantitative investment appraisal and how flexibility can help to cope with those impacts. Flexibility characteristics of some existing heat and power generation technologies are discussed. I find that climate policy increases the value of flexibility in energy investments. There are structural differences in flexibility between heat and power generation technologies. Whereas some technologies provide managerial flexibility through the options to alter operating scale and through the option to switch between fuels or products, others provide passive flexibility (robustness).

Keywords: Investment, heat and power generation, climate policy instruments, flexibility

Acknowledgement: I would like to thank Pekka Pirilä, Peter Letmathe, Ilkka Keppo and Frank Gagelmann for valuable comments. All remaining errors are those of the author. The financial support of National Technology Agency of Finland and the Environment Pool of the Finnish Energy Industries Federation is gratefully acknowledged. The views expressed in this paper are those of the author and do not necessarily reflect the opinions of the financiers.

1 Introduction

Climate policy and the regulation of greenhouse gas (GHG) emissions have become a new factor affecting strategy and investment decisions. Heat and power producers are among those facing the largest changes due to their high emission intensity (emissions / turnover).

Globally, 4.800 GW of new electric capacity and investments of about 4.6 trillion USD are projected to be required until 2030 (IEA 2004). In addition, other final energy use in residential, services, industrial and agricultural sectors is expected to grow by 16.000 TWh (37%)[1] until 2030 (our own calculation based on IEA 2004), implying an increasing demand for heat only plants. All in all, it is recognized that "*increased investment in the energy sector, from both public and private sources, is necessary*" (EDC 2004). Also, large numbers of power plants will change hands as a result of *acquisition processes* in deregulated markets.

The economic lifetime of an investment in heat and power capacity typically ranges from 20-40 years (OECD NEA/IEA 1998). During the lifetime, various policy instruments intended to regulate GHG emissions can influence the cash flows of the plant and thus its viability. Such instruments can range from problem-specific tradable emission permits[2] to more general policy instruments, such as taxation (fuel tax, emission tax) and subsidies (investment subsidy, fixed feed-in tariff)[3].

Climate change related decision-making is essentially a sequential process under uncertainty (IPCC 2001). The development of climate science, climate policy goals and business-as-usual emissions is uncertain for an investor ("how big should the emissions reduction cake be?"). Moreover, the negotiation results between various parties, i.e. countries, sectors, and companies, create uncertainty ("how big is our piece of the emissions reduction cake?").

Assuming that the future of the energy sector is to a significant extent "carbon-constrained" - which seems to be the case, in Europe in particular - any investment or valuation should take into account the financial impacts of climate change mitigation (e.g. Vrolijk 2002; de Leyva and Lekander 2003; IEA 2003b). The European Union emissions trading scheme (EU ETS) will be one of the key climate policy instruments until 2012 (EC 2003). In the medium-term, European utilities are likely to consider abatement options based on traditional technologies, and be keen on emissions trading due to the high long-term technological uncertainty (Söderholm and Strömberg 2003).

Climate policy instruments are often overlapping, i.e. the introduction of a new instrument (e.g. tradable permits) causes a need to change the earlier instruments (e.g. taxation). In most European countries several climate policy instruments are likely to co-exist in the future for many reasons. First, the structure of climate pol-

---

[1] Includes petrochemical feedstocks.

[2] In this paper, I apply the term *tradable emission permit* as a generic concept including emission *credits* (e.g. within the project-based Kyoto mechanisms) and emission *allowances* (e.g. within the EU ETS).

[3] For a good overview of climate policy instruments, see IPCC (2001).

icy instruments is already very heterogeneous in many countries (Vrolijk 2002; OECD/EEA 2003). Second, the United Nations Framework Convention on Climate Change (UNFCCC), the EU ETS, and the plans for tradable green certificate systems in several countries will enhance the role of currently less influential market-based climate policy instruments in the coming years. Third, climate policy instruments have complementary goals, which implies that they would not replace each other completely (Johnstone 2003). Finally, a policy mix consisting of several instruments can help to reduce abatement cost uncertainty, overcome technology market failures and increase behavioural responsiveness (Johnstone 2003).

It is, therefore, important for individual enterprises and/or investors to understand the logic of how different climate policy instruments affect their investment decisions, in order to recognize the structural differences between investment alternatives regarding climate policy instruments, and to be able to make informed predictions about the impacts of climate policy. At present, companies are not always fully informed about the quality of information and application of decision-support technologies (IPCC 2001).

This paper explores the mechanisms through which climate policy instruments affect heat and power capacity investment decisions. First, I analyze the role of climate policy instruments in an investment decision process. Second, I examine how climate policy instruments affect the key components of a quantitative investment appraisal and how flexibility can help to cope with those impacts. Flexibility characteristics of some existing heat and power generation technologies are discussed and compared.

2 Investment decision process

Three kinds of factors affect strategic decision processes such as investments: (1) organization's operating environment (e.g. uncertainty and complexity), (2) organizational conditions (e.g. internal power structure, past performance, past strategies and the extent of organizational slack) and (3) decision-specific factors (e.g. impetus for the decision, the urgency associated with the decision, the degree of outcome uncertainty, and the extent of resource commitment) (Rajagopalan et al. 1993). Environmental factors are macro level variables: they are equal to all investment decisions within the observed investment environment. Organizational and decision-specific factors influence at micro level and vary between decisions.

The growing importance of climate policy instruments is reflected in all three categories (Figure 1). Climate policy changes stochastic environmental factors, such as electricity market price, fuel prices and allowance market price[4]. Organizations perceive these future changes differently. This implies not only differences in price forecasts, but also, more profoundly, differences in perceptions of the problem character: some managers and board members believe that climate policy is "here to stay", whereas others expect it to be "a short-term fashion". The differ-

[4] This classification assumes that the investor cannot exert market power.

ent perceptions most likely result from diverse evaluations and beliefs regarding both the scientific foundation of climate change research[5] and the expected reactions of governments and companies.

Decision-specific factors are linked to location-specific risks, e.g. whether certain climate policy instruments, such as investment subsidies or emission credits, are relevant in a specific investment context. Within the EU ETS, allocation of free emission allowances can also depend on the situation on the planned site: e.g. whether the investment is considered a new entrant or a modification of an existing installation. Climate policy can also change other decision-specific factors affecting the process, such as heat demand and price. Some fuels, such as biomass, have more location-specific prices than others, such as coal. Climate policy is likely to improve the availability of bio-fuels for energy generation.

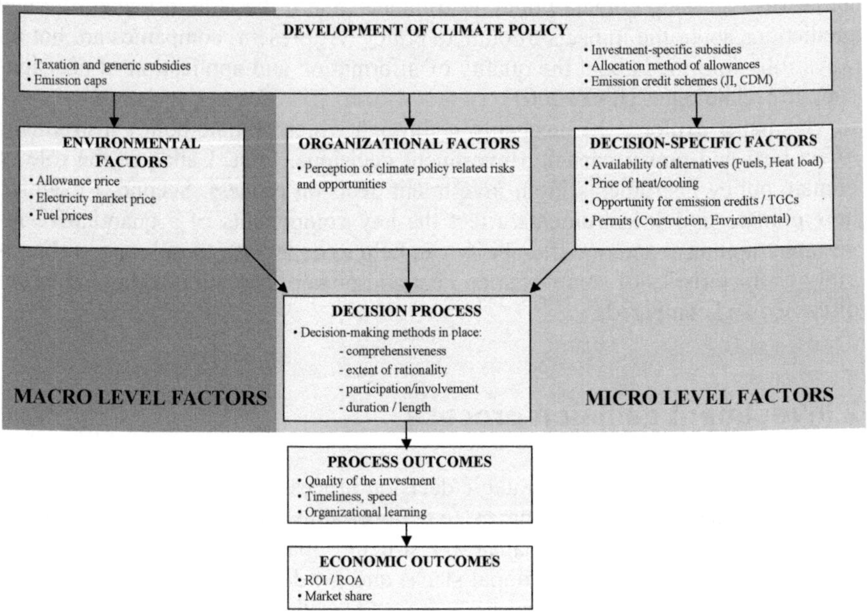

Fig. 1. Climate policy instruments in the investment decision process (modified from Rajagopalan et al. 1993).

The resulting decision process characteristics, such as the duration of the process and the degree of rationality, help determine process outcomes, such as the timeliness of the decision and the extent of organizational learning. Process characteristics as well as process outcomes influence economic outcomes.

Decision processes involving evaluation of capital expenditure typically consist of at least four steps: (1) Identification of spending proposals; (2) Quantitative analysis of the incremental cash flows; (3) Qualitative issues which cannot be fit-

[5] This is well illustrated by the existence of industrial organizations like the Global Climate Coalition and the Business Environmental Leadership Council in the US. See also Innovest (2003).

ted into the cash-flow calculus; and (4) Making the decision (Shank 1996). The importance of the identification of values and objectives as generators of creative alternatives for phase 1 has also been highlighted (Keeney 1994; Clemen and Reilly 2001). In this paper I assume, for simplicity, that decision-makers aim at maximizing their long-term financial performance without taking into account any other objectives they might have.

Decision processes vary according to their extent of *rationality* and *comprehensiveness*. Human behaviour in organizations is intentionally rational, but only boundedly so, which means that human decision-making is non-optimizing in practice (Simon 1997; Selten 2002). An organizational decision is therefore deemed rational if it is orientated towards the organization's goals (Simon 1997). Three types of rationality have been proposed as drivers of investments in energy capacity: (1) economic viability, (2) understanding of the business model of the investment, and its fit to the corporate strategy, and (3) the impact on the operational performance (Sandoff 2003). Comprehensiveness has been defined as the extent to which organizations attempt to be exhaustive or inclusive in making and integrating strategic decisions (Fredrickson and Mitchell 1984). In heat and power capacity planning, comprehensiveness has been pursued through large energy models in order to overcome the deficiencies of individual techniques, such as optimization, simulation and decision analysis. This "model synthesis" has been criticized as an elusive and ultimately impractical objective (Ku 1995). In line with this argument, for example, the record of US model-based energy forecasting yields evidence that the models have given biased estimates (Laitner et al. 2003).

The results of energy models, such as energy prices, have been used as inputs in the strategic investment appraisal. Standard approaches to strategic investment appraisal include quantitative methods such as payback, internal rate of return (IRR), return on investment (ROI), and discounted cash-flow (DCF) methods. In addition to these, an expanded DCF-based framework and a number of other analytical techniques, such as strategic cost management (SCM), the multi-attribute decision model (MADM), value analysis, the analytical hierarchy model, and the uncertainty method, have been developed in literature[6].

Perhaps the most important complement to the standard approaches is the *real option theory* and its applications, which have been strengthened by a substantial body of research during the last two decades[7]. The key reason for this interest has been the notion that standard approaches are inadequate in that they cannot properly capture management's flexibility to adapt and revise later decisions in response to market developments. Standard approaches are based on "*decisioneering*", which assumes that the future is probabilistic and an optimal course of action can be found by comparing payoffs in scenarios (Lessard and Miller 2001). The standard approaches implicitly assume "expected scenarios" of cash flows and presume management's passive commitment to the selected operating strategies.

---

[6] For a review, see Adler (2000).

[7] For an overview, see e.g. Trigeorgis (1995) or Schwartz and Trigeorgis (2001).

The proponents of real options or *"managerial approaches"* assume the future indeterminate, where outcomes are not only difficult to assess but depend on exogenous events or endogenous processes that can lead to multiple possible future states (Lessard and Miller 2001). Hence, the standard approaches should be extended to take the value of new information into account by modelling flexibility for contingency (Ku 1995; Trigeorgis 1995). Two kinds of flexibility can be identified corresponding to two ways of responding to uncertainty: *active* flexibility, a state of readiness such as the ability to react to change, and *passive* flexibility (often referred to as „robustness") as a state of being, such as a resistance or an immunity to change (Ku 1995).

A key impact of the real option theory in the context of this paper is that it opens up a new perspective to the evaluation of an investment appraisal. The traditional valuation tools do not work well, if there is a high uncertainty about the future and options available (Amram and Kulatilaka 1999). Climate policy has significantly increased uncertainty in the business environment through the value of tradable emission permits, the number of free emission allowances, and the impacts on subsidies and taxation. This results in a higher value for flexibility. The decision context and various energy technologies hold characteristics that need to be taken into account in order to end up in (organizationally) rational decision outcomes.

3 Quantitative investment appraisal

A quantitative investment appraisal can be broken down into sub-problems from the Net Present Value (NPV) framework: the initial investment, the annual cash flows and the discount rate. I will go through them one by one and discuss the role of climate policy instruments in each case. The following analysis is limited to financial implications.

3.1 Initial investment

The initial investment cost (I_c) of a plant with an output capacity P_{out} is given by

$$I_c = P_{out} \cdot i_c = P_{unit} \cdot n \cdot i_c(P_{unit}, t) \quad (1)$$

with P_{unit} being the characteristic unit size (in MW) of the technology in question; n the number of units required to generate P_{out}; and i_c the specific investment cost (in €/kW) for a given level of P_{unit} at time period t. P_{unit} differs significantly between technologies: in particular, nuclear and coal plants tend to have large unit sizes, whereas renewable energy plants have small unit sizes. Typical unit sizes range from 455 - 1.460 MW_e for nuclear power, 120 - 800 MW_e for coal and 250 - 750 MW_e for a combined cycle gas turbine (OECD NEA/IEA 1998). The average wind turbine size in the largest single market in Germany was 1.4 MW in 2002 (EurObserv'ER 2003).

The specific investment costs i_c for energy production technologies vary by technology, and typically show fairly strong economies of scale. i_c also tends to decrease in time as a result of technological innovation, if quality requirements remain the same. This is generally modelled in a formulation, in which each doubling of experience (e.g. cumulative capacity) reduces i_c by a certain percentage called the *learning rate*. Importantly, both changes in experience and learning rates differ by technology (McDonald and Schrattenholzer 2001).

Climate policy can affect the rate of experience accumulation for technologies and thus change i_c. Additionally, climate policy may change any regional investment subsidies available for different technologies. Let us denote this investment subsidy as a percentage (k) of the total investment. Then the effective specific investment cost i_{eff} at time t for the investor is[8]:

$$i_{eff,t} = (1-k_t)\cdot i_{c,t} = (1-k_t)\cdot i_{c,o}\cdot(1-r\frac{\ln(\frac{C_t}{C_o})}{\ln 2}) \tag{2}$$

with $i_{c,t}$ being the specific investment cost at time t,r being the learning rate, and C_t being the experience measure (e.g. the cumulative capacity) at time t. Combining Equations 1 and 2, we get:

$$I_{c,t} = P_{unit}\cdot n\cdot(1-k_t)\cdot i_{c,o}\cdot(1-r\frac{\ln(\frac{C_t}{C_o})}{\ln 2}) \tag{3}$$

Equations 1 to 3 formalize the concept of *modularity*: in an uncertain environment, modularity enables the operator to implement the investment in several phases, i.e. it provides an *option to staged investment* (e.g. Dixit and Pindyck 1994, p. 51-55; IEA 2003a, p. 29). In addition, each investor also has an *option to wait*: either to invest in one of available alternatives or to postpone and to wait for more favourable conditions (e.g. McDonald and Siegel 1986). An exception might be a situation in which the existing thermal capacity is running out (e.g. in a district heating system). Combined, these two options seem to create a situation, whereby a higher uncertainty delays capacity additions, but also makes those additions larger when they do occur (Pindyck 2001; Kort et al. 2004).

3.2 Annual cash flows

When the regulator applies economic climate policy instruments, cash flows of energy projects during a period can be simplified as follows:

[8] Equation 2 is deterministic. See Murto (2003) for a stochastic case.

$$R - C = P_{tot} \sum Sx + R_{f,CO2} - C_f$$
$$= P_{tot} \sum \left[\alpha\eta(\gamma)\frac{\gamma}{\gamma+1} p_e + \eta(\gamma)\frac{1}{\gamma+1} p_h + r_{CO2} - \mu c_v - c_{CO2} \right] \cdot x + R_{f,CO2} - C_f \tag{4}$$

with R being the revenue (in €) and C the total cost (in €). P_{tot} denotes the thermal capacity of the plant and S the *spark spread* i.e. the difference between the unit price of energy and the variable production cost of the plant (in €/MWh$_{th}$). x denotes the number of full-load operating hours during the period. $R_{f,CO2}$ (in €) is the corresponding climate-policy related fixed revenue, such as the value of free emission allowances, and C_f is the corresponding fixed cost (in €). p_e is the average market price of electricity (in €/MWh$_e$), γ is the *power-to-heat ratio* and $\eta(\gamma)$ is the thermal efficiency of the plant for a given power-to-heat ratio. α is the *production profile factor* of the technology in question during the period: if the technology can systematically benefit from output price fluctuations during the period, then the production profile factor exceeds one ($\alpha > 1$) and vice versa[9]. The price of heat delivered is denoted by p_h (in €/MWh$_h$).

Revenue per unit from any climate-policy related products, such as tradable green certificates, is given by r_{CO2} (in €/MWh$_{th}$) and c_{CO2} is the climate-policy related variable cost (in €/MWh$_{th}$), such as an emission tax or the market price of a surrendered emission allowance. c_v is the average "normal" variable cost including fuel costs (in €/MWh$_{th}$).

Corresponding to α, μ is the *consumption profile factor*. If the technology can systematically benefit from fuel price fluctuations during the period, then the consumption profile factor undercuts one ($\mu < 1$) and vice versa.

There is no multiplier for r_{CO2} or for c_{CO2} in Equation 4. Tradable emission permit markets do not have a similar basis for a systematic daily and monthly price fluctuation as compared to the electricity market, since compliance is required only on an annual basis[10]. In power exchanges, the market needs to be settled every hour.

Variables of Equation 4 may all be affected by climate policy. I will first take a look at revenues and then move to the cost side.

Market price of energy products (p_e, p_h): Economic theory suggests that market price of energy products will increase due to emissions trading or taxes. This results from increased production costs in utilities (see below), which are reflected in output prices. The effect is similar in heat production. Heat suppliers are even more likely to transfer increased production costs into prices due to the monopolistic character of the business.

In the power sector, theory and studies suggest that the price increase in p_e reflects the emission factor of the power plant on the margin, which often is a con-

[9] Note that α is not necessarily constant in time.

[10] Note that prices can still fluctuate systematically for other reasons.

densing power plant due to the merit order of technologies. For this reason, the projected price increases can be significant. It has been projected that the EU ETS can raise wholesale power prices from 15% to 60% (de Leyva and Lekander 2003; Ernst&Young 2004; Reinaud 2003).

Combined heat and power (CHP) extraction plants with simultaneous excess capacity in heat production have a *switching option* due to *product flexibility*. This means that during a period of high power prices, a CHP extraction plant can produce more power and shift the corresponding heat production into heat only plants and vice versa. Thus the switching option due to product flexibility is reflected in the operator's ability to optimize the power-to-heat ratio (γ) between [γ_{min}, γ_{max}] in Equation 4. Backpressure plants do not typically have this option: the power-to-heat ratio is constant.

Revenue from climate policy related products (r_{CO2}): Investments can gain additional revenues due to new climate policy instruments in several forms:

- Tradable green certificates: in some countries, such as the Netherlands and Sweden, national trading schemes for green certificates have been established.
- Higher feed-in tariffs or tax subsidies for green electricity are used in many countries, such as Germany and Austria.
- GHG emission reduction credits: heat and power capacity investments in one of the signatory states of the Kyoto Protocol may obtain additional financing through joint implementation (JI). A host country approval is always required, and in the EU countries, a JI project affecting installations within the EU ETS may be implemented only, if an equal amount of allowances is cancelled from the registry of the host country (EC 2004). In the developing countries investments complying with the relevant rules and modalities and with approval from a host country may similarly obtain additional financing through the clean development mechanism (CDM).

Fixed revenues related to climate policy ($R_{f,CO2}$): Within the EU ETS, utilities obtain at least 95% of the emission allowances for the period 2005-2007 and at least 90% for the period 2008-2012 free of charge. In an investment analysis, this asset transfer may be considered a fixed - but uncertain - annual cash in-flow for the plant. The transfer is linked to a simultaneous *obligation* to surrender emission allowances during the same period, i.e. R_{CO2} is linked to the introduction of c_{CO2}. Free allowances are obtained regardless of whether the plant is used during *that* period or not. The fixed revenue for *subsequent* periods may change due to the selected operating strategy in the *preceding* periods, i.e. $R_{CO2,\ t+j}$ may depend on x_t, with j being, for example, between 1 and 5 periods.

Full-load operating hours (x): of a technology depend on the characteristics of demand and the technology in question. Operators of thermal plants have an *option to alter their operating scale* if the market turns out to be unfavourable, i.e. if the spark spread S is negative, the plant can be turned down taking into account the physical constraints, such as lead times, and switching costs (McDonald and Siegel 1985; Tseng and Barz 2002). This implies that:

$$R - C = P_{tot} \sum MAX \left[Sx - C_{sw,u}, -C_{sw,d} \right] + R_{f,CO2} - C_f \quad (5)$$

where $C_{sw,u}$ and $C_{sw,d}$ are the (potential) switching costs in start-up and shutdown (in €/MW). Power plants that rely on flowing resources (e.g. wind and run-of-river hydro power) or base load plants with heavy initial investments and low fuel costs (e.g. nuclear power) also have this option, but its value is negligible: in normal circumstances the optimal decision is to run the plant as much as possible. For wind and run-of-river hydro power plants, the number of operating hours is a stochastic variable depending on weather conditions and technical availability – not on managerial judgment.

Conventional variable cost (c_v): In thermal power plants, fuel costs constitute the bulk of the variable costs. Climate policy instruments can indirectly change fuel prices. The magnitude of this impact is not necessarily defined within the fuel market context only, since coordination with the tradable emission permit market might occur (Hagem and Mæstad 2002; Holtsmark 2003). It has been estimated that the producer price of coal would decrease by 7-10% and the price of oil by 2 % within an (unreachable) 100% implementation[11] of the Kyoto Protocol (Holtsmark and Mæstad 2002). Most analysts project an increase in natural gas prices due to the EU ETS (e.g. Ernst & Young 2004; de Leyva and Lekander 2003; Reinaud, 2003).

Many thermal heat and power producers have a *switching option* due to *process flexibility*. They are able to burn two or more different fuels and switch between them, or ex-post conversions are relatively straightforward and inexpensive (Kulatilaka 1993; Söderholm 2000). For this reason, Equation 5 can be modified as follows for multi-fuel plants:

$$R - C = P_{tot} \sum MAX \left[MAX_f (S_f x - C_{sw,f}) - C_{sw,u}, -C_{sw,d} \right] + R_{f,CO2} - C_f \quad (6)$$

with $MAX_f (S_f x - C_{sw,f})$ referring to the opportunity of the producer to optimize the operating cost through fuel selection taking into account the potential related switching costs ($C_{sw,f}$).

Climate policy related variable costs (c_{CO2}): Similarly to revenues, climate policy related costs potentially accrue in many forms, such as in taxes and in the value of surrendered allowances. Emissions are determined by the energy output, the emission factors of the fuels used and the heat rate of the plant[12]. Emission factors can generally be considered constant in a long-term investment analysis, although some uncertainty concerning global warming potentials (GWPs) of different greenhouse gases, for example, can be taken into account.

---

[11] Including the United States and Australia.

[12] Methane and nitrous oxide emissions are also affected by the combustion technology in question.

Hoff and Herig (1996) point out that *"the decision to build any polluting generation source includes the plant owner's decision to give a valuable option to the government. The option gives the government the right (but not the obligation) to change emission standards or impose externality costs (i.e. environmental taxes) associated with environmental damages at any time."* In fact, this option applies for all plants, not only for directly polluting plants. In the context of climate policy, it is, however, more likely that the government will raise standards for emission intensive plants.

In terms of Equation 4, having an emission free plant 1) *determines* c_{CO2} *to zero* and 2) gives the holder of the plant an *opportunity for climate-policy related revenues* (r_{CO2}) for the entire lifetime of the plant. Interestingly, the option hold back consisting of having a less emission-intensive plant is comparable to the notion of "robustness" or passive flexibility (see Chapter 2 or Ku 1995). A robust investment can also be regarded as an *"insurance investment"*. An insurance investment reduces exposure to uncertainty, but with a cost or "insurance premium" (Amram and Kulatilaka 1999).

Fixed costs (C_f): Some climate policy instruments will increase the fixed costs of power plant investments through *transaction costs*. For example, in order to obtain climate policy related revenues from JI and the CDM, companies need to pay for preparations up-front (i.e. r_{CO2} increases C_f). These costs are not strongly dependent on project size (e.g. Fichtner et al. 2003; Krey 2004). Empirical data on 15 Indian CDM projects suggest transaction costs ranging from USD 60.000 to USD 480.000 (without monitoring and verification) (Krey 2004). The Prototype Carbon Fund (2003) reports transaction costs of USD 265,000 for JI and CDM projects (without verification). It is estimated that even with simplified procedures and modalities the annual value of emission reduction must exceed USD 180.000 in order to keep the relative transaction cost at a sustainable level (CDCF 2004).

Within the EU ETS, installations are also subject to annual monitoring and verification. This will similarly cause additional transaction costs.

3.3 Discount rate

In standard financial theory, a dollar today is worth more than a dollar tomorrow. This implies that the future expected payoffs should be discounted by the rate of return offered by comparable investment alternatives, i.e. the opportunity cost of capital. In practical applications, the opportunity cost of capital is often either the cost of equity or the after-tax weighted average cost of capital (WACC), which additionally takes into account the cost of debt and the marginal corporate tax rate.

The cost of equity has been commonly estimated through the Capital Asset Pricing Model (CAPM) by Sharpe (1964) and Lintner (1965)[13]. The model suggests that in a market equilibrium, the value-weight asset market portfolio is

[13] CAPM has been a simple and attractive tool for practitioners, although model tests have shown it to be insufficient. Beta has been found unable to explain average returns alone (see e.g. Fama and French 1993; Wang 2003).

mean-variance efficient. This implies, firstly, that beta (β), the slope in the regression of a security's return on the market return, is the only risk needed to explain expected return. Secondly, there is a positive expected premium for β risk. Beta measures the systematic risk of the asset and it can be estimated as the co-variance of the asset value with the value of the market portfolio divided by the variance of the market portfolio.

According to many authors (e.g. Beaver et al. 1970; Ismail and Kim 1989; Young et al. 1991) the volatility of a company's earnings or cash flows compared to the volatility of the overall economy-wide earnings/cash-flows, its *accounting/cash flow beta*, is an applicable concept for predicting stock betas. If the annual cash flows between different technologies differ and have low mutual correlations, their stock betas should consequently be different as well. In oil projects, a high volatility in oil prices has been empirically found to be associated with a higher cost of capital (IEA 2003a, p. 53).

Risk sources of annual cash flows of various energy technologies differ significantly (IEA 2003b). For example, cash flows of thermal power plants depend on spark spreads, whereas those of hydropower and wind power mainly depend on electricity prices and hydrological / meteorological conditions. In addition, fuel prices do not correlate fully.

In the evaluation of heat and power capacity investments, a single discount rate (with a potential sensitivity analysis) is often selected for *all* investment alternatives, implying that they share the same systematic risk (e.g. OECD NEA/IEA 1998). If discount rates differ, high discount rates are typically used for new technologies, due to their inherent uncertainty (IPCC 2001).

Taking into account the discussion above, the existing practices may turn out to be too simplified. As different technologies have different characteristic emission factors, climate policy and, in particular, the market-based instruments seem to modify the risk related to annual cash flows. However, it is not self-evident what exactly the impact is on discount rates for different technologies. GHG-emission-free power plants, such as renewable energy and nuclear plants, are insured against changes in climate policy related *costs*, whereas plants using fossil fuels remain either fully or partly exposed to climate policy risks. However, even renewable energy plants remain exposed to climate-policy-related volatility in *revenues*.

3.4 Comparison of technologies

As discussed above, energy production technologies provide different kinds of flexibility regarding the financial impacts due to climate policy instruments. These are roughly summarized in Table 1.

Thermal power investments provide an option to alter the operating scale, if the market conditions turn out to be poor. Multi-fired power plants have an option to switch between fuels, whereas CHP extraction plants can provide an option to switch between products. The rate up to which a technology is insured against changes in climate policy related *costs* is reflected through the relative emission factor (in $kgCO_2/MWh$) of the plant. Fossil fuels remain either fully or partly exposed to climate policy cost risk.

Table 1. Flexibility regarding climate policy: comparison of technologies (+ = option available/significant).

Source of flexibility		Technology								
		Thermal plants						Other		
		Multi-fired	CHP extraction	Coal	Gas	Oil	Bio-mass	Hydro with reservoir	Hydro (run-off-river), wind	Nuclear
Active flexibility	Option to wait	+	+	+	+	+	+	+	+	+
	Option to alter operating scale	+	+	+	+	+	+	+	-	-
	Option to switch between fuels	+	-	-	-	-	-	-	-	-
	Option to switch between products	-	+	-	-	-	-	-	-	-
Passive flexibility	Robustness to changes in climate policy related costs[1]	Fuel (mix)-dependent	Fuel (mix)-dependent	0%	40%	17%	100%	100%	100%	100%

[1]Defined as the relative emission intensity of the fuel (Coal = 0%)

4 Discussion and conclusions

This paper has explored the impact of climate policy instruments on heat and power capacity investment decisions. Climate policy changes environmental, organizational and decision-specific factors in capacity investment decision processes. The quantitative investment appraisal is modified through an influence on the initial investment, the annual cash flows and, potentially, the discount rate. Climate policy increases uncertainty in the business environment and, as a consequence, the value of flexibility, which is not fully captured with standard methods of quantitative investment appraisal.

Heat and power generation technologies show significant structural differences in flexibility to stochastic changes in climate policy instruments. Whereas some technologies provide managerial flexibility through the options to alter operating scale and to switch between fuels or products, others provide passive flexibility (robustness).

Managerial flexibility can be taken into account in the investment decision process through a real-options analysis using partial differential equations, dynamic programming or simulation (Amram and Kulatilaka 1999). Robustness, on the other hand, either creates upside potential in annual cash flows or provides insurance against the downside risk. In many cases, there are multiple, interactive options *and* robustness present in the investment problem, which requires a joint valuation of different sources of flexibility.

This setting is interesting for the quantitative investment appraisal that is more or less comprehensively applied in all investment decision processes. While it is obvious that the value of climate policy related robustness depends on the prospects of climate policy, we may ask how large the climate policy related uncertainty must be in order to compensate for (in many cases) higher investment costs on the one hand, and worse managerial flexibility (in comparison to thermal plants) on the other hand. Also, the value of robustness can, in some circumstances, be reduced by the risk of windfall taxation: if it is perceived that some technologies receive sudden increased profits "without effort", the regulator may pursue "not to distort the market" and to prohibit this phenomenon through an additional tax. From the flexibility perspective, biomass-fired co-generation projects with a multi-fuel option seem ideal, since they provide both robustness (if the emission factor of the fuel-mix is significantly lower than with fossil fuels only) and managerial flexibility.

References

Adler RW (2000) Strategic Investment Decision Appraisal Techniques: The Old and the New. Business Horizons, November-December

Amram M, Kulatilaka N (1999) Real Options – Managing Strategic Investment in an Uncertain World. Harvard Business School Press, Boston, Massachusetts

Beaver W, Kettler P, Scholes M (1970) The association between market determined and accounting determined risk measures. The Accounting Review 45, October: 654-682

CDCF (Community Development Carbon Fund) (2004) Annual Report 2004

Clemen RT, Reilly T (2001) Making Hard Decisions with DecisionTools®. Duxbury. Pacific Grove.

de Leyva E, Lekander PA (2003) Climate change for Europe's utilities. The McKinsey Quarterly. New York

Dixit AK, Pindyck RS (1994) Investment under uncertainty. Princeton University Press, Princeton, New Jersey

EC (European Commission) (2003) Directive of the European Parliament and of the Coun1cil establishing a scheme for greenhouse gas emission allowance trading within the Community and amending Council Directive 96/61/EC.

EC (European Commission) (2004) Directive of the European Parliament and of the Council amending the Directive establishing a scheme for greenhouse gas emission allowance trading within the Community, in respect of the Kyoto Protocol's project mechanisms.

EDC (Energy for Development Conference) (2004) Chairman's conclusions. December 12 - 14, Noordwijk

Ernst&Young (2004) Press release 17.06.2004
EurObserv'ER (2003) Wind Energy Barometer
Fama EF, French KR (1993) Common risk factors in the returns on stocks and bonds, Journal of Financial Economics 33: 3-56
Fama EF, French KR (1997) Industry costs of equity, Journal of Financial Economics 43: 153-193
Fichtner W, Graehl S, Rentz O (2003) The impact of private investor's transaction costs on the cost effectiveness of project-based Kyoto mechanisms. Climate Policy No. 3: 189-235
Fredrickson JW, Mitchell TR (1984) Strategic Decision Processes: Comprehensiveness and Performance in an Industry with an Unstable Environment. Academy of Management Journal 27 (2): 399-423
Hagem C, Mæstad, O (2002) Market power in the market for greenhouse gas emission permits – the interplay with the fossil fuel markets. CICERO working paper 2002-08
Hoff TE, Herig C (1996) Strategic Planning in Electric Utilities: Using Renewable Energy Technologies as Risk Management Tools. 58th Annual Meeting of the American Power Conference. Chicago, Illinois, USA
Holtsmark B, Mæstad O (2002) Emission trading under the Kyoto Protocol - effects on fossil fuel markets under alternative regimes. Energy Policy 30: 207-218
Holtsmark B (2003) Russian Behaviour in the market for permits under the Kyoto Protocol. Climate Policy, 3: 399-415
IEA (International Energy Agency) (2001a) Natural Gas Information 2001. Paris
IEA (International Energy Agency) (2001b) Coal Information 2001. Paris
IEA (International Energy Agency) (2003a) World Energy Investment Outlook 2003. Paris
IEA (International Energy Agency) (2003b) Power Generation Investment in Electricity Markets. Paris
IEA (International Energy Agency) (2004) World Energy Outlook 2004. Paris
Innovest (2003) Carbon Finance and the Global Equity Markets. Carbon Disclosure Project 2003. Innovest Strategic Value Advisors
IPCC (Intergovernmental Panel on Climate Change) (2001) Third Assessment Report - Climate Change 2001: Mitigation. EarthPrint.com
Ismail BE, Kim MK (1989) On the Association of Cash Flow Variables with Market Risk: Further Evidence. The Accounting Review, 64, January: 125-136
Johnstone N. (2003) The Use of Tradable Permits in Combination with Other Policy Instruments: a Scoping Paper. ENV/EPOC/WPNEP(2002)28. OECD
Keeney RL (1994) Creativity in Decision Making with Value-Focused Thinking. Sloan Management Review, Summer: 33-41
Kort P, Murto P, Pawlina G (2004) The Value of Flexibility in Sequencing Irreversible Investment. Working Paper, September
Krey M (2004) Transaction costs of unilateral CDM projects in India - results from an empirical survey. Energy Policy (In press)
Ku A (1995) Modelling Uncertainty in Electricity Capacity Planning. Ph.D. thesis, London Business School
Kulatilaka (1993) The Value of Flexibility: The Case of a Dual-Fuel Industrial Steam Boiler, Financial Management 22 (3): 271-280
Laitner JA, DeCanio SJ, Koomey JG, Sanstad AH (2003) Room for improvement: increasing the value of energy modelling for policy analysis. Utilities policy, 11: 87-94
Lessard D, Miller R (2001) Mapping and Facing the Landscape of Risks. In: Lessard, D (2001) Strategic Management of Large Engineering Projects: Shaping Institutions, Risks and Governance. MIT Press. Cambridge, MA, USA

Lintner J (1965) The Valuation of Risk Assets and the Selection of Risky Investments in Stock Portfolios and Capital Budgets, Review of Economics and Statistics 47: 13-37

McDonald A, Schrattenholzer L (2001) Learning rates for energy technologies. Energy Policy 29: 255-261

McDonald RL, Siegel DR (1985) Investment and the Valuation of Firms when there is an option to shut down. International Economic Review vol 26, issue 2, June: 331-349

McDonald RL, Siegel DR (1986) The value of waiting to invest. The Quarterly journal of Economics vol 101, issue 4, November: 707-728

Murto P (2003) Timing of investment under technological and revenue related uncertainties, Systems Analysis Laboratory Research Report E 11. Helsinki University of Technology

OECD NEA/IEA (Nuclear Energy Agency/International Energy Agency) (1998) Projected Costs of Generating Electricity - Update 1998. Paris

OECD/EEA (2003) Database on economic instruments and voluntary approaches used in environmental policy and natural resources management. http://www1.oecd.org /scripts/env/ecoInst/

PCF (Prototype Carbon Fund) (2003) PCF Annual Report 2002 www.prototypecarbon fund.org

Pindyck RS (2001) Irreversibility, uncertainty and investment. in Schwartz, E.S., Trigeorgis L (2001) Real Options and Investment under Uncertainty - Classical Readings and Recent Contributions. The MIT Press. Cambridge, Massachusetts

Rajagopalan N, Rasheed MA, Datta DK (1993) Strategic Decision Processes: Critical Review and Future Directions. Journal of Management 19(2): 349-384

Reinaud J (2003) Emissions trading and its possible impacts on investment decisions in the power sector. IEA information paper. Paris

Sandoff A (2003) Teoretiskt ramverk för analys av aktörernas drivkrafter för investeringar. in Rydén B (ed) Nordleden – Slutrapport för etapp 2. Gothenburg (in Swedish)

Schwartz E Trigeorgis L (2001): Real options and investment under uncertainty: classical readings and recent contributions. MIT press., Cambridge, Massachusetts

Selten R (2002) What is Bounded Rationality? in Gigerenzer, G., Selten, R. (ed) Bounded Rationality – the Adaptive Toolbox. Dahlem Workshop Reports. The MIT Press, Massachusetts

Shank JK (1996) Analysing technology investments - from NPV to Strategic Cost Management (SCM). Management Accounting Research 7: 185-97

Sharpe WF (1964) Capital Asset Prices: A Theory of Market Equilibrium under Conditions of Risk. Journal of Finance 19: 425-442

Simon HA (1997) Administrative behavior – a study of decision-making processes in administrative organizations. Fourth Edition. Free Press, New York

Söderholm P (2000) Fuel flexibility in the West European power sector. Resources Policy 26: 157-170

Söderholm P, Strömberg L (2003) A utility-eye view of the CO2 compliance decision process in the European power-sector. Applied Energy, 75: 183-192

Trigeorgis L (1995) Real options: an overview in Trigeorgis L (ed) Real options in capital investment - models, strategies and applications. Praeger, Westport, Connecticut

Tseng C, Barz G (2002) Short-term generation asset valuation: a real options approach. Operations Research, 50(2), March-April: 297-310

Vrolijk C (ed) (2002) Climate Change and Power - Economic Instruments for European Electricity. Earthscan Publications Ltd, London, p. 300

Wang, K.Q. (2003) Asset pricing with conditioning information: a new test. Journal of Finance 58(1): 161-196

Young SD, Berry MA, Harvey DW, Page JR (1991) Macroeconomic Forces, Systematic Risk, and Financial Variables: an Empirical Investigation. The Journal of Financial and Quantitative Analysis 26(4): 559-564

Implications of the European emissions trading scheme for strategic energy management in small and medium enterprises

Anja Pauksztat[I], Martin Kruska[II]

[I] Rheinisch-Westfälische Technische Hochschule Aachen
Lehrstuhl für Technische Thermodynamik
Schinkelstraße 8, 52062 Aachen, Germany
pauksztat@LTT.RWTH-Aachen.de

[II] EUtech Energie & Management GmbH
Dennewartstraße 25-27, 52068 Aachen, Germany
kruska@EUtech.de

Abstract

This paper analyses the relevance of the European emissions trading scheme for the strategic energy management of those enterprises covered by the scheme. The administrative requirements are discussed and a short estimation of the prospective financial burden is given, based on different scenarios.

Since January 2005, companies are operating in a "carbon restrained market". To make adequate "make-or-buy" decisions in this market, the internal options and the internal reduction potentials need to be known, the dynamic development of the emission trading market should be continuously studied and analysed and an internal operation strategy has to be developed. Regarding the financial burden, different scenarios are analysed for an industrial boiler as well as for a cement production plant. The results show that the impact brought about by the limitation of emissions allowances within the European emissions trading scheme depends largely on the circumstances considered. However, the financial impact on the considered industrial installations is assessed as rather insignificant.

Keywords: Carbon restrained market, make-or-buy, internal emissions reduction measures, marginal costs curve

1 Introduction

The introduction of an emissions trading scheme in Europe is, as of now, the most prominent signal that the commitments made in the framework of the Kyoto protocol by Governments as legal parties are passed on to those who are actually responsible for the emissions. The European emissions trading scheme was implemented in January 2005. The companies covered by this scheme are now operating in a "carbon restrained market" and face new challenges – and additional administrative requirements – which are discussed on the following pages.

The general advantage of an emissions trading scheme consists in the realisation of the – at least theoretically – economically most favourable emissions reduction measures within the considered system. This implies the cost minimization of environmental protection. From an external point of view, an additional advantage is that the system implicitly guarantees the reduction of emissions to a pre-defined degree. Furthermore, the emissions become traceable through emissions trading, if the appropriate monitoring, validation and verifying process is ensured.

This paper analyses the relevance of the European emissions trading scheme for the strategic energy management of those enterprises covered by the scheme. The administrative requirements are discussed and a short estimate of the prospective financial burden is given, based on different scenarios.

2 Relevance of the European emissions trading scheme for small and medium enterprises

The European Parliament and the Council are established a European-wide emissions trading scheme. The Community scheme also includes small and medium enterprises, if their installations fall under the selected activities and exceed certain threshold values given in the directive. The categories of activities listed in annex I of the directive and the design characteristics will be discussed regarding their implications for small and medium enterprises within this chapter.

2.1 Categories of activities

An overview of the categories of activities referred to in annex I of the European directive on greenhouse gas emission allowance trading is given in Table 1.

Table 1. Categories of activities listed in annex I of the European directive on greenhouse gas emission allowance trading, *EU Commission (2003)*

Activities
Energy activities: including combustion installations with a thermal input exceeding 20 MW, mineral oil refineries, coke ovens
Production and processing of ferrous metals
Mineral industry: including installations for the production of cement clinker or lime, as well as for the manufacture of glass and ceramic products
Other activities: including industrial plants for the production of pulp and paper

Regarding these activities, especially within the categories "energy activities", "mineral industry" and "industrial plants for the production of pulp and paper" and through the possibility of accumulating installations within the plant boundaries, small and medium enterprises are affected by the Community scheme.

For example, if a food-processing firm has three industrial boilers at its site with a thermal input of 7 MW_{th} each, the accumulated thermal input adds up to 21 MW_{th}, thus the installations fall under the Community scheme.

2.2 Design characteristics[1]

The Community scheme is compulsory and will started with a three-year period in January 2005. From 2008 on there are five-year periods. At first the Community scheme is restricted to carbon dioxide emissions and the emissions of activities referred to in annex I of the directive, but the extension to further activities and emissions is intended as stated in article 24 "procedures for unilateral inclusion of additional activities and gases" of the directive. However, alternative policies and measures are demanded for the sectors and installations not mentioned in annex I of the directive.

The operators of installations are allowed under certain conditions to form a pool of installations from the same activity. This may be applied by some firms who need to report several installations to different competent authorities. On the other hand, the temporary exclusion of certain installations is restricted to the first three-year period and was not applied extensively.

Since January 2005 on, every installation listed in annex I of the directive needs a greenhouse gas emissions permit to operate. The permit grants the authorisation to emit greenhouse gases from the installation if the operator is capable of monitoring and reporting emissions. However, the permit only grants authorisation to the operation of the installation. For the actual emissions of this installation the

[1] The following information is taken from EU Commission (2003), referred to in this document as the "directive". In accordance with this directive, the European scheme for greenhouse gas emission allowance trading is referred to as the "Community scheme".

operator needs to hold the necessary amount of allowances in accordance with the annual emissions of the installation.

For each period, the member states need to develop a national allocation plan wherein the emission targets and allocation method are stated. The allocation of allowances is generally considered as one of the major issues of this trading scheme. The directive sets basic criteria for national allocation plans in its annex III. During the three-year period from 2005 to 2007 the allocation of emissions allowances by member states needs to be at least 95% free of charge. For the five-year period starting January 2008 at least 90% of the allowances need to be allocated free of charge. This is stated in article 10 of the directive.

For the monitoring and reporting of emissions, guidelines are being established. Generally, high demands are made on both and verification of the reported data is necessary. The monitoring and reporting as referred to in article 14 "guidelines for monitoring and reporting of emissions" and in annex IV of the directive require the operator of an installation to report their emissions annually including an assessment of the uncertainty of the reported data. Additionally, in article 15 "verification" and in annex V of the directive, a verification of the credibility, reliability and accuracy of the monitoring, as well as of the reported data, is demanded.

In article 16 "penalties" the Community scheme also provides rules on penalties which must be effective, proportionate and dissuasive. For non-compliance the payment of an excess emissions penalty is fixed at 40 €/t carbon dioxide within the first period and at 100 €/t CO_2 for each successive period.

3 Implications of the community scheme for small and medium enterprises

In the framework of the Community scheme several new and additional implications for enterprises need to be taken into account. These include operational as well as strategic implications which will be discussed in more detail in this chapter.

3.1 Operational implications

The operational tasks in the context of the Community scheme concern foremost the cost accounting and controlling departments as well as the tax and purchasing departments.

For the annual inventory the carbon dioxide emissions of the concerned installations need to be measured or calculated and an annual emissions account needs to be drawn up. The requirements of monitoring and reporting are detailed by guidelines of the European Commission as mentioned above. It is clear that the internal data base needs to be complemented by carbon dioxide emissions. For this there are already tools available, e.g. the *Bavarian greenhouse gas monitoring sys-*

tem (2003). But it may also be necessary to improve the data collection by installing additional meters.

It may be the task of the tax or purchasing departments to coordinate the annual accounting of emissions allowances with the competent authority including the transfer, surrender and cancellation of the number of allowances equal to the total emissions from the firm's installation.

Additionally, the trade of emissions allowances needs to be organised, possibly as over-the-counter-transactions at the end of each year (or at the beginning of a year for the passed one) to balance the emissions allowance account or as continual trade in the course of the year. The continual trade offers the opportunity to buy emissions allowances cost effectively in periods of low prices and to sell one‘s own allowances where appropriate in times of high prices.

Finally, all mentioned tasks imply an enlargement of the already existing tasks of the different departments involved. Considering a functioning energy management, especially the task of inventory, monitoring and reporting should be rather simple to accomplish. Usually, it will be unnecessary to create a new department for emissions trading, but the new tasks can be comprised within existing departments. However, the provision of personnel demands a certain amount of resources that should not be neglected. Thus the responsibility may lie e.g. with the purchasing department or with the environmental management.

3.2 Strategic implications

The strategic implications of the emissions trading scheme include above all the detailed analysis of the internal emissions reduction potential, the continual follow-up and analysis of the dynamics of the emissions market as well as the development of an internal operation strategy taking into account all possible actions. The follow-up of the market development is already customary for energy purchase and treaty negotiations, this will also be the case for the emissions allowance market and its consequences for the financial situation of the enterprise.

In the framework of the Community scheme, the carbon dioxide intensity needs to be considered within the demand side management, which is part of the firm's energy management. This means that an enterprise which can employ different fuel types, e.g. in a two-fuel-type burner, has to consider not only the supply costs but also the emissions intensity of the utilized fuel and the costs connected herewith.

3.2.1 Determination of the internal emissions reduction potentials

When considering the internal reduction potential of emissions, or, generally speaking, every possible measure a company can take in a carbon restrained market, the evaluation includes the additional expense influencing factor of CO_2 emission. This means that measures which are considered as not economically feasible should also be taken into account as possible internal reduction measures. Energy analyses in different companies often show that even economically advantageous measures are not yet implemented because of a lack of financial liquidity of the company. But there are also measures that are profitable, but result in an increase in emissions, e.g. a fuel switch from gas or oil to coal. In a carbon restrained market the expense influence of this increase in emissions needs to be evaluated as well.

Identification of the internal direct and indirect emissions reduction potentials results in the knowledge of the specific internal emissions reduction costs. The different measures can be presented in a marginal costs curve which shows the specific reduction costs in correlation with the amount of emissions reduction per year, as shown in Figure 1. In this figure the emissions reduction measures of several companies in Northern Germany are presented as identified in *ESSH (2003)*.

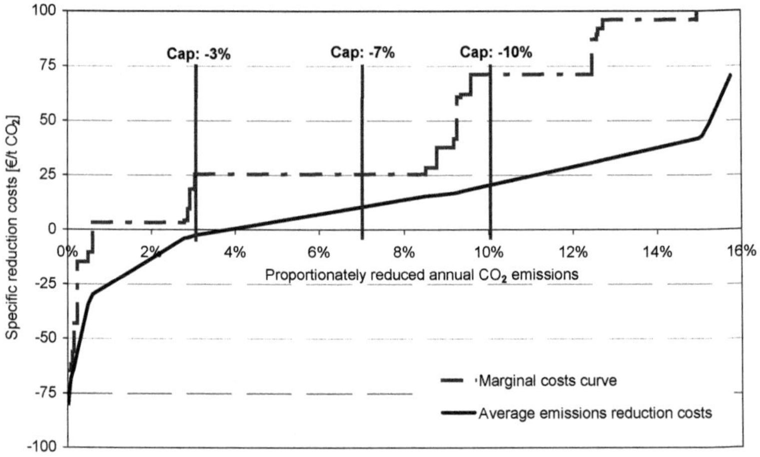

Fig. 1. Marginal costs curve and average emissions reduction costs, *ESSH (2003)*

For the 29 internal measures considered, the specific emissions reduction costs range between -80 €/t CO_2 (thus being already economically feasible) and over 100 €/t CO_2. Subject to the market price, each company will decide between realizing their own measures or buying emissions allowances from the market. It is interesting to consider the average emissions reduction costs presented in the marginal costs curve, when realizing more than one measure. In the example of Figure 1 the cap of -3% emissions reduction can be realized for all companies at a negative average price of -2.6 €/t carbon dioxide. For a cap of -10% the average price reaches about 18 €/t CO_2.

Concerning the implementation of their own measures, it is important to note that emissions reduction measures are not infinitely variable. Thus the cap of -7% emissions reduction could not be realized with the identified measures of the selected firms as shown in Figure 1. It is also imperative to take into account that the particular amounts, as well as the specific reduction costs, are liable to incertitude. This is illustrated for the example given in Figure 2. For the measure marked in the figure, the quantitative risk describes the uncertainty concerning the amount of emissions this measure will actually reduce. The price risk describes the uncertainty with regard to the assessed specific reduction costs implied by the measure.

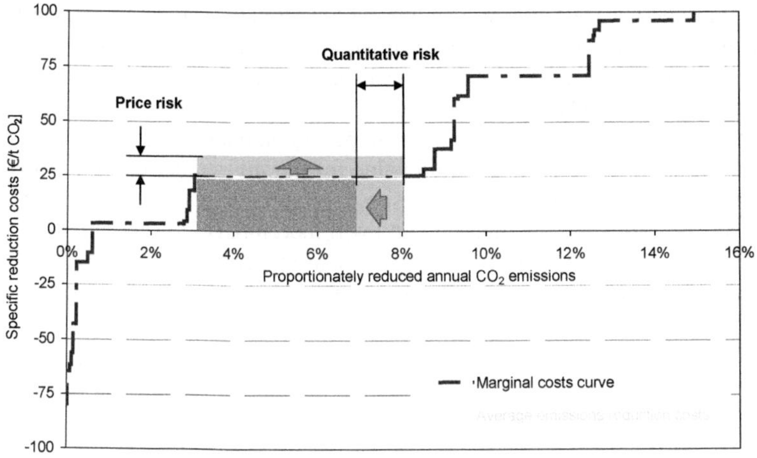

Fig. 2. Risk assessment considering the marginal costs curve, *ESSH (2003)*

3.2.2 Continuous analysis of the dynamics of the emissions market

The market of emissions allowances proved to be volatile if not chaotic in the beginning. Therefore it is difficult to determine the price development from the market dynamics. But this is crucial, as the potential internal measures have a distinct period necessary for the implementation. The influence of the realization period on the additional costs for a company is illustrated in Figure 3. When the market price for emissions allowances exceeds the emissions reduction costs of an internal measure, an investment decision will be made. However, the measure will need its realization period before being put into operation, thus causing additional costs for the company.

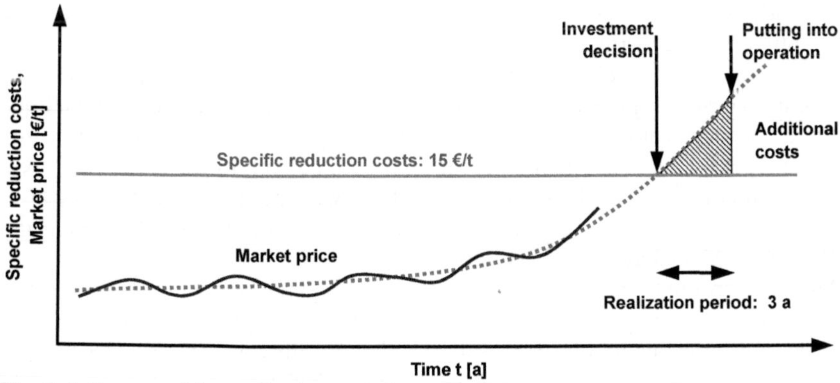

Fig. 3. Influence of the realization period

To adjust to the market development and to prevent, or at least minimize, additional costs for the company, assessments of the dynamics of the emissions market as well as of the implementation time of the emissions reduction measures are of utmost importance. To determine a sound market strategy it is, therefore, crucial to consider the sensitivity of the market price and the internal costs to minimize the risks, also including the quantitative risks connected to the internal measures and the realization period of each measure.

In Figure 4 the influence of the realization period on the additional costs for the company is illustrated when a threshold value for the investment decision of each internal measure is considered. This threshold value is below the specific reduction costs of the corresponding measure and will therefore (partly) compensate the additional costs caused by the necessary realization period of the measure. The additional costs are then considerably dependent on the actual development of the market price.

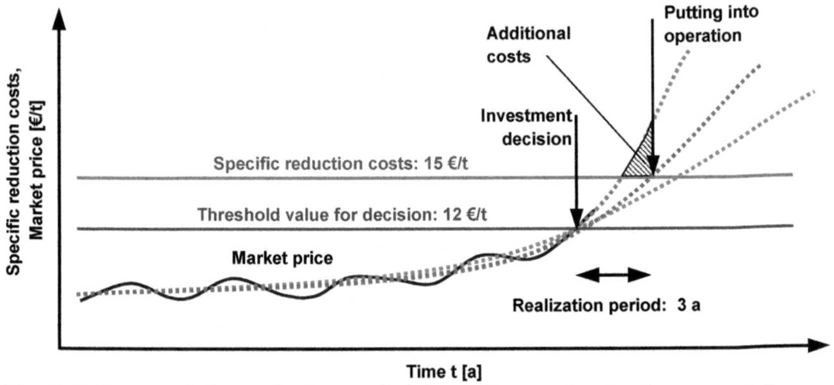

Fig. 4. Influence of the realization period depending on the development of the market price when considering a threshold value

3.2.3 Development of an internal operation strategy

For the development of the internal operation strategy of the firm within the framework of the European emissions trading scheme short, medium and long term options can be distinguished.

Short term options include the use of the alternative compliance options and instruments. These comprise e.g. emissions certificates from "joint implementation" (JI) and "clean development mechanism" (CDM) projects, as the necessary link with the Kyoto scheme is established within the framework of the European directive. A company then has the option to directly invest into a JI or CDM project or to buy the generated certificates of a realized project. Additionally, the use of derivatives for risk management is possible.

Medium term options are, for example, treaty negotiations with energy suppliers, neighbouring companies, etc. Conceivable are the cooperation in planning, financing and implementation of emissions reduction measures as well as integration of the energy supply and demand. An enhanced energy exploitation leads also to a reduced energy demand and therefore to decreasing emissions.

Finally, long term options encompass inclusion of the cost factor emissions allowances within the production process. This also involves product and process innovations, which may achieve emissions reductions.

4 Scenario analysis involving dependence on the emissions allowance price

Since the beginning of the discussion about the implementation of an EU-wide emissions trading scheme, different data have been published concerning the financial implications of such a scheme on the enterprises. However, the actual market prices of emission allowances depend on the size and liquidity of the Community market, on the possibilities to use alternative instruments like JI and CDM as well as on the actual costs for emissions reduction measures.

As discussed in *Kruska et al. (2003)*, the assessed prices range from 1 €/t CO_2 to over 30 €/t CO_2. In existing trading schemes the prices lie between 2 €/t CO_2 and 15 €/t CO_2. The prices for EU emission allowances started at about 10 €/t CO_2. In this price discussion the emissions reduction potentials within industrial sites are often neglected. As mentioned above, from an economic point of view, these potentials are feasible and would thus imply no or even negative emission reduction costs.

To determine the economical implications of an emissions trading scheme for enterprises, different factors of influence need to be taken into account including, of course, the specific market price of allowances, but also the emissions intensity of the installation and the employed fuel, the fuel price as well as the consideration of process related emissions. For an objective assessment of the possible implications for industrial plants, a scenario analysis for two examples of installations is discussed by *Kruska et al. (2003)* analysing the CO_2 emissions as well as the relative additional costs through a limitation of emissions allowances. Three different scenarios are considered:

- Worst case
  The supply of the total demand of allowances on the market as worst case scenario is actually not considered for existing installations (at least in Germany), but might be applied to new entrants depending on the national allocation plan.
- Reduction obligation of 20% in 5 years
  The primary allocation of emission allowances is 100% free of charge, but the emission allowances are successively reduced by 4 percentage points per year. The amount of emissions remains unchanged. This scenario implies excessive requirements of the firms and is assessed as rather unlikely.
- Increase of production output by 10%
  E.g. through production fluctuations or plant extension the demand for emis-

sions allowances rises above the allocated amount. The additionally needed emissions allowances must be bought on the market.

As a typical installation within small and medium enterprises, an industrial fossil fuelled boiler as well as a cement production plant is analysed. Considering the possible price range discussed above, the scenarios consider prices between 5 and 30 €/t CO_2.

4.1 Industrial fossil fuelled boiler

An industrial fossil fuelled boiler can be found in most industrial plants, so that the analysis of this example will be relevant for a large number of enterprises taking part in the European emissions trading scheme. The cost relevance for the industrial boiler needs to be analysed regarding energy costs which are usually indirect costs.

The technical details and the emissions data of a (group of) boiler(s) with a thermal power of 30 MW_{th} and about 6.000 h/a production time as considered by *Kruska et al. (2003)* are listed in Table 2 and the results of the scenario analysis are given in Table 3.

Table 2. Technical details and emission data of the industrial boiler, Kruska et al. (2003)

Indicator	Value
Technical data	
thermal power	30 MW (approx. 37 t steam/h)
annual efficiency	85%
annual load period	6.000 h/a
fuel price (natural gas)	25 €/MWh_{th}
emissions factor (natural gas)	0.2016 t CO_2/MWh_{Hu}
Emissions data	
fuel input	212.000 MWh_{Hu}/a
total fuel costs	5.300.000 €/a
CO_2 emissions	42.700 t CO_2/a

Table 3. Results of the scenario analysis as surcharge on the net fuel costs, Kruska et al. (2003)

Price in €/t CO_2	5	10	20	30
Worst case	4.0%	8.1%	16.1%	24.2%
Reduction obligation of 20% in 5 years	0.5%	1.0%	1.9%	2.9%
Increase of production output by 10%	0.4%	0.8%	1.6%	2.4%

The analyses show that in the worst case scenario a surcharge on the net fuel costs between 4% at about 5 €/t CO_2 and 24% at 30 €/t CO_2 can be assessed resulting from the supply of allowances on the market. Considering a reduction obligation of 20% over 5 years, which is still evaluated as rather unlikely, the surcharge ranges between 0.5 and 3% depending on the allowance price. The additional costs for an increase in production output by 10% come to the same range of about 0.4 to 2.4%.

As the third scenario is rated as the most realistic and if prices reach around 10 to 15 €/t CO_2, the enterprises need to take into account a surcharge on the net gas costs of about 0.8 to 1.6%. This value is still within the normal fluctuations of the fuel price.

4.2 Cement production plant

The cement industry is an energy intensive industrial branch with a high amount of process emissions. When analysing the cement production plant energy related emissions and process emissions need to be treated separately. The energy related emissions depend on the plant operation and especially on the fuels, as there are usually a number of different fuels employed, including e.g. coal, alternative fuels and biomass. The process emissions depend on the calcinations of the raw materials and can be calculated to about 0.53 t CO_2/t cement clinker. The cement industry has the possibility – within a certain range – to reduce the clinker portion in the final product through the addition of materials that are neutral regarding carbon dioxide emissions. But this alternative only influences the emissions balance of the firm, if the clinker production is actually reduced.

The cost relevance needs to be analysed regarding direct production costs. The technical details and the emissions data of a cement production plant with an annual production of 400.000 t/a cement clinker as considered by *Kruska et al. (2003)* are listed in Table 4. The results of the scenario analysis for the energy related emissions are given in Table 5 and the results for the process emissions are shown in Table 6.

Table 4. Technical details and emissions data of a cement plant, *Kruska et al. (2003)*

Indicator	Value
Technical data	
annual production	400.000 t/a cement clinker
clinker portion	85%
fuel price	6.1 €/MWh_{th} (50 €/t SKE)
average cement price	53 €/t cement
emissions factor (coal)	93 g CO_2/MJ_{Hu}
emissions factor (others)	80 g CO_2/MJ_{Hu}
emissions factor (calcinations)	0.5 t CO_2/t clinker
Emissions data	
fuel input (coal)	1.500 TJ/a
fuel input (others)	190 TJ/a
total fuel costs	2.800.000 €/a
CO_2 emissions (fuel)	155.000 t CO_2/a
CO_2 emissions (calcinations)	200.000 t CO_2/a

Table 5. Results of the scenario analysis of the energy related emissions as surcharge on the net fuel costs, based on the annual fuel demand, *Kruska et al. (2003)*

Price in €/t CO_2	5	10	20	30
Worst case	27.2%	54.4%	108.7%	163.1%
Reduction obligation of 20% in 5 years	3.3%	6.5%	13.0%	19.6%
Increase of production output by 10%	2.7%	5.4%	10.9%	16.3%

Table 6. Results of the scenario analysis of the process emissions as surcharge on the product price, based on the annual cement production, *Kruska et al. (2003) and own calculations*

Price in €/t CO_2	5	10	20	30
Worst case	4.0%	8.0%	16.0%	24.0%
Reduction obligation of 20% in 5 years	0.5%	1.0%	1.9%	2.9%
Increase of production output by 10%	0.4%	0.8%	1.6%	2.4%

The cost relevance of the energy related emissions has been determined, based on the annual net fuel costs of the cement production plant. The analysis shows that the energy related surcharge on the net fuel costs ranges from 27% at about 5 €/t CO_2 up to over 160% at 30 €/t CO_2 for the worst case scenario. Considering a reduction obligation of 20% over 5 years, which is still assessed as rather unlikely, the energy related surcharge on the net fuel costs is between 3.3% at 5 €/t CO_2 and about 20% at 30 €/t CO_2. The additional net fuel costs for an increase in production output by 10% come to the same scale of about 2.7 to 16.3% depending on the allowance price per ton of carbon dioxide.

In addition there is a surcharge on the product price caused by the process emissions. This surcharge is based on annual cement production considering an average cement price of 53 €/t cement and a clinker portion of 85%. The analysis shows that in the worst case scenario a surcharge on the product price ranging from 4% at 5 €/t CO_2 to 24% at 30 €/t CO_2 can be assessed resulting from the supply of allowances on the market. For a reduction obligation of 20% over 5 years the surcharge on the product price is between 0.5% at 5 €/t CO_2 and 2.9% at 30 €/t CO_2. Finally, for the third scenario, the additional costs for cement production caused by the process emissions come to about 0.4 to 2.4%, depending on the allowance price per ton of carbon dioxide.

In conclusion regarding the given data for a cement production plant, taking into account only the third scenario, as it is the most realistically assessed, and allowing for prices to reach around 10 to 15 €/t CO_2, the cement industry needs to consider a surcharge caused by the energy related emissions of about 5 to 10% on the net fuel costs and, additionally, a surcharge caused by the process emissions of about 0.8 to 1.6% on the cement price.

5 Conclusion

Since January 2005, companies are operating in a "carbon restrained market". The inclusion of the emissions within the energy management of the plant is therefore of utmost importance. To make adequate "make-or-buy" decisions in this market and thus reduce the additional costs for the company, the internal options and the internal emissions reduction potentials need to be known. The dynamical development of the emissions trading market should be continuously studied and analysed and an internal operation strategy has to be developed.

The emissions trading scheme allows participating firms the advantage of a high degree of flexibility for whether to reduce emissions at plant level or to buy any additionally needed emissions allowances on the market. Furthermore, within the framework of the emissions trading scheme the investment cycles of a company can be taken into account, as there is no constraint to actively implement internal measures. Negative employment incentives will be avoided; on the contrary, the European emissions trading scheme may even have positive effects through innovation incentives. It can be concluded that through the European emissions trading scheme, there will be no additional costs except for transaction costs. Financial surcharges result exclusively from the politically motivated emissions reduction obligation.

However, the examples in chapter 4 illustrate that the impact through the limitation of emissions allowances within the European emissions trading scheme depends largely on the circumstances considered. Regarding the more realistic framework conditions of an increase of production output by 10%, the financial impact on the considered industrial installations will be rather insignificant. Therefore, dislocation of production sites to non-European countries only on the basis of the European emissions trading scheme seems unlikely.

References

EU Commission (2003) Directive 2003/87/EC of the European Parliament and of the Council of 13 October 2003 establishing a scheme for greenhouse gas emission allowance trading within the Community and amending Council Directive 96/61/EC, Official Journal of the European Union, L 275, pp. 32-46

Kruska M, Meyer J, Pauksztat A, Schubert A (2003) Die Bedeutung des Europäischen Emissionshandels für das strategische Energiemanagement in der Industrie, In: VDI-Gesellschaft Energietechnik (ed): Betriebliches Energiemanagment, Düsseldorf, Germany, pp. 167-180

ESSH (2003) Pilot project "Emissions Trading North", Project P1553, Energiestiftung Schleswig-Holstein, Germany

Bavarian greenhouse gas monitoring system (2003) Freiwilliges Treibhausgas-Monitoring-system für Betriebe, http://www.vbw-bayern.de/pdf/021129_Rahmstorf_Leitfaden_ CO_2-Monitoring.pdf.

Management and optimization of environmental data within emissions trading markets – VEREGISTER and TEMPI

Bernhard Grimm[I], Stefan Pickl[II], Alan Reed[III]

[I] TÜV Süddeutschland, Carbon Management Service
Westendstraße 199, 80686 München, Germany
bernhard.grimm@tuev-sued.de

[II] Universität der Bundeswehr München, Fakultät für Informatik
85577 Neubiberg, Germany
stefan.pickl@unibw.de

[III] University of New Mexico, Veregister Corporation
Albuquerque, NM 87104, USA
areed@veregister.net

Abstract

The conferences of Rio de Janeiro 1992 and Kyoto 1997 expressed the world-wide demand for new economic instruments which focus on environmental management in both macro and micro economies. In this contribution we describe two main software tools which help to establish such an optimal energy management within future emissions trading markets: TEMPI and VEREGISTER. The tools are characterized and their features, as they are related to emissions trading markets, are discussed.

Keywords: Environmental management, emissions trading markets, software, energy management

Acknowledgement: The authors want to thank Peter Letmathe for his careful reading of the manuscript and fruitful discussions which gave us useful insights to our approach.

1 Introduction

The conferences of Rio de Janeiro 1992 and Kyoto 1997 expressed the world-wide demand for new economic instruments which focus on environmental protection in both macro and micro economies. An important economic tool in that area is Joint-Implementation (JI) which is defined in Art. 6 of the Kyoto Protocol. It is an international instrument which intends to strengthen international co-operation between enterprises on reducing CO_2-emissions. A sustainable development can only be guaranteed if the instrument is embedded in an optimal energy management (Bültmann 2002). In this article we describe two main tools which help to establish such an optimal energy management within emissions trading markets, which are described in Gagelmann 2002: TEMPI and VEREGISTER. Veregister is constructed to assist companies and institutions in addressing the complex process of managing greenhouse gas emissions in a manner consistent with applicable rules and regulations. TEMPI is a forecasting tool to estimate investments in greenhouse gas reduction programs.

2 Management of greenhouse gas emissions[1]

The Kyoto Protocol of the UN Framework Convention on Climate Change (UNFCCC) calls for a complex new structure for greenhouse gas management that implies both responsibilities and opportunities for countries, companies, and financial markets (Böhringer 2003). The standard measurement unit for GHG is the CO_2equivalent (CO_2e) metric tonne that takes into account the physical effects of different gases. There are four types of emission units:

- **Assigned Amounts (AAU)**
  National emissions rights for the industrial nations based on targets set by the Protocol for the period 2008-2012
- **Certified Emission Reduction (CER)**
  Tradable emission rights originating from Clean Development Mechanism projects in developing nations
- **Emission Reduction Unit (ERU)**
  Emission rights originating from Joint Implementation projects between industrial countries
- **Removal Unit (RMU)**
  Emission rights originating from absorption of GHG in industrial countries (e.g. sinks such as forests).

[1] This introductory part is a summary of REED2001.

Achieving the goals of the Kyoto Protocol, or of other regimes for reducing greenhouse gas emissions, will require multidimensional approaches, especially

- technological
- social
- economical

which involve complex, long-term, transparent, and verifiable interactions among governments, agencies, and the private sector, (see for example Hansjürgens 2000, 2003).

The Protocol calls for strict and detailed accounting and reporting of production of GHG emissions and holdings of emission rights. The Protocol includes penalties for procedural mistakes or failures to comply with the requirements of the Protocol. Companies, operating within national registry requirements, must have accurate and up-to-date inventories and tools for tracking, analyzing, and reporting the data. In addition, markets are emerging, both regionally and nationally, for trading emission credits as a new commodity, similar to the long-existing markets for emissions of sulphur and nitrogen oxides. Such markets provide economic incentives for better management of GHG emissions. Tradable credits are derived from certified reductions recorded over a period of time and must be tracked and properly documented. Thus, a successful trading system calls for accurate and verifiable records of all transactions and data such as serial numbers, transacting entities, account balances, and certification status.

Proper underpinning and support for these various interactions and mechanisms require specialized data/information management software that includes a registry system for tracking holdings of metric tonnes of CO_2e in various Protocol-based categories *and* a transaction logging system that records and updates transfers, sales and purchases of CO_2e that communicates with official registries. At the moment, several distinguished registries exist; one of them is VEREGISTER (http://www.veregister.com). As described below, Veregister provides tools for the following purposes:

- maintain communication between the applications
- provide appropriate documentation and reports
- meet all United Nations requirements

In addition, the system contains software for GIS-based geospatial characterization and analysis of GHG sources and sinks, a valuable tool for planning and policy development. In the following we present parts of a reduced PC-version in order to illustrate the structure of the program:

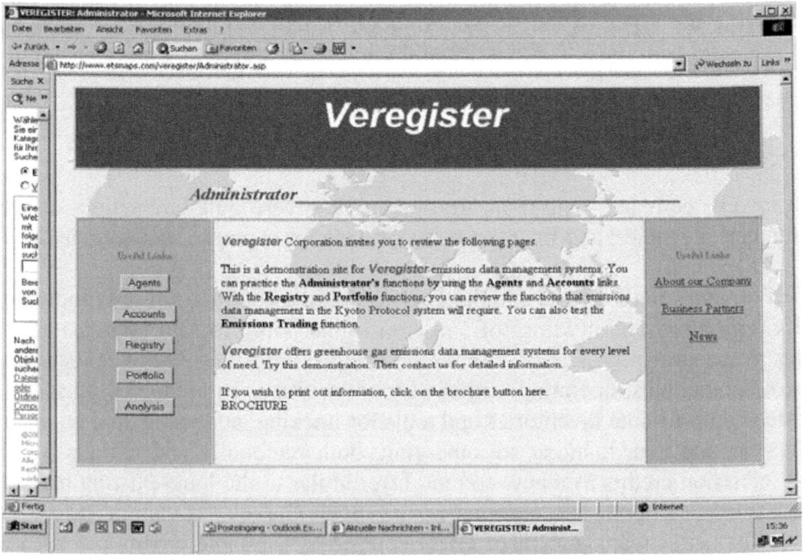

Fig. 1. VEREGISTER[2] - a tool for planning and policy development

Veregister is designed to assist companies and institutions in addressing the complex process of managing greenhouse gas emissions in a manner consistent with applicable rules and regulations and is, at the same time, cost effective in the long-term. The analytic part of VEREGISTER which distinguishes the five parts provides Agents, Accounts, Registry, Portfolio and Analysis:

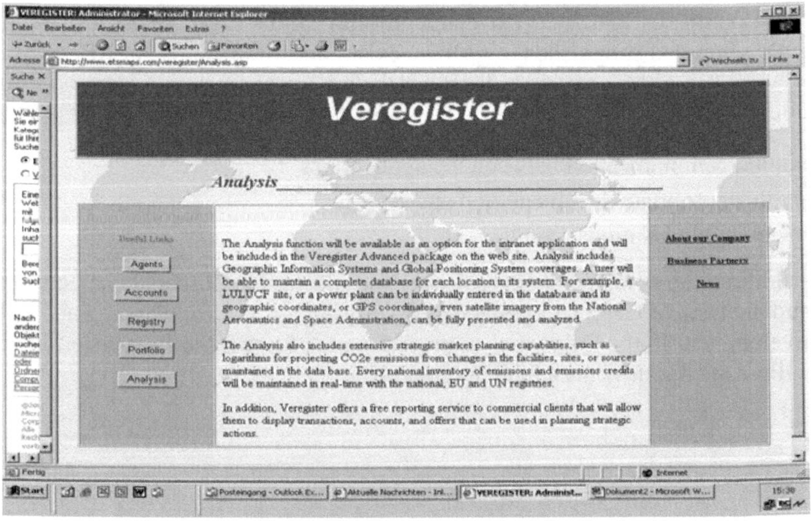

Fig. 2. Complex process of managing greenhouse gas emissions

[2] VEREGISTER is a trademark of Annex I Corporation, see http:www.veregister.com.

3 Forecasting and econo-mathematics

These five functions allow management of all kinds of CO_2e units. Several agents can define their profile, can administer their accounts and registry and can combine certain portfolios. Optimal energy management according to such processes demand further analytic tools. The reporting service should be expanded to a simulation and forecasting unit. There are only a few approaches which try to simulate and forecast investments in the field of emissions trading markets. In the following we want to describe the analytic tool TEMPI (Technology Emissions Means Process Identification) and a possible combination of TEMPI and VEREGISTER in the future.

TEMPI is based on the so-called TEM model (Technology-Emissions-Means model) which was developed by PICKL1998a, providing the opportunity to simulate such an economic behaviour. It is a model in the field of econo-mathematic models which integrates economic and technical investments in a coupled time-discrete nonlinear system of equations. The Framework Convention on Climate Change (FCCC) makes demands for reductions in greenhouse gas emissions by the industrialized countries. On the other hand, developing countries are expanding their energy consumption, which leads to increased levels of greenhouse gas emissions. The preparation of an optimal management tool in that field requires the possibility to identify, assess, compare and forecast several technological options. For that reason, TEMPI (Technology Emissions Means Process Identification) was developed as a comfortable software. TEMPI is based on the TEM model which will be explained in the following.

According to the FCCC (Article 4, paragraph 2(a)), additional control parameters are incorporated which have to be determined iteratively, according to a game theoretic negotiation process. The iterative solution to the nonlinear time-discrete TEM model is an approach to initiate cooperative behaviour in the realization of joint CO_2-reduction initiatives and programs . The cooperative aspect is coming more and more into the centre of interest. Hansjürgens 2003 gives a very good overview of this field.

TEMPI offers several possibilities within an emissions data management:

- Documentation for the user
- Interactive learning system for the user
- Data bank access via the internet (ORACLE)

In the following we give a short description of the underlying TEM model.

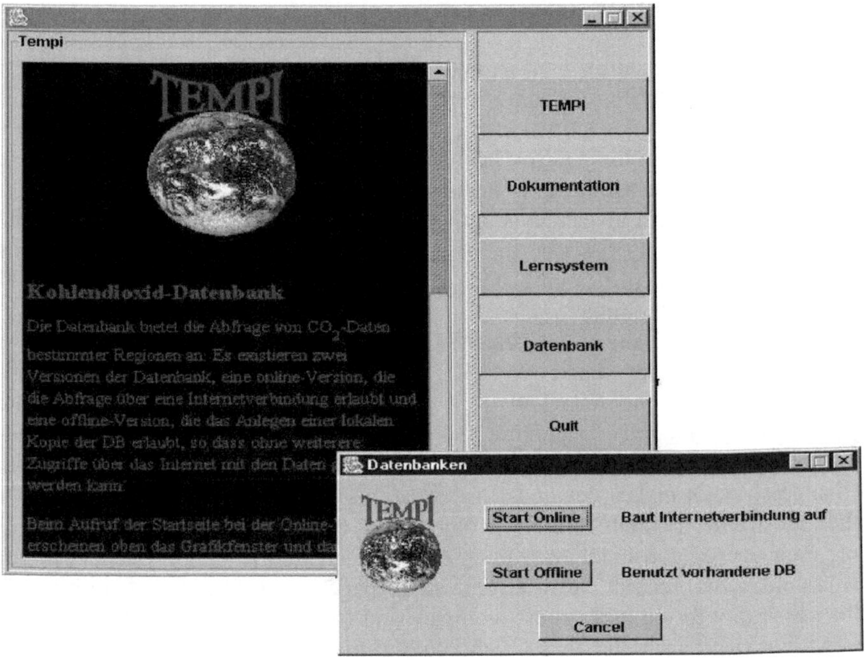

Fig. 3. TEMPI-Analytic instrument to simulate greenhouse gas emissions and investments

3.1 Short description of the underlying TEM model

This part is concerned with a mathematical derivation of the nonlinear time-discrete Technology-Emissions Means (TEM-)model. A detailed introduction to the dynamics modelling a Joint Implementation Program concerning Kyoto Protocol is given in Pickl 1998a, 1998b. As the nonlinear time-discrete dynamics tends towards chaotic behaviour, the necessary introduction of control parameters in the dynamics of the TEM model leads to new results in the field of time-discrete control systems (Krabs 1997). Furthermore, the results give new insights into a Joint Implementation Program and hereby, they may improve this important economic tool. The presented TEM model describes the economic interaction between several actors (players) who intend to maximize their emissions reduced E_i caused by technologies T_i using expenditures of money M_i or financial means, respectively. The index i stands for the i-th player. The players are linked by technical co-operations and the market, which expresses itself in the nonlinear time-discrete dynamics of the TEM model. The TEM model is based on a general conflict model of Jürgen Scheffran which was refined by Krabs 1997, Pickl 1998a, 1998b:

$$\Delta E_i(t) = \sum_{j=1}^{n} em_{ij} M_i(t) \tag{1}$$

$$\Delta M_i(t) = -\lambda_i \cdot M_i(t)[M_i \cdot - M_i(t)][E_i(t) - \varphi_i \Delta E_i(t)] \tag{2}$$

The first equation describes the time-dependent behaviour of the emissions reduced so far by each player (nations or companies). These levels E_i, are influenced by financial investments M_i, which are restricted by the second equation. The em_{ij}-parameters determine the effect on the emissions of the i-th actor, if the j-th actor invests money. We can say that it expresses how effective technology co-operations are, which represents the essence of a Joint Implementation Program. Joint Implementation is mentioned explicitly in the Kyoto Protocol (for details and a very good survey, see Stronzick 2001). The parameter φ_i, is called the memory parameter. The consideration of the term E_i, multiplied with φ_i, can be regarded as a memory effect. The expression stands for the influence of earlier investments. The first part of the second equation is similar to a logistic difference equation, where the proportional factor λ_i can be seen as a growth parameter.

The TEM model can be regarded as a mathematical model which supports the development of a management tool in the area of the international climate change convention, namely in the creation of a Joint Implementation Program. Joint Implementation projects intend to strengthen technical co-operations in order to fulfill the Kyoto Protocol. The different em_{ij} stand for different technical relationships. Their economic interpretation can lead to case-studies, in which the range of relevant data can be gained. These data sets might be a good basis for the iterative solution and game theoretic bargaining approach. In the next section we want to describe the integration of several technologies.

3.2 Derivation of the model – technology vector

First of all, we have to introduce for each actor i a vector of his possible technologies

$T_i(t) = (T_{i1}(t), T_{i2}(t), \ldots, T_{im}(t))$

making an energy supply available to the actors. This vector describes the energy-mix, which means that each component expresses the share of the possible energy sources e.g. (in suitable dimensions), for example

T_{i1} ~ petroleum
T_{i2} ~ natural gas
...
T_{im} ~ solar energy

In order to fulfil the climate change conventions, every actor intends to maximize reduction of his emissions, which are caused by these technologies. Multi-step investments (time horizon 1997-2012) in several strategies are possible. We motivate that approach in the next part.

3.3 Multi-step investments

In order to get a sense of the problem, we start with a very simple case in which we have only two players. Two players have two alternatives to invest. The origin of the coordinate system (see figure 4) is the starting point for the two players. Each actor tries to reach the black square which stands for the level of reductions of emissions mentioned in Kyoto Protocol, KYOTO1997. For that reason, we have a limited time-horizon. After the first time-step one of the squares with a grey square will be reached. The players make their choice independently and simultaneously. The directions are attached to the small diagram. The first player goes to the right. If he reduces one unit of CO_2-emissions, then he has to invest 3 financial units. If he wants to attain a reduction of two units he has to invest 5 financial units. One of the main problems in the area of Joint Implementation is the question of an optimal schedule of technical innovations. Therefore, in Pickl 2001 the control problem is solved from a purely mathematical point of view. Here, we model such a situation with a time-discrete approach in which we have three investment units. Each actor can choose between two alternatives (2-step-1-step or 1-step-2-step). The strategy 2-step-1-step stands for a great reduction at the beginning and a smaller investment at the end of the period. The costs are lower than in a 1-step-2-step strategy. We can transfer this simple model with two players and two time-steps to an easy matrix game, which we call the Kyoto-game which will be described in the next part:

3.4 The Kyoto Game

The Kyoto Game is a game-theoretic model to simulate investments in CO_2 reduction programs. Different paths indicate different strategies of the players. Weights are put upon the paths. They can be interpreted as additional sanctions.
Nash-equilibria can be determined. Their characteristics may lead to cooperative behaviour.

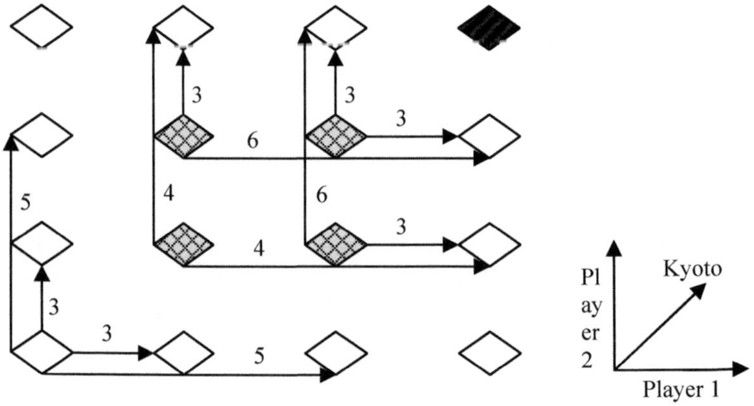

Fig. 4 . Kyoto Game

The pairs (7,7) and (8,8) are Nash-equilibria. A derivation from that chosen strategy is not favoured by one player acting alone. Nevertheless, the tuple ((1,2),(1,2)) is favoured by every actor, yielding a minimum of financial costs. If we want to support a special path or technology - within our *technology mix* -, then we can introduce an additive tax to a special energy path. It is possible to regard the tax as a Pigou tax.

It is possible to prove that for a special tax, only one Nash-equilibrium exists (Meyer-Nieberg 2003; Pickl 2001, 2002). This value can be seen as the maximal trading interval, or as a necessary technical effectiveness measure for guaranteeing uniqueness. Furthermore, as the necessary data are given to the Clearing House, we are able to compare the obtained results with real world phenomena. An extension to an n-player situation with various time-steps may lead to new insights in that economic field and support an improvement of such important energy management tools. In this paper, we have provided the three main issues for optimal management of environmental data within emissions trading markets, namely

1. VEREGISTER as a software package for the management of greenhouse gas emissions
2. TEMPI as analytic forecasting tool
3. The model of the Kyoto Game as a multi-step investment model

In the future, a combination and integration is in the centre of interest in order to optimize the management of greenhouse gas emissions in a very comfortable way according to actual requirements.

4 Emissions trading market

The international effort to reduce atmospheric effects of Greenhouse Gas emissions (GHG) generates complex problems for information generation and management. Information in all forms such as data, calculations, quantities, dates, reports must be created, recorded, compiled, maintained, exchanged and analysed. This is especially crucial for the initiation of an emissions trading market because the market will demand certified units, as close to stock share certificates as possible (see Grimm 2003).

Although the international business world has not agreed on the financial standards and data format for emissions trading in detail, contracts for emissions units are already being traded. Regional pilot projects have, of course, been operating for several years. The Dutch and Danish markets, and in the U.S., the Chicago Climate Exchange, are in the midst of full implementation. At international meetings, participants in these embryonic emissions trading markets admit that they are "learning by doing". Their business plans include calculated risks connected with their limited inventories of sequestered metric tonnes of carbon, or futures contracts for allocated amounts of carbon dioxide. The global stakes involved in the generation and management of information about GHG are enormous. Certified metric tonnes of carbon promise to be one of the most precious commodities in the world's economy for generations to come (Reed 2001). For further aspects we refer to Löschel 2001.

The financial considerations will control the requirements for information management which we presented here in this paper. Reductions in emissions will produce units to be traded. So, the focus is on both to reduce GHG and to establish information management that will provide the necessary administrative framework of the reductions that serve as the instruments for gaining wealth in the marketplace. Even at this elementary stage of the system, it is clear that the 21st Century will be characterized by the creation of billions of dollars worth of new tradable certificates, representing not a share of a producing entity, but rather a number of metric tonnes of an ephemeral by-product of modern industry, tonnes which no longer exist. With so much at stake, the emissions trading market will rely on the institution of new means for originating information about the output of GHG emissions and the location, control and ownership of those emissions. The information required will be found in several places. The emissions will start at the source. The credits generated and account balance for every entity or party will be maintained in inventories. Ownership will be established in registries, which will also capture changes in ownership through emissions trading markets. Reporting accounts can summarize this information (Reed 2001):

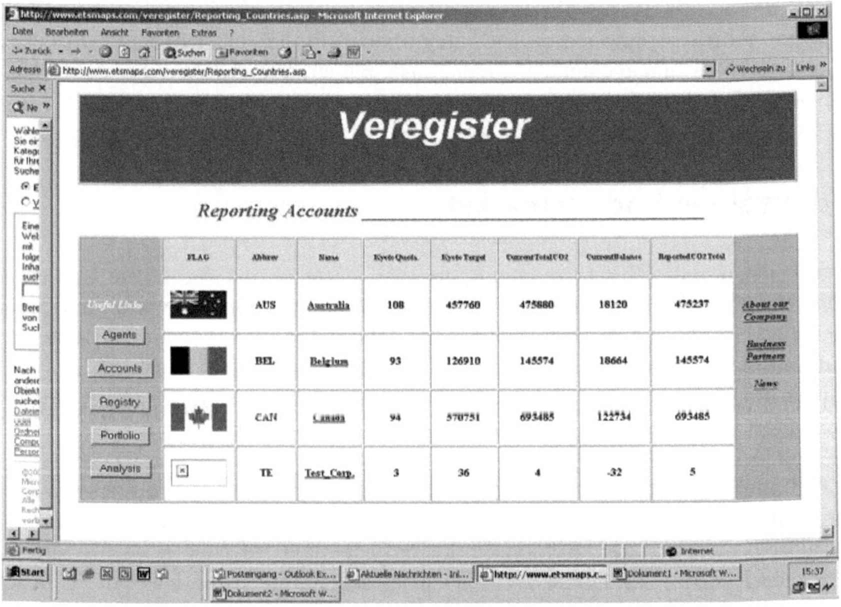

Fig. 5. Reporting accounts and changes in ownership within emissions trading markets

Brokerage and financial organizations have already adapted standard software to record the small volume of speculative trades completed to date. These will not satisfy future requirements for certifiable units, especially when the international regime under the Kyoto Protocol or a similar treaty, could impose damaging penalties on national governments unable to maintain proper inventories and registries. The United Nations Conference on Trade and Development (UNCTAD) financed the prototype software for maintaining registries (updated through trading), inventories, and reporting data. In any case, certified trades must embody several stages of data management (Reed 2001):

1. The source emissions must be recorded through adequate monitoring. This might be done "internally", if instruments like the GHG Protocol are used, but could (should) involve independent consulting firms.
2. The continuous stream of data from the monitoring stage will need verification from a third party. It is an open question if this is necessary for the whole stream of data at this stage.
3. The record of the verified emissions will be the basis for a certification, (probably a consulting environmental or quality assurance organization). Only at this stage will certified credits (here, credits of the Flex Mex after verification through UNFCCC) be available for trading.

All of the stages of this process must be logged through an inventory software that can show real-time balances for any level of organization, from individual source sites, to corporate entities, to the regional or national governmental party:

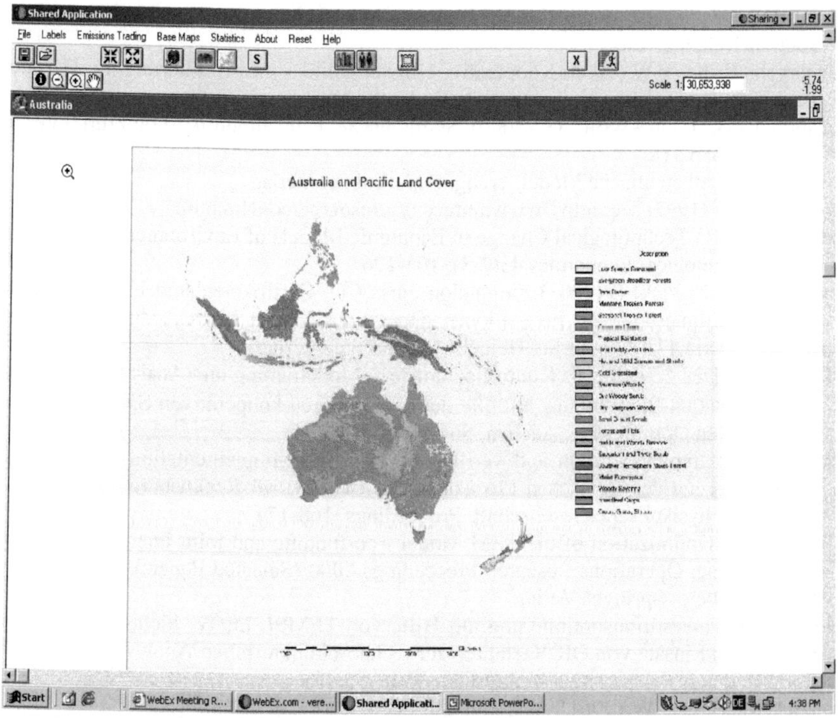

Fig. 6. Regional and national data processing within VEREGISTER

National inventories based on this process will be reported to the UN Climate Change Secretariat and will provide the basis for compliance reviews by the global enforcement organization.

5 Conclusion

Software like VEREGISTER and TEMPI can provide integrated information management origination to final submittal to the Climate Change Secretariat, even to forecast investments in the future (2005-2012).
Options being incorporated into the first version include geographic information modules, optimization algorithms for performing alternative scenarios and trading registries for updating inventories in real time. Information technology often encourages the creation of data.

References

Bültmann A, Hansjürgens B (2002) Umweltpolitik und Umweltmanagementsysteme, in: Zabel, H.-U. (ed) Betriebliches Umweltmanagement - nachhaltig und interdisziplinär, pp. 195-210

Gagelmann F, Hansjürgens B (2002) Der neue CO_2-Emissionshandel in der EU, Wirtschaftsdienst. Zeitschrift für Wirtschaftspolitik, 4: 226-234

Grimm B (2003) Wandel durch Handel, Umweltschutz nach Angebot und Nachfrage, TÜV Journal 1.Quartal 2003

Hansjürgens B, Köck W (2003) Kooperative Umweltpolitik - eine Einführung, In: Hansjürgens (ed): Kooperative Umweltpolitik, Baden-Baden

Hansjürgens B, Lübbe-Wolff G (2000) Symbolische Umweltpolitik, Frankfurt (Verlag Suhrkamp), p.318

Krabs (1997) Mathematische Modellierung, B G Teubner Verlag

Kyoto-Contract (1997) see: http://www.unfccc.org/resource/convkp.html

Löschel A (2001) Technological Change in Economic Models of Environmental Policy - A Survey. Ecological Economics 43(2-3): 105-126

Meyer-Nieberg S, Pickl S (2003) Simulation eines CO_2-Zertifikatenhandels und algorithmische Optimierung von Investitionen. Operations Research Proceedings 2002 (Selected Papers) 471-474 Berlin, Heidelberg: Springer Verlag

Pickl S (1998) Der τ -value als Kontrollparameter - Modellierung und Analyse eines Joint-Implementation Programmes mithilfe der dynamischen kooperativen Spieltheorie und der diskreten Optimierung. Aachen, Shaker Verlag

Pickl S (1998) Implementation and verification of a Joint-Implementation program. 62. Physikertagung der Deutschen Physikalischen Gesellschaft Regensburg 1998, DPG - Deutsche Physikalische Gesellschaft, Proceedings, 168-174

Pickl S (2001) Optimization of the TEM Model - co-funding and joint international emissions trading. Operations Research Proceedings 2000 (Selected Papers) 113-118. Berlin, Heidelberg: Springer Verlag

Pickl S (2000) Investitionsoptimierung mit Hilfe von TEMPI. In: W. Fichtner, J. Geldermann (eds) Einsatz von OR-Verfahren zur techno-ökonomischen Analyse von Produktionssystemen. Peter Lang. Frankfurt am Main, 95-109

Stronzick M, Bräuer W (2001) Accreditation, Verification & Monitoring of Joint Implementtion. JOINT Working Group 4 Position Paper, Mannheim

Reed A (2001) Delivering Data, Environmental Finance, no. 9

VEREGISTER http://www.veregister.com

Emissions trading with changing future commitments – some initial thoughts

Marcus Stronzik

Scientific Institute for Infrastructure and Communications Services (WIK)
Rhoendorfer Str. 68, 53604 Bad Honnef, Germany
m.stronzik@wik.org

Abstract

In climate policy a shift from traditional command and control regulations to more market-oriented approaches such as emissions trading can be observed. Concerning the European context, the Directive for a European-wide emissions trading regime that started in 2005 is currently one of the major topics in European climate policy debates. We used contingent claims analysis in order to shed some light on the investment behaviour of companies covered by the EU Directive. Two different approaches to design an allocation free of charge were looked at: grandfathering without updating the base year and a rolling base year. It is shown that in the first case a considerable option value exists, whereas in the second case the option will more or less expire worthless.

Keywords: Climate policy, emissions trading, investment behaviour, real options, allocation

1 Introduction

In climate policy a shift from traditional command and control regulations to more market-oriented approaches such as emissions trading can be observed. Concerning the European context, the Directive for a European-wide emissions trading regime that started in 2005 is currently one of the major topics in European climate policy debates.

This change is welcomed by most of the economic research community while still a great part of industry is sceptical concerning this shift in regulation as they are not familiar with the new policy approach. Major concerns brought forward by industry are the uncertainties associated with the introduction of an emissions trading scheme. Generally, two types of risks can be identified. On the one hand, the price for the emission permits (allowances) will fluctuate. Therefore, investors will face a price risk when deciding about abatement options. On the other hand, the market is set up by politics and driven by their regulations. It might be that regulators will change or adjust the overall framework (e.g. according to the upcoming results of the ongoing international climate negotiations) while the scheme is already up and running. As this second type of uncertainty is due to changes in regulation it can be classified as regulatory risk. Furthermore, investments in the sectors covered by the EU Directive are usually quite capital intensive, e.g. setting up a new power plant involves payments of some hundred million Euro. The installations usually can't be sold on a secondary market, payments are sunk which leads to irreversibility. Last but not least, emissions trading is offering greater flexibility than most of the other policy instruments. In order to comply with the regulations, the investor can either invest in abatement options or buy a corresponding amount of emission allowances. Only the overall emissions are fixed. How the companies will comply with the given target is up to them. Therefore, while investments in abatement technologies incorporate irreversibility, holding certificates gives the investor the option to invest later.

Uncertainty, irreversibility and some flexibility concerning the investment decision build the basis for the application of real option theory when evaluating investment options (Dixit and Pindyck 1994; Trigeorgis 1995). Economic literature is full of studies analyzing the economic benefits of emissions trading compared to other policy approaches (see e.g. Böhringer 2001; Böhringer and Löschel 2002). Surprisingly, not much has been contributed by economic science to tackle the concerns of industry. So far, not much advice has been provided regarding how companies might have to adjust their investment behaviour, or how it will be affected under an emissions trading regime. Only a few studies exist – at least to our knowledge – that consider emissions trading under a real option approach. Herbelot (1994a and 1994b) as well as Edleson and Reinhardt (1995) had a look at the US Acid Rain Program for SO_2 allowances. Lembie (2003) analyses the effect of two different allocation methods on the investment decision for the case of CO_2 abatement measures. He compares auctioning vs. the allocation free of charge. This paper attempts to broaden the scope by accounting for one peculiarity of the discussion around the EU Directive. The Directive is more or less silent regarding what the actual initial allocation has to look like. Only some guidance is given (e.g. auctioning of up to 5% of the overall number of permits in the first phase and up to 10% in the second phase). Member States have to substantiate this by providing national allocation plans in which they have to specify how the permits are actually allotted to the affected installations. So, it is quite obvious that grandfathering will be the dominant procedure – at least at the beginning. Some Member States – like Germany – have considered a grandfathering scheme with a rolling base year. Therefore, our special focus is on the case where the abatement decisions in the first phase will have an impact on the allocation in the second phase, e.g. through the usage of a rolling base year or period as a means for determining the amount of permits allocated to the corresponding plants. The question is how this specific design will affect investment decisions. We use a contingent claims approach considering a very simple two period model in order to demonstrate the basic effects.

The remainder of the paper is organised as follows. We will start with an introduction of the example of an investment decision under uncertainty where traditional discounted cash flow (DCF) approach will be applied. Thereafter, real option theory will be used in order to show the main differences between these two approaches. In doing so, we distinguish two possible allocation designs, one with a never-ending grandfathering, the other with an adjustment of the allocated allowances (rolling base year). Some basic implications are derived. In the last section we draw conclusions.

2 Application of the discounted cash flow approach[1]

A power company is running an old coal-fired plant of 800 MW that produces electricity only. The utility faces an emission constraint due to the implementation of an emissions trading scheme. The government allocates 90% of the base year emissions (year 0 in Figure 1) free of charge. The company is now considering an investment in a new and more efficient plant with a combined cycle gas turbine (CCGT) (for details see Table 1).

Table 1. Basic data for power plants

Plant type {i}	Fuel	Capacity C_i (MW)	Operating hours l_i	Efficiency η_i (%)	CO_2 e_i (kg/MWh)
Old plant	coal	800	7.500	31	1.093
New plant	gas	800	7.500	57	352

Sources: Söderholm and Ströberg (2003a and 2003b)

In order to evaluate the costs and benefits incurred by replacing an old plant with a new less carbon intensive one, we have to make some assumptions. First of

[1] The example is based on Trigeorgis (1995: 151 ff.)

all, the only benefit that is considered is the number of permits saved through the less carbon intensive plant. The cash flow – the value of the project – for the different states of nature can be calculated as follows:

$$V_t^j = C_i \times l_i \times \Delta \times P_t^j \quad (1)$$ [2]

with V_t^j: value of the project in year t and state j, C_i: capacity of plant i, l_i: operating hours p.a. of plant i, $\Delta = (e_{coal} - e_{CCGT})/1000$; e_i specific CO_2 emissions of plant i, P_t^j: allowance price in year t and state j.

Therefore, we abstract from any other advantages or disadvantages of the new installation.[3] The cash accrues only in the year after the decision has been made.

The only risk the investor faces is the price risk of the allowances. The price development is shown in Figure 1. There is a 50% probability for an upward move. The magnitudes of the price shifts are u=1.8 (upward) and d=0.6 (downward)[4]. The starting price in year 0 is 20 €/t_{CO2}.

In order to use the DCF we need a discount rate. In NPV analysis the existence of a twin security is implicitly assumed for purposes of estimating the required rate of return on a project. In our example, the allowance price can be regarded as such a twin security as it is perfectly correlated with the cash flows of the project. The required return i is

$$i = \frac{E_0(P_1)}{P_0} - 1 = \frac{0.5 \times 36 - 0.5 \times 12}{20} - 1 = 0.2$$

The value of the investment V_0 is:

$$V_0 = \frac{E_0(V_1)}{1+i} = \frac{qV_1^u + (1-q)V_1^d}{1+i}$$

which amounts to around 89 Million Euro. Let us assume – just for demonstration purposes – that the investment costs I are 90 Million Euro[5]. This would mean that management would cancel the project and continue running the old plant.

[2] For l and C we will skip the subscripts in the remainder of the paper as they are the same for both plant types.

[3] Because of the higher fuel to electricity efficiency, the new plant also saves fuel (in MWh_{fuel}). On the other hand, gas is substituted for coal with higher fuel costs for gas (in €/MWh_{fuel}, see Söderholm and Strömberg (2003a)).

[4] The idea behind this price development is the so-called Brownian motion. The general process is characterised through $dP = \mu P dt + \sigma P dz$ with drift rate μ, volatility σ and dz as a Wiener Process. In our discrete time model the price shifts amount to a volatility of roughly 55% ($u = e^{\sigma} \Leftrightarrow \sigma = \ln u$) (Hull 2003).

[5] Investment costs for CCGT are about 560 USD/kW (Söderholm and Strömberg 2003a). As we only consider cash flows in the period after the investment decision, the set-up costs are arbitrarily chosen in order to demonstrate the main effects in the remainder of the paper.

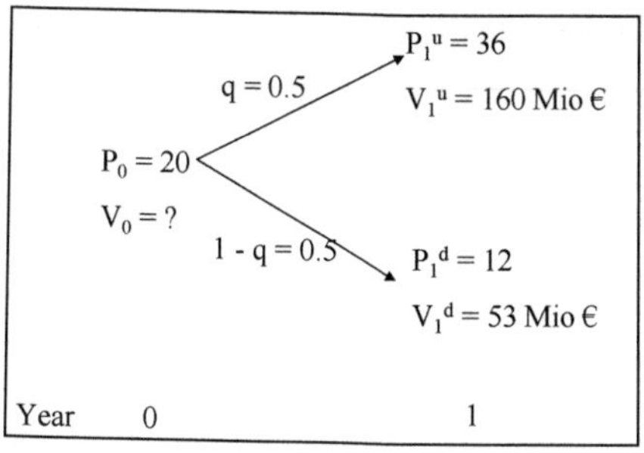

Fig. 1. Development of allowance prices

Note that the results for the DCF method remain unchanged if we assume the same up- and downward shifts for allowance prices for a second period with the same probabilities. In the case of DCF, only expected values, which stay the same, matter.

3 Emissions trading and real options

Now, let us turn to a real option approach using contingent claims analysis. Beforehand, analysing the above mentioned problem, we briefly introduce here the basic concept of contingent claims analysis.

3.1 Basics of application

Real option theory considers investments as investment options E, i.e. the investor has the right to invest, but not the obligation. Therefore, investment options (or real options) are evaluated using the same standard option-pricing hedging strategies as for financial options. This requires an underlying asset, upon which the valuation of the option is based. If this is the case – as in our example, the investment option E is perfectly correlated with the permit price P-, a portfolio can be constructed consisting of N shares of the underlying asset P partly financed by borrowing an amount of B at the risk-free rate r. In doing so, this portfolio can be chosen in such a way that it will exactly replicate the payoffs of the option whether the state of nature in the following year will be d or u (see Figure 2).

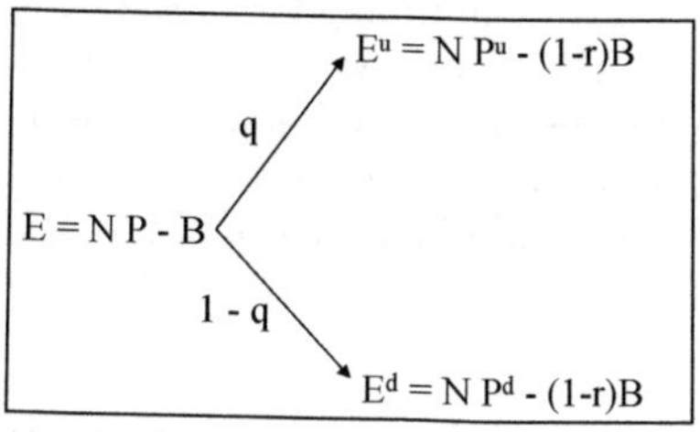

Fig. 2. Portfolio design

Using the two equations for the two different states we can determine the two unknown values of N and B:

$$N = \frac{E^u - E^d}{P^u - P^d}$$

and

$$B = \frac{NP^d - E^d}{1+r}.$$

Assuming perfect markets, no arbitrage opportunities will exist. Therefore, in year 0 the value of the investment opportunity must equal the value of the portfolio with:

$$E = \frac{pE^u + (1-p)E^d}{1+r} \tag{2}$$

with

$$p = \frac{(1+r)-d}{u-d} \tag{3}$$

Assuming r = 0.08 and inserting the corresponding values for u and d we get p=0.4. The value of the investment opportunity does not explicitly involve the actual probabilities q. Instead, it is expressed in terms of adjusted or risk-neutral probabilities, p, which allow expected values to be discounted at the risk-free rate. This constitutes the main principle evaluating investments using contingent claims analysis, the so-called risk-neutral valuation principle.

Applying the adjusted probabilities, we obtain

$$V = \frac{pV^u + (1-p)V^d}{1+r} = \frac{0.4 \times 160 + 0.6 \times 53}{1.08} \approx 89 \quad \text{Mio €}.$$

The result is identical with the gross project value obtained earlier using DCF with the actual probability q and the discount rate i. In the absence of any managerial flexibility, both methods come to the same results.

3.2 Case 1: Grandfathering without adjustment

Let us now assume that management has some flexibility in timing the investment. The investment can be deferred (option to defer). The plant owner can either invest in t=0 or in t=1. In the case of pursuing the investment, the initial allocation will not be adjusted. The plant owner will still get the number of permits for which the determination was based on the emissions of the old plant.[6] We will also assume that the investment costs I increase with the risk-free interest rate r. This as-

[6] We abstract from the discussion about new or modified plants. The base year is not changed (year 0).

sumption is intended to make the analysis somewhat more realistic and is not crucial to the analysis.

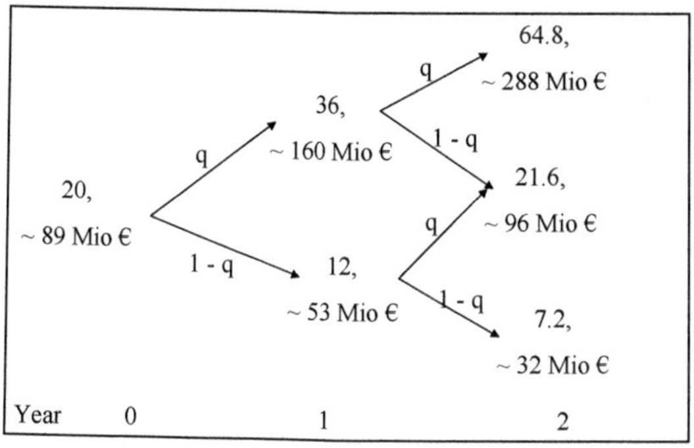

Fig. 3. Price movement and project value

It is expected that permit prices will develop the same way as they do during the first year, i.e. u=1.8 with q=0.5 and d=0.6 with (1-q)=0.5 also for the second year. The resulting binomial tree for prices as well as for the gross project value V is shown in Figure 3.

Because the option to defer the project for a year gives managers the right, but not the obligation, to make investment by next year, they will wait and make the investment if the project value next year turns out to exceed the necessary investment at that time. The option to wait can be regarded as a call option on the gross project value V with an exercise price equal to the required expenditures next year. This translates into the right to choose the maximum of the project value minus the required investment costs or zero, since management will simply allow the option to expire worthless (not invest) if project value turns out not to cover the necessary costs. That is,

$$E^u = \max(V^u - I_1, 0) \approx \max(160 - 97.2) = 62.8 \tag{4}$$
$$E^d = \max(V^d - I_1, 0) \approx \max(53 - 97.2) = 0$$

with $I_1 = 1.08 * I_0$.

Thus, with the option to wait, the payoff structure will be as shown in Figure 4.

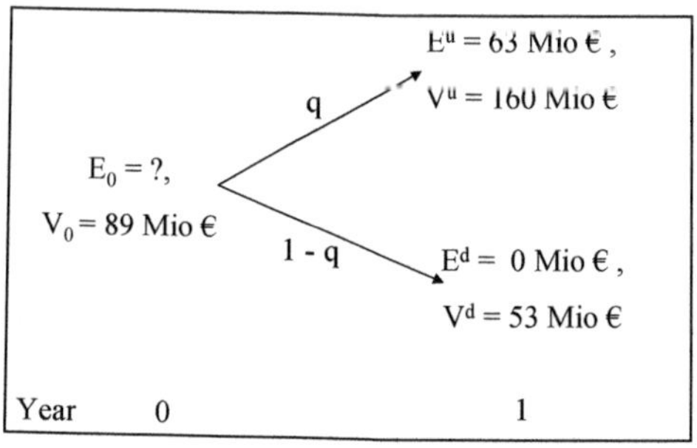

Fig. 4. Option pricing with grandfathering without adjustment

The option to defer has asymmetrically altered the structure of payoffs. Instead of paying 90 million Euro immediately to receive either 160 million Euro or 53 million Euro next period, the investor is now able to wait and observe if the outcome is favourable, in which case he would go ahead with investing for a net payoff of 62.8 million Euro; or he could decide not to proceed if permit prices go down, with a payoff of zero.

The total value of the investment opportunity can be obtained by

$$E_0 = \frac{pE^u + (1-p)E^d}{1+r} \tag{5}$$

Inserting the corresponding values we get E_0=23.3 million Euro. Although, the project has per se a negative (passive) NPV if taken immediately, the investment opportunity should not be rejected because the option to invest in the new plant within a year is actually worth a positive amount. The value of the option to wait is thus given by

Option premium = Expected NPV – Passive NPV
= 24.2 million Euro (6)

3.3 Case 2: Rolling base year

A rolling base year for the determination of the initially allotted allowances means that the base year will be regularly updated. In our simple example we assume an immediate update with still an allocation of 90% of base year emissions. Therefore, the gross project value is affected by such a regulation. It is now given by

$$V_t^j = 0.1 \times C \times l \times \Delta \times P_t^j \tag{1a}$$

In our second case the gross project value is reduced by 90% compared to the first case. The investor still saves 10% of the emissions of the old plant for which he does not have to buy allowances. But also, he has now to buy permits representing 10% of the emissions of the new plant. The resulting gross project values are presented in Figure 5.

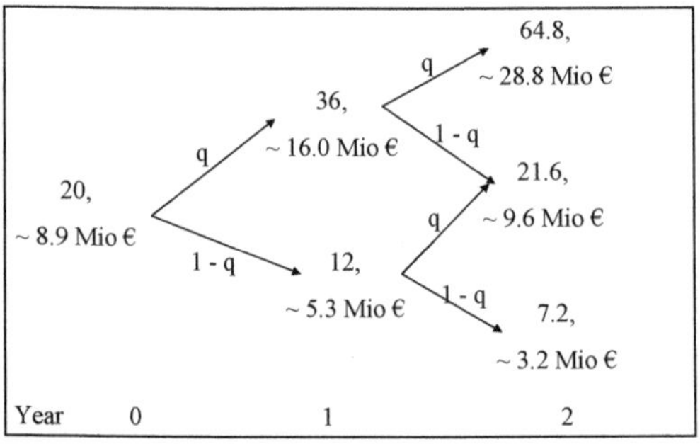

Fig. 5. Gross project values for case 2

It can easily be seen that in none of the states the investor will be willing to invest in the new plant. The option will never be exercised, it will expire worthless (E=0 in all states of nature). In such a case, traditional DCF would give accurate results for the investment decision.

The results are not really surprising as the regulation of a rolling base year punishes outriders. Such a passage will lead to delays of investments in new and less carbon-intensive technologies.

4 Conclusions

Traditional DCF is unable to capture the value of operating options properly, because of their discretionary asymmetric nature and their dependence on future events that are uncertain at the time of the initial decision. The DCF technique is mapping a "now or never" decision. Managerial flexibility can be better captured through the use of decision tree analysis that helps management to structure the decision problem by mapping out all feasible alternative actions contingent on the possible states of nature in a hierarchical manner. Decision tree analysis can actually be seen as an advanced version of DCF – one that correctly computes unconditional expected cash flows by properly taking account of their conditional probabilities given each state of nature. Its main drawback is the problem of determining the appropriate discount rate to be used working back through the decision tree. The fundamental problem of both traditional approaches lies in the valuation of investment opportunities whose claims are not symmetric or proportional. The asymmetry resulting from operating flexibility options and other strategic aspects can properly be analysed by thinking of investment opportunities as options on real assets. The option based approach enables management to quantify the additional value of a project's operating flexibility. In the absence of such flexibility, the contingent claims analysis gives results identical to those of traditional DCF (as seen in sub-section 3.1).

We used contingent claims analysis in order to shed some light on the investment behaviour of companies covered by the EU Directive for a European-wide emissions trading scheme. Two different approaches to design an allocation free of charge were looked at, grandfathering without updating the base year and a rolling base year. It was shown that in the first case a considerable option value exists, whereas in the second case the option will more or less expire worthless.

References

Böhringer C (2001) Industry-level emission trading between power producers in the EU, Applied Economics 34(4): 523-533

Böhringer C, Löschel A (2002), Assessing the Costs of Compliance: The Kyoto Protocol, European Environment 12(1): 1-16

Dixit AK, Pindyck RS (1994), Investment under Uncertainty, Princeton

Edleson ME, Reinhardt FL (1995) Investment in Pollution Compliance Options: The Case of Georgia Power, In: Trigeorgis L. (ed) Real Options in Capital Investment, Westport, pp. 243-264
Herbelot O (1994a) Option Valuation of Flexible Investments: The Case of a Coal Gasifier, Working Paper MIT-CEEPR 94-002WP, Massachusetts Institute of Technology
Herbelot O (1994b) Option Valuation of Flexible Investments: The Case of a Scrubber for Coal-Fired Power Plant, Working Paper MIT-CEEPR 94-001WP, Massachusetts Institute of Technology
Hull JC (2003) Options, Futures, and Other Derivatives, 5th Edition, New Jersey
Lambie NR (2003) Analysing the Effect of a Distribution of Carbon Permits on Firm Investment, Mimeo, The Australian National University, Canberra
Söderholm P, Strömberg L (2003a) Options, Costs and Strategies for CO_2 Reductions in the European Power Sector, Energy Studies Review 11(2): 171-204
Söderholm P, Strömberg L (2003b) A Utility-eye View of the CO_2 Compliance-decision in the European Power Sector, Applied Energy 75: 183-192
Trigeorgis L (1995) Real Options - Managerial Flexibility and Strategy in Resource Allocation, Cambridge, MA

Part C

Emissions trading and business administration

Emission Trading North – important findings from a business perspective

Katja Barzantny

EUtech Energie & Management GmbH
Dennewartstraße 25-27, 52068 Aachen, Germany
barzantny@eutech.de

Abstract

In co-operation with the Association of the Chambers of Industry and Commerce in Schleswig-Holstein and the Union of Employers' Associations in Hamburg and Schleswig-Holstein, the Energy Foundation Schleswig-Holstein carried out the pilot project Emission Trading North - Benefits for the Economy and the Environment (May 2002 - May 2003). It focused on capacity building, communication of important findings into the national political decision-making process (based on detailed company focused case studies) and the positioning of the Energy Foundation in a future emissions trading system. The opportunity to prepare for EU emissions trading at an early stage was taken by nine companies: three from the power sector, three from the pulp and paper sector, one from the cement industry and non ferrous metal sector each and one from the renewable energy sector. Special attention was paid to questions relating to internal requirements within the corporations, the identification and evaluation of emission reduction measures, the development of an emissions trading strategy, including an internet based trading simulation and the allocation of emission allowances in the context of the national allocation plan. Emission Trading North revealed a number of important findings on information, reporting and decision-making processes and on an optimal emission management strategy in the short, medium and long term and it, finally, allowed the identification of research questions for future work. Detailed information on the project can be obtained from www.emisssionshandel-nord.de.

Keywords: Emissions trading, pilot project, emission reduction measures, trading simulation, business strategies

1 Introduction

On October 25th 2003 Directive 2003/87/EC of the European Parliament and of the Council establishing a scheme of greenhouse gas emission allowance trading within the community and amending Council Directive 96/61/EC became effective. The EU member states were asked to set up their national allocation plans by March 2004 and to decide on the different design options. Emissions trading is no longer just a theoretical concept, but is a central element of the environmental policy mix of the European Union. It aims to improve cost efficiency of climate protection and to increase the flexibility of action from a business point of view - the two essential advantages of emissions trading frequently mentioned in the economic literature. However, to let these advantages become reality business needs to know how emissions trading works. For that reason, the Energy Foundation Schleswig-Holstein in co-operation with the Association of the Chambers of Industry and Commerce in Schleswig-Holstein and the Union of Employers' Associations in Hamburg and Schleswig-Holstein initiated the one-year pilot project Emission Trading North - Benefits for the Economy and the Environment starting in May 2002.[1]

2 The pilot project

Targets and participants

Primary target of the pilot project was the regional application of an instrument developed in an international context together with the industry of Schleswig-Holstein and Hamburg. Specific targets included:

1. Capacity building of the different stakeholders involved to realise first mover advantages, especially with respect to the participating companies.
2. Communication of important findings into the national political decision making process.
3. Positioning of the Energy Foundation as a regional promoter of renewable energy, energy efficiency and energy saving measures in a future emissions trading system.

The opportunity to prepare for EU emissions trading at an early stage was taken by nine companies: Three from the power sector with very different production structures, three from the pulp and paper sector, an internationally operating company from the cement industry, one company from the non ferrous metal sector and one from the renewable energy sector.

---

[1] The project itself was carried out by the advisory firms ERM Lahmeyer International, 500 PPM and EUtech.

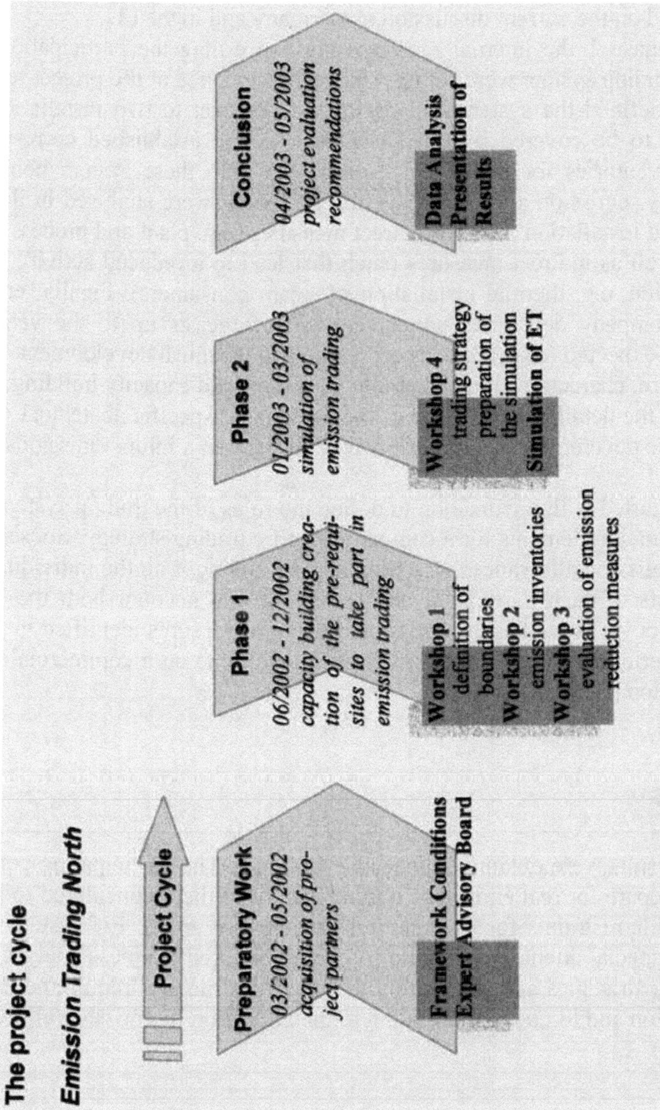

Fig. 1. The project cycle consisted of three phases (see Figure 1).

The preparatory work of phase 0 included the acquisition of the companies mentioned above, the constitution of an expert advisory board with a regional focus but also including members of the European Commission and the German Environment Ministry and the definition of realistic framework conditions for the pilot project based on the current discussion in Germany and in the EU.

During phase 1 the internal requirements concerning the participation in an emissions trading system were set up. Under the guidance of the project team, the companies defined the system boundaries with respect to two installations that were likely to be covered by the EU Directive and established corresponding emission inventories for direct CO_2-emissions within these system boundaries. Subsequently, emission and cost reduction potentials were analysed in detail for each selected installation including direct measures (e.g. plant and process optimisation) as well as indirect measures (such that lead to a reduced activity level of the installation, e.g. thermal installation of steam consumers). Finally, each participating company developed annual emission forecasts up to the year 2012, based on its expected (or presumed) economic and technical development. Phase 1 was, therefore, characterized by "learning by doing" and capacity building. As one main result, the detailed knowledge of the companies' specific abatement costs allowed for the development of a profit driven strategy in a future emissions trading system.

Phase 2 started with a workshop to define the rules of the trading system and to develop the major elements for a company specific trading strategy. Subsequently, trading of emission allowances was simulated by basing it on the individual emission forecasts of each participant and by taking into account both the strategy mentioned above, as well as the emission reduction measures identified in phase 1. The multi-period-trading simulation was realised online on a commercially available simulation platform.

3 Results

Important results were obtained in capacity building. The participating companies started to prepare for real emissions trading and all of them considered to have reduced possible restraints for a successful participation in the EU system. During the project, special attention was paid to energy oriented business analyses, emissions trading strategies and simulation, the internal structural requirements within the corporation and to questions relating to the allocation of emission allowances.[2]

[2] The last issue will not be dealt with in this article but is analyzed in detail in Hahn et al. (2003) and Kruska et al. (2003).

3.1 Energy oriented business analyses

The framework conditions with respect to the energy oriented business analyses were characterized by the choice of individual, company specific investment and decision criteria, e.g. with respect to the investment period, the internal interest rate of return and the lead-time of investments to guarantee the consistency with the general business strategy and the prevailing investment process. The companies were explicitly asked to analyse all potential emission reduction measures including the (so far) unprofitable ones, because in an emissions trading system profitability is a function of the current or rather expected allowance price. The de-tailed plant analyses showed that even within the small sample of participating enterprises a broad range of energy saving and emission reduction potentials could be realised on an economic basis, i.e. at negative or zero specific abatement cost ("no-regret potentials"). A large majority of emission reduction potentials was found to have specific abatement costs between 0 and 100 € per ton of CO_2 avoided annually. Up to 15% of the overall emissions of all participating enterprises could be avoided at these costs. Figure 2 shows the abatement cost curve of the emission reduction potentials identified within the project, not taking into account the lead-time, price risks and volume risks of emission reduction measures.

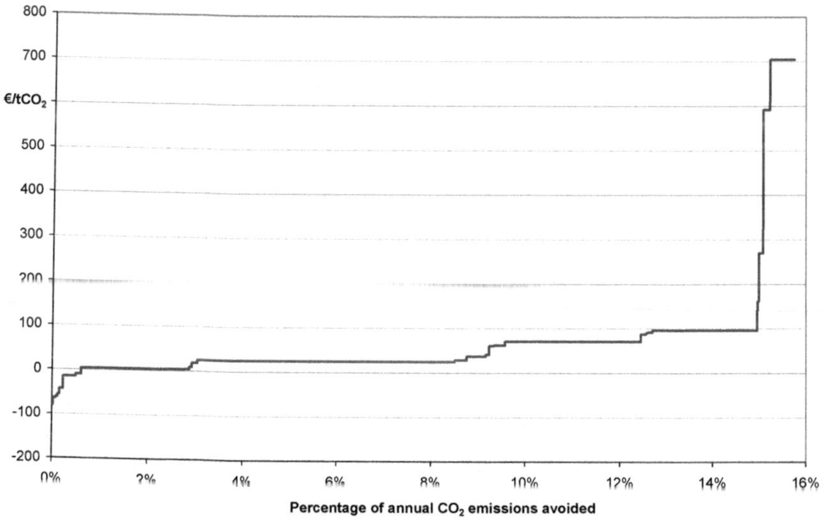

Fig. 2. Specific abatement cost curve of the potential emission reduction measures

3.2 Trading strategy and simulation

For each installation, emission forecasts for the years 2005 to 2012 were made based on projected activity levels, planned implementations of reduction measures and investments. By opposing these emission forecasts to the predefined allocation scheme, emission balances were drawn to determine each company's compliance situation (shortfall or surplus) of each year. Figure 3 gives an example for such an emission balance.

Emission inventory (period x)		Compliance instruments (period x)	
baseline emissions	659.905 t	allocated emission allowances	685.395 t
emission reductions by emission reduction measures	66.557 t	+ allowances bought	+ 28.322 t
		- allowances sold	- 42.483 t
		+ CER's (CDM)	0 t
		+ ERU's (JI)	+ 70.806 t
		BALANCE: surplus/shortfall	- 15.577 t
sum	726.452 t	sum	726.452 t

Fig. 3. Example for an emission balance

Companies were then invited to develop a trading strategy taking into account investment cycles, the legal, financial and technological framework and the marginal abatement costs compared to market price forecasts. The possible strategies were tested in an internet based trading simulation and included "make-or-buy" decisions, forward transactions, banking according to the allocation scheme, and the choice between three types of compliance instruments (emission allowances, JI credits, CDM credits).

In most cases, such a trading strategy will mainly consist of a compliance strategy which has to focus on the following three elements:

- strategic evaluation of emission reduction measures
- definition of a long term trading strategy
- cover against (short term) technical and economic risks of non-compliance.

A strategic evaluation of emission reduction measures has to consider the specific abatement costs (€/t CO_2), the avoided emissions in the course of time, the lead-time of investments and absolute investment costs. It will be based on the definition of a company specific risk attitude and has to be integrated into the general business strategy. Furthermore, the company needs to define a long term trading strategy by analysing price movements, the overall characteristics of the emission trading market and policy developments. Finally, it has to cover against technical and economic risks of non-compliance in the short term which requires the use of a portfolio of compliance instruments and derivates as well as the resort to internal know-how, external experts and consultancy services.

3.3 Internal structural requirements

By pricing CO_2-emissions technical issues are transformed into monetary issues and a new input factor is introduced into production and investment decisions. This implies that a wide range of departments of a company will, on the one hand, be affected by emissions trading and on the other hand, their decisions will affect the compliance situation of the company. Therefore, all the relevant departments need to communicate with each other and to co-ordinate their internal and external activities to allow for a sufficient transfer of information and an optimal reporting and decision making process. A transparent and effective assignment of tasks is required, which has to fit to the individual internal structures of each company. In this context, the management will most likely be in charge of the risk management, "make-or-buy" decisions with respect to emission reduction measures and a company specific emission strategy. The production department has to analyse the effects of the current production on the actual emissions. Besides this, it has to develop production and emission scenarios considering potential emission reduction measures, technological aspects and volume risks of different kinds. These activities will closely be linked to the research and development department dealing with product and process innovations that might reduce CO_2-emissions. Concerning the trading transactions, a decision has to be made whether the company will use or build up internal know-how (e.g. of the purchasing or electricity trading department) or whether it will resort to external services. Finally, the reporting department could possibly be responsible for the establishment and maintenance of an emission inventory and an adequate monitoring and reporting system. They have to communicate with the controlling and production department on a regular basis and could also be in charge of co-ordination tasks.

In addition to the establishment of effective internal business structures, the company should take advantage of the inherent flexibility of emissions trading to ensure cost efficiency. This flexibility is not fully described by the term "make-or-buy" decision and trading activities between companies. What is more, it encloses the possibility of synchronizing emission reduction measures with respect to investment plans, i.e. investment cycles, financial market conditions and production targets. Besides this, a transfer of emission allowances between different installations, product lines and sites is another interesting option to allow for a balancing of shortfalls and surpluses within companies. Internal trading can be accompanied by shifting production between installations, sites and/or countries. However, there is not always a need for such "physical" measures.

All the aspects mentioned above should enter into a profound business strategy for the short term, the medium term and the long term. In the short term, an optimal preparation for the EU emissions trading system was in the centre of interest. Emission Trading North focused on steps essential in this context (see phase 1 and 2) including the documentation of historic emissions and early action, the implementation of a monitoring and reporting system, the determination of all potential internal emission reduction measures and their costs and the realisation of no-regret potentials. The medium term will be characterized by an optimal positioning in the emissions trading system based on the definition of a company specific risk attitude and the use of a diversified portfolio of compliance instruments. Finally in the long term, a decarbonisation of business activities might be necessary to ensure competitiveness which requires a different focus of research and development projects resulting in new products and scopes of business.

4 Final remarks

The pilot project Emission Trading North illustrated that projects of capacity building are important and necessary to increase the acceptance of emissions trading and to allow for business friendly solutions. With respect to the review process required by the EU Directive until June 2006 this finding implies that a constructive dialogue between business and politics on the range of experiences made should be organized. At present, the allocation of allowances and the emission reduction commitments of the individual installations are the actual challenge for industry. Therefore, distributional aspects and fundamental discussions on future energy strategies are currently dominating the national debate, which will most likely also be the case for the upcoming commitment periods. Nevertheless, business and politics are called upon to secure cost efficiency and the inherent flexibility of emissions trading by minimizing transaction costs, maximizing market liquidity and establishing a purposeful incentive structure for emission reductions, especially with respect to new entrants and their long term investments. In this context, the reduction of transaction costs for small and medium sized companies - necessary to increase market liquidity and to avoid market distortions - turned out to be an important issue for future work. Business is now called upon to collect and verify emission data and to establish a monitoring and reporting system on the one hand and to determine the internal reduction options and their related costs on the other hand. While doing this, companies have to be aware of the fact that emissions trading is an overlapping task and will result in new internal business structures. The inclusion of emissions trading into production and investment decisions demands a permanent consideration of this new input factor in all operational and strategic processes. Further research should focus on the design of these internal business structures to stimulate climate friendly products and/or process innovations within corporations. By simultaneously looking at the static and dynamic efficiency of different allocation methods, it would be possible to combine the macroeconomic and microeconomic aspects of cost efficiency. Last but not least, we have to remember that emissions trading will only be one element of an optimal future environmental policy mix. Further research is, therefore, needed to define this optimal environmental policy mix, taking into account that there are numerous installations currently not covered by the EU Directive and that there exists the possibility for these installations to "opt-in" the EU trading system.

References

European Parliament, European Council (2003) Directive 2003/87/EC of the European Parliament and of the Council of 13 October 2003 establishing a scheme of greenhouse gas emission allowance trading within the community and amending Council Directive 96/61/EC

Hahn M, Klein M, Kruska M, Barzantny K (2003) Zwischenbericht Emissionshandel Nord - Anforderungen an einen nationalen Allokationsplan. For the Energy Foundation

Schleswig-Holstein, February 2003. In: http://www.emissionshandel-nord.de, download

Kruska M, Hahn M, Klein M, Barzantny K (2003) 10 Forderungen an den nationalen Allokationsplan - Schlussfolgerungen aus dem Pilotprojekt "Emissionshandel Nord". In: Energiewirtschaftliche Tagesfragen 53, 2003

Corporate greenhouse gas management in the context of emissions trading regimes[1]

Ralf Antes

Martin-Luther-University Halle-Wittenberg
Faculty of Economics
Department of Corporate Environmental Management
Große Steinstraße 73, 06099 Halle, Germany
ralf.antes@wiwi.uni-halle.de

Carl von Ossietzky-University Oldenburg
Faculty of Economics
CENTOS – Oldenburg Center for Sustainibility Economics and Management
Uhlhornsweg, 26129 Oldenburg, Germany
ralf.antes@uni-oldenburg.de

Abstract

The article analyses the impact greenhouse gas emissions trading (GHG-ET) regimes have on companies from the angle of ecological involvement, the stakeholder approach and the organizational field. It concludes that some form of GHG management is advisable. The main consequences of the European Union's emissions trading scheme (which, albeit limited to CO_2 emissions, has come into effect in January 2005) are discussed, in particular strategy options for procurement management for GHG certificates and the scheme's impact on corporate environmental management.

Keywords: Certificates procurement management, ecological involvement, emerging organizational fields, organizational impacts, stakeholder analysis

---

[1] The article is a result of a research visitorship at the University of East Anglia, Norwich, School of Environmental Sciences, Centre for Social and Economic Research on the Global Environment (CSERGE). I am indebted to Andy Jordan, Tim O'Riordan, Kerry Turner, and others for their fruitful discussions and invaluable support. The visitorship was kindly funded by Stiftungsfonds Dresdner Bank im Stifterverband für die Deutsche Wissenschaft.

1 Introduction

International climate protection policy has two innovations in store for the business community, indeed as early as 2005 for companies based in the EU. Firstly, the common right to emit CO_2 (and other GHGs) will be transformed into a property right with GHG emissions being capped. As a result, previously externalized costs of GHG emissions will become partly internalized. Secondly, this transformation will be effected by means of an approach which is novel for most corporate environmental management and indeed for business administration research – namely through trading in emissions certificates Below I examine how companies will be affected by the problem of GHGs themselves (ecological involvement; Section 2), external effects on business (emerging organizational fields of a system of GHG management; Section 3) and internal effects for corporate management (procurement management of emissions certificates and environmental management; Section 4.). The study focuses on GHGs in general, even though the EU emissions trading regime will initially only cover CO_2. Then again, CO_2 is the main GHG (accounting for 80% of overall emissions); moreover, both the Kyoto Protocol (Art. 3 (1)) and the EU Directive (Art. 24) make provision for the regime to be expanded to other GHGs.

2 How GHGs affect companies

2.1 Ways in which companies are affected by GHGs

Companies can be affected by the emission of GHGs in various ways. The concept of general ecological 'involvement' (Antes 1992, pp. 490-93, 1996, pp. 84-88) serves as a heuristic explanation and combines subjective constructivist and realistic objective viewpoints and findings. It allows three phenomena to be integrated and interrelated: (1) the causing of ecological effects (in this case the emission of GHGs through economic activity); (2) the inevitable interdependencies in the production process resulting from the division of labor among different companies, especially interdependencies with a bearing on the emission of GHGs; (3) recursive stakeholder relations influencing the emergence of GHG-related organizational fields and corporate GHG management. The ecological (GHG) involvement of a company can accordingly be described in five pairs of terms: self-produced (directly/indirectly), externally acquired (ecological effects/ecological demands of others), objective-realistic versus constructivistically-subjectively perceived, and in temporal terms acute/latent/potential and temporary/permanent.

Direct involvement through the emission of GHGs: First of all, a company can produce GHGs itself. Private business causes well over half the emissions of CO_2 and other GHGs.[2] The main 'culprits' are power generators along with energy-

[2] In the respective statistics, companies are represented not just by the sector 'industry' but also by 'energy', 'transportation', 'tourism' and 'farming'. For the GHG emissions of the

intensive processes such as aluminium and cement production, transport and the chemical industry.

Indirect involvement by influencing the GHG emissions of others or their scope for GHG management: A business can (intentionally or unintentionally) influence or even predetermine the ecological performance of other economic actors (competitors, suppliers, buyers, etc.). Ways of reducing or avoiding GHGs can be encouraged or discouraged in preceding and following steps due to the interdependencies within the production process ("the next process is your customer"). For instance, the design of a piece of equipment determines its range of energy efficiency and hence its GHG emissions. As a result, the operator's ecological performance (in terms of minimizing emissions of GHGs) is restricted to a certain predefined range. Another example is decisions on aspects such as location, production and logistics (e.g. in connection with just-in-time operations), which have a considerable bearing on the volume and the means of transport needed (Antes/Prätorius/Steger 1992).

GHG involvement imposed by other economic actors: Just as a business can cause GHG involvement among other economic actors, it can in turn also be ecologically affected by the activities of others. For example, a company's own production may be affected by other companies' GHG emissions (resulting for example in increased retention and reprocessing costs or lower quality). However, damage caused directly by local concentrations of GHGs can be ruled out because the sinks of GHGs arise not locally but instead globally in the atmosphere. The long duration of this process (dispersion and residence times)[3] makes the consequences or 'global feedback' of GHG emissions by other actors more or less irrelevant for company employees (who, even if they spend a lifetime with the same employer, will still fluctuate from one department to another) and may even exceed the company's lifetime. The more relevant GHG involvement imposed by others therefore seems to be the pressure that stakeholders and the organizational field could (depending on their awareness of climate change and the GHG issue) bring to bear on the GHG-emitting company now in anticipation of possible harm. The legal institutions and institutional regimes introduced by climate policy ranging from the global to the national level (such as the Climate Change Convention, the Kyoto Protocol, the EU Emissions Trading Directive, the EU linking Directive, National Allocation Plans, and monitoring and reporting systems as yet to be implemented) are of outstanding importance in this respect. As well as internalizing the minimum standard(s) for companies' GHG management, they also induce additional expectations and demands on the part of other stakeholders (for more details see the next section).

Temporal GHG involvement: The temporal dimension emerges because firstly a company may be affected by the GHG issue temporarily or permanently. Both cases are covered by the EU Directive. For example, permanent involvement can be assumed for installations defined in Annex I of the EU Directive. If a company

OECD member states in 1998 measured in CO_2 equivalents (OECD 2001, pp. 160 f.; for Europe EEA 2001, p. 62).

[3] CO_2 about 100 years; CH_4 about 10 years, N_2O about 150 years (Graßl 1993, pp. 76-79).

succeeds in the technological substitution of its Annex I installation, it can escape legal regulation and hence automatically sidestep other requirements entailed by installations of this type. The possibility of opt-out (i.e. the national exclusion of installations and activities from the EU emissions trading scheme like currently permitted for the Netherlands and the UK) is only permitted until 31 December 2007 (Article 27 (1) EU Directive).[4] Secondly, a distinction is drawn between acute, latent and potential involvement depending on the level of urgency (although the differences between them are not hard and fast). Before the EU Directive was passed, all operators of the activities and installations plants defined in Annex I were latently affected due to the residual uncertainty surrounding it; now they are acutely affected. By the same token, installations currently excluded such as combustion installations with a rated thermal input falling below 20 MW may be included in subsequent revisions of the Directive. Moreover, owing to the EU members states' opt-in possibilities, i.e. for the "unilateral inclusion of additional activities and gases" as of 2008 (Article 24 EU Directive), other CO_2-emitting non-Annex I activities and installations are potentially affected. Areas that have already come under consideration include transport and private households (RSU 2002, p. 235f.; Cames, Deuber, Rath 2004; Deuber 2002). Besides, the set of rules may be expanded to the other GHGs listed in Annex II of the EU Directive,[5] meaning that other activities/installations in the same company (and indeed previously unaffected companies) could be affected for the first time.

Objective/subjective GHG involvement: One aspect that is crucial for the maintenance of an ecosystem is the objective, actual emissions load ('critical load'), i.e. pollution by GHGs (= direct GHG involvement). Meanwhile, the motivation for an emitting company to reduce or even eliminate GHGs (as well as for stakeholders to exert pressure on a GHG-emitting company) depends on the subjective perception and shape of objective involvement. Accordingly, the objective load and the load subjectively perceived need not be the same and indeed is normally not. One particular aspect of GHGs is that although emissions of GHGs cannot be directly perceived by the human senses, their consequences (climate change) can, albeit much later (i.e. in the form of regular and catastrophic flooding and periods of drought). Consequently, these ecological effects and their perceivability are often externalized, i.e. they are perceived by others than the polluter (because they occur much later or in a different place). This is a pitty, because it has been shown that being able to perceive consequences and attributing them to one's own behaviour are vital if actors are to take the problem seriously and be motivated to act in a more ecologically sound manner (Pawlik 1991; Antes 1996, pp. 134-143; Hinding 2002, p. 30 f.). If at all possible it requires highly complex models to attribute temporally and spatially remote effects to particular (economic) activities.

4 The possibility of pooling (Article 28 EU Directive) is a temporarily granted option. However, rather than excluding the company from the GHG regime of the EU Directive, it merely places it under a special form of this regime.

5 Article 3, Annex A of the Kyoto Protocol (1997) covers those gases obligatory for the period 2008-12.

2.2 Degrees of externally acquired GHG involvement

The stakeholders' importance for the GHG management of a company was mentioned above on the basis of the involvement passed on by others. Table 1 contains an overview of (a) the benefits that could be provided by stakeholders to a manufacturing company emitting GHGs, (b) the sanctionable demands they could press, and (c) the basis of their power and influence.[6] Readers are assumed to be sufficiently familiar with the stakeholder approach from the literature, and so below we concentrate on specific aspects of GHG management. The fact that for example shareholders could exercise their ecological preferences via capital fluctuation or the company organs is regarded as a general aspect and is not emphasized here. Stakeholders are assumed to be keen to reduce/avoid GHGs.[7] Contrary attitudes and strategies are of course possible and have been empirically observed, but can easily be imagined by reversing the content of the cells in the table.

Compared to the traditional understanding of the stakeholder approach, one important difference that needs to be borne in mind is that a company is by no means merely just a recipient of its stakeholders' demands that tries to adapt accordingly. On the one hand, this stems from the variety of possible demands, which are of course not limited to GHGs. The GHG problem generates (inter alia) a management task. It is hardly likely that all the numerous demands will complement or be neutrally related to each other. Instead, conflicts exist between the demands of different stakeholders. Consequently, it will be impossible for a company to meet all its stakeholders' (GHG-related) demands. Scope for strategic behaviour inevitably emerges both in favour and at the expense of the reduction/avoidance of GHG, despite stakeholders' contrasting expectations (Meyer, Rowan 1991, 1977; Ortmann, Zimmer 1998; Antes 2004). On the other hand companies actively influence the design of the behaviour framework or of the organizational field for GHG management. Empirically this can indeed be observed in the current development of the GHG legal standards at UN, EU and national level.[8] Companies must therefore be regarded as structural and cultural actors of the GHG orientation of their environment.[9]

---

[6] Of course, a stakeholder analysis is for its part feasible for every type of stakeholder potentially relevant for the GHG management of an emitting company.

[7] The restrictions to which motivation for a global ecological problem is subject apply to all protagonists essentially similarly; regarding subjective involvement see above, also Preuss 1991; Antes 1996, Chapter. 3.

[8] Examples include the options of pooling and opt-out now granted for a certain period in the EU Emissions Trading Directive, the negotiation of the National Allocation Plans, the set-up of the Emissions Trading Group within the UK emissions trading scheme, and the establishment of a discussion group between the Ministry of Environment and trade associations in Germany prior to the national guidelines coming into effect (Vorholz 2003; Wurzel et al. 2003).

[9] Schneidewind 1998; Pfriem 2000; Seo and Creed 2002. And depending on their standardization capacity to encourage GHG-reducing/avoiding management in their own organizational field, a standardization responsibility for their organizational field also falls

Table 1. Stakeholder relations of GHG emitters in the context of emissions trading

Stakeholder	Benefits	Demands = f (preferences)	Basis of power & influence
Internal			
Management	GHG-specific expertise and commitment	Sustainable jobs/tasks → Esteem and legitimation (self-, by others) of own activity, identification with a socially responsible company	Withdrawal of GHG-specific expertise and commitment
Other employees		→ Income and job security (affected by ET, GHG management, pollution, climate change risks)	
Works committee	GHG co-management	Income and job security (affected by ET, GHG management, pollution, climate change risks)	Formal and informal industrial relations e.g. worker participation laws (e.g. Germany Betriebsverfassungsgesetz/ Industrial Constitution Law), wage agreements
External			
Investors of equity and external capital (e. g. shareholders)	Financial resources (capital) for GHG projects and management	(Ethical?) ROI, low-loss risks → Seizing the chances and preventing the risks of climate change through GHG management → Transparent reporting about GHG involvement and management	Capital outflow Rating of GHG pollution and management
Suppliers			
- Manufacturers of GHG-emitting means of production (plants, materials) and substitutes	GHG-reducing/avoiding substitutes (energy efficient plants, renewable sources of energy)	Emerging markets for climate protection Buyers' compliance with GHG legal regulations	GHG-specific market position/know-how monopoly
- Producers of certificates: JI-/CDM project partners	Production of emissions certificates (CERs, ERUs)	Transfer of GHG-reducing technology as a contribution to regional/national development	Currently high (only few certified projects), but expected to drop (multitude of potential projects and hot air-problems)
- Suppliers/agents of certificates (AAUs, CERs, ERUs): exchanges, brokers, funds, companies/nations directly	Provision of GHG emission rights, Reduction of transaction costs of GHG management	Emerging markets for climate protection	Contingent: pricing for GHG transactions
- Risk management-products	Reduction of risks from GHGs and GHG re-	Emerging markets for climate protection	Know-how monopoly

to them for ethical reasons. This is not the place to examine this aspect in detail; see Zabel 2001, p. 208 f.; Antes 2004.

	gimes, and transaction costs of GHG management		
Buyers			
- Products	Honouring of GHG managment/performance (revenues, legitimation)	Products: goods reducing GHGs and the risk of GHGs (means of production, services) Processes: GHG management, ethical contract partner Transparent reporting about GHG involvement and management	Withdrawal of demand, substitution of the contract partner
- Certificates (all legal and private entities)	Revenues increasing ROI of GHG-reducing projects	Valid certificates (own compliance, ethical, speculative)	Pricing Shortage of certificates
Competitors	Promotion of an organizational field supporting GHG management	Compliance with the rules of the GHG regime	Shifting the competitive position through more effective and efficient GHG management Shortage of certificates through purchase Standard setting (BAT)
		Pooling (Art. 28 EU-Directive), voluntary agreement	Power of the sector to impose sanctions
Community of nations (ratifying states)/nation state and authorities	Compulsory institutional regime(s) Competition chances for ethical and innovative GHG management/products/services	= f (ratifying Kyoto Protocol) Holding GHG emissions allowances at the level of emissions Observing the other rules of the GHG regime	Sanctions/penalties if actual emissions exceed allowances held (Art. 16 EU Directive: until 31 Dec 2007 €40/t CO_2; from 1 Jan 2008 €100/t CO_2 + surrender of allowances additionally required)
Public (media, industrial associations, neighbours, other nation states)	Legitimation of climate protection policy/GHG management	Transparent reporting on GHG involvement and management	Bad publicity stemming from a negative climate protection image Shortage of certificates through purchase

AAU: Assigned Amount Unit; BAT: Best Available Technology; CER: Certified Emission Reduction; ERU: Emission Reduction Unit ET: emissions trading; GHG: Greenhouse Gas; NAP: National Allocation Plan

3 Emerging organizational fields of GHG management

Stakeholders' sanctionable expectations (Picot, Dietl, Franck 1999, p. 11; similarly North 1990, p. 4; Richter and Furubotn 1996, p. 7) can be regarded as institutions. Moreover, by considering the mutual interconnections of the sanctionable expectations of all akeholders with those of the company (rather than the relationship between an individual stakeholder and the company in isolation), the result is now an analysis heuristic referred to by organization-sociological neoinstitutionalism as an 'organizational field': "By organizational field we mean those organizations that, in the aggregate, constitute a recognized area of institutional life: key suppliers, resource and product consumers, regulatory agencies, and other organizations that produce similar services or products." (DiMaggio, Powell 1991, 1983, pp. 64 f.) The actors are hence comparable with those in the stakeholder approach, although in addition a dynamic perspective is introduced and enabled by virtue of questions being asked regarding how co-evolution and institutional alignment processes take place between the actors of an organizational field and what the driving forces are. The driving forces are the institutions: a company is dependent on the influx of resources from its organizational field and will therefore be keen for its activities to meet with the organization field's expectations. Therefore, it will either adopt the institutional structures of the organizational field ('institutional isomorphism') or alternatively try to create the impression that its activities meet the organizational field's expectations (institutional façades). Ultimately it can even use its authority (based on a powerful competitive position or a strong reputation) to mould the criteria of approval and hence its organizational field.

The topic of GHG management is still too new to be able to empirically identify such processes or to reject these propositions. In fact, numerous legal institutions are still only being set up (e.g. the national rules and international verification and monitoring standards). However, what can be observed in advance of the global and EU-wide start of GHG emissions trading systems is the co-evolution of new areas of business and markets ('emerging markets'), i.e. the material core of organizational fields. At present, numerous actors/stakeholders of GHG management are currently emerging:

- The operators of the industrial installations and activities explicitly affected by the rules (see Kyoto Protocol 1997, Annex A; Emissions Trading Directive 2003/87/EC, Annex I). Previous European Commission estimates put their number in the range of four or five thousand (see COM 2001 581 final), although more recent estimates since the adoption of the guidelines in July 2003 are in the order of 12-15.000 (Carbon Market Europe, 27.06.2003) – a volume that could easily increase given the opt-in clause. The responses observed among GHG emitters (regardless of whether they are acutely, latently or potentially affected) range from open minded to defensive (see Hove, Menestrel Bettignies 2002). Even so, despite various efforts (such as pilot systems in the UK and Denmark and simulation studies in Germany), a large number of companies are assumed to be badly informed or even not at all;[10]
- Suppliers of low-GHG and GHG-free technologies (i.e. the companies in the supply chain subject to the changing considerations of the regulated GHG emitters, especially direct suppliers who have themselves come under competitive

[10] For an indication of this, cf. the self-assessment of a study carried out back in October/November 2001 on a relatively small sample by Santarius and Ott (2002, pp. 16-18).

pressure from producers of more energy-efficient systems and suppliers of substitute fuels (such as gas and renewable energies to replace coal and oil);[11]

- Direct JI/CDM project organizers or funds promoting assistance in GHG emission reduction projects such as the World Bank's Prototype Carbon Fund (PCF 2001; 2002);
- Financial service providers acting in various capacities: by investing in companies whose creditworthiness may have been reduced owing to their GHG involvement, by directly financing JI/CDM projects, by acting as fund/asset managers, 'stock brokers' or risk management service-providers (e.g. with portfolios of projects providing GHG allowances for sale as well as liability and insurance products; see Sandhövel 2003; UNEP FI 2002; Innovest 2002; Whittaker, Kiernan, Dickinson 2003);
- Exchanges (such as the Chicago Climate Exchange) and specialized brokers (Rosenzweig et al. 2002, p. 19);
- Service-providers handling monitoring, reporting, verification and validation (see CCAP 2001).

In this connection, particular attention needs to be drawn to the special role of competitors. Although described in Porter's sector structural analysis as (almost) solely a threat to a company's own competitive position, this is in fact by no means the case[12] (even though the basic possibility to purchase emissions allowances will result in new strategies to drive others out of the market). Instead, competitors who invest their resources in GHG management, products or services will in fact be helping to prepare the ground for the adequate development of fields of business and markets.[13] Their activities will encourage the growth of sustainable organizational fields beyond niche status – which as well as being in the interests of sustainability is also essential if they are to exploit their own potential strategic success.[14] This may still be the case even if a company's market share slightly decreases compared to that of its competitor, provided this drop is more than made up for by (permanent) growth in market volume.

---

[11] See Fichtner and Enzensberger and Rentz (2003). In addition, the World Bank's Prototype Carbon Fund emphasizes that even the comparatively low price of emissions reduction of \$3/t CO_2 it calculated for its projects in 2002 (more energy-efficient technologies and renewable fuels) enabled internal rates of return to be raised by 0.5-4%, making them competitive (PCF 2001, pp. 25 f.).

[12] Although Porter does not completely rule out the possibility of mutual advantage, he limits its likelihood to isolated cases and so deems it to be of marginal importance. Categories such as 'threat' and 'rivalry' systematically and explicitly predominate in his analysis of relations between competitors ("driving forces of sectoral competition") (see Porter 1998a, 1998b).

[13] Not dissimilar to the old adage "competition is good for business"!

[14] Regarding co-evolution between "greening Goliaths vs. multiplying Davids" using the example of developing a mass market for green electricity see Wüstenhagen (2000, pp. 129-95), and for a generalized view Villiger and Wüstenhagen and Meyer (2000, pp. 303-05).

4 Internal effects of GHG emissions trading regimes on companies

Below we shall narrow our gaze in two respects: after primarily addressing the territory of the EU, we shall move on to a company running an installation of the type specified in Annex I of the EU Emissions Trading Directive. Once again we should bear in mind that emissions trading in the EU will initially be restricted to CO_2 (which by itself accounts for 80% of all GHGs), although its coverage may be extended to other GHGs at a later date. Ultimately, the impact of GHG emissions trading on companies' internal management will largely depend on the actual form of the legal institutions – an area which is still shrouded in uncertainty.

4.1 Procurement management strategies for GHG certificates

The EU Emissions Trading Directive requires "that, by 30 April each year at the latest, the operator of each installation surrenders a number of allowances equal to the total emissions from that installation during the preceding calendar year" (2003/87/EC, Art. 12 (3)). Consequently, a procurement management system for GHG allowances will have to be set up handling not only emissions allowances but also emissions credits. A number of strategic options arise within the context of the EU Directive's overall institutional regime (cf. Fig. 1; and also Scharte 2002, pp. 194-211; Betz et. al. 2002, pp. 82-89). Planning needs to start by the initial allocation of allowances to each regulated installation. At least 95% of these allowances must be free of charge for the first phase (2005-07), a volume which afterwards will be reduced to least 90%.[15] The allowances are to be distributed on the basis of the respective National Allocation Plans. Companies will then have either too many or too few allowances:

(a) Companies starting off with a surplus of emissions allowances will in particular include those whose emissions reductions achieved since 1990 ("early action") have not yet been formally recognized. The surplus can be sold and/or stored for future use ('banked'). The total amounts involved may be considerable. Vorholz (2003) cites the (currently politically contentious) example of Vattenfall Europe, which while rehabilitating the electricity sector in eastern Germany has slashed emissions of CO_2 from 120 to 70 million tonnes.

(b) Companies that do not receive enough allowances can choose between two basic strategies: the (internal) creation or (external) buying of allowances from the market. Their choice will depend on the marginal costs of producing their own allowances and the price at which allowances are available. Four types of strategies can be distinguished:

[15] The term "free of charge" in the guidelines means a price of €0. Transaction costs are of course also incurred for the administrative process of allocation and any exertion of political influence beforehand.

1. *Reduction/avoidance of GHGs at installation X to reduce the quantity of allowances needed.* This is basically the approach already generally practised in response to classical environmental legislation, albeit with the exclusion of the retention of CO_2 ('end-of-pipe' technology). It amounts to integrated environmental protection (more efficient processes and substitution by for example natural gas or renewable energies). In contrast to current environmental legislation, this strategy is only one of a number.
2. *Internal creation of emissions credits through projects abroad.* This option boils down to the basic idea of market instruments, namely accomplishing emissions reductions at the place where the marginal avoidance costs are the lowest. This usually means in threshold and developing countries (CDM projects), although this is not to say that industrialized and transformation countries ("parties" in the UNFCCC and Kyoto Protocol; JI projects) are ruled out. Under the linking directive adopted by the European Parliament on 20 April 2004, EU member states can allow installation operators to use credits from both CDM projects (as of 2005) and JI projects (as of 2008) to meet their obligations. However, these mechanisms may only be employed to supplement domestic action. Starting in the second phase (2008-12), the member states will therefore have to set upper limits in their National Allocation Plans for each trading period capping the extent to which installation operators may use JI/CDM credits.
3. *Purchase of emissions allowances and credits.* If a company's emissions exceed the level of allowances allocated, it can buy allowances on the market to cover the shortfall. CO_2 emissions allowances can be traded within the EU as of 2005. This market looks set to successively expand by 2008, since as of 2005 every EU member state will have the option of recognizing CDM credits and other GHGs could also be included (Art. 30 (1), 24 (1) 2003/87/EC). Hence by the time the Kyoto Protocol enters into force, the allowance market will have expanded in three respects as of 2008: extensively due to JI/CDM projects, and also because of other GHGs (Annex A: CO_2, CH_4, N_2O, HFCs, PFCs, SF_6) and to cover all the parties (Annex B countries) in the protocol. Transaction partners can include not only all classical brokers but also funds as well as countries and companies acting directly.
4. *Company-internal creation and trading.* The rules imply that companies structured into a number of divisions etc. can set up their own internal market for allowances created internally. BP Amoco and the Royal Dutch/Shell Group have already introduced their own internal trading systems in anticipation of environmental trading regimes.[16] Just as with the general requirements, the more heterogeneous the emissions levels (i.e. the potential for reduction) and the divisions' marginal avoidance/creation costs are, the more favourable are the conditions for internal trading systems. This situation is most likely to apply to multinationals with branches in threshold and devel-

[16] Regarding Amoco cf. Knoedel (2000), Trautwein (2002), pp. 137-41 and the article by Langrock in this volume; regarding Shell cf. Royal Dutch/Shell Group (2003a); for more on both see Nichols (2003).

oping countries. Since all of a company's business units in the territory covered by the Kyoto Protocol can also conduct their transactions with each other via external markets, the economic advantage of an internal trading system depends on whether transaction costs are likely to be permanently lower than in external trading;[17] although possible, this is by no means certain, as Scharte assumes (2002, pp. 209 f.). After all, by adopting an opportunistic attitude, principal–agent problems for instance are almost bound to occur, including in connection with internal transactions. True, corporate culture is generally thought to be more confidence-building than market coordination and hence to reduce transaction costs. But on the other hand, an internal trading system in fact means the institutionalization of market coordination. Moreover, increasing multinationality exacerbates the differences attributable to different management cultures,[18] meaning that coordinating activities and transactions costs will increase too. Moreover, the compatibility of the internal with the external trading system is also relevant for transaction costs. Shell's experience in this respect is not encouraging (Royal Dutch/Shell Group 2003b).

Allowance surpluses resulting from the various strategies can of course (like those stemming from surplus allocation) be retained for future use (banking) or sold on the market.

[17] By contrast, the possibility of achieving lower allowance prices can be ruled out because this would go against the interests of the internal seller.

[18] In the context of environmental management cf. Baumast (2002).

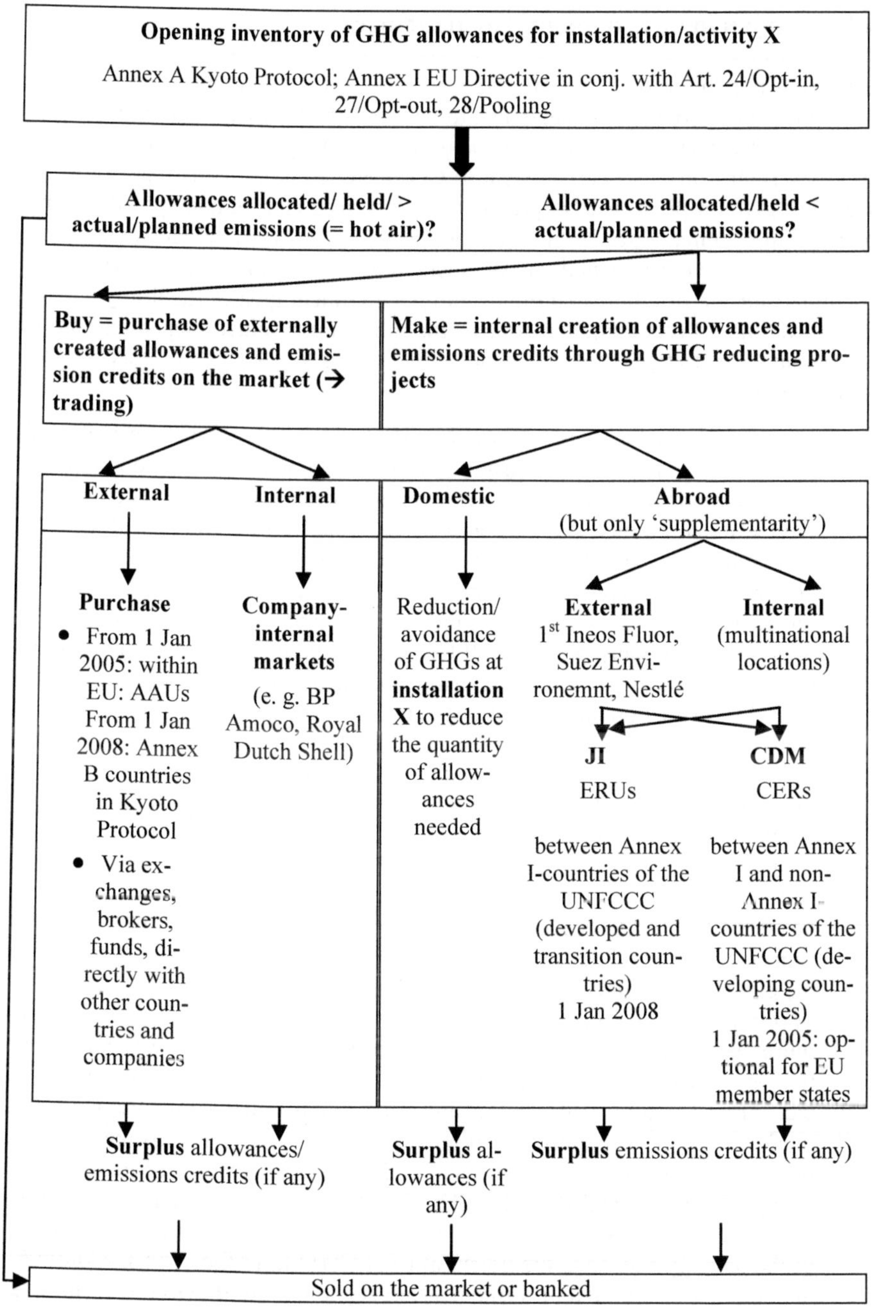

Fig. 1. Procurement management strategies for GHG certificates in the context of the UNFCCC, the Kyoto Protocol and the EU emissions trading scheme

4.2 Changes in environmental management

The legal regulation of CO_2/GHG emissions extends the minimum expenditure[19] of corporate environment management to decisions and activities which directly or indirectly cause such emissions. This is by itself nothing new: a previously public property right – in this case the right to CO_2/GHG emissions – is (partly) privatized and given a price, and so previous factor prices change and spread to other areas of the company due to the integrated nature of its activity. However, a new dimension for environmental management arises from the strategy types now permitted for the first time and the resulting subsequent greater flexibility. Generally speaking, all related administrative functions are revalued, depending on the combination of strategies selected. The main changes to coordinating functions are listed below:

1. *Investment and financing.* Action to reduce or avoid emissions at the installation itself is no longer necessary. And as long as there are enough certificates on the market, there is not even any need for an individual company to reduce its emissions. As the company can itself create and sell allowances and credits, the investment calculus is extended in two directions: firstly in connection with investment alternatives ranging from technical alterations to the regulated installation X to compensatory and overcompensatory measures (other company branches, JI/CDM projects, trading), and secondly because simple cost analysis (devising the cheapest way of performing required measures on an installation) is transformed into a portfolio and profit analysis of the entire combination of strategies. For example, if investing to reduce GHGs increases the internal rate of return to the extend of the sale of allowances less the transaction costs incurred,[20] this may well alter the advantageousnes of investments (rating and ranking) towards the company or business units (mergers and acquisitions) (Scharte 2002, pp. 184-191). The role of financing will also increase (i.e. not simply as a correlate of the investment function) because emissions certificates will be akin to stocks and shares (Sandhövel 2003, p. 41). As a result, environmental management will have to cover a range of previously irrelevant areas.
2. *Information.* GHG management in connection with emissions trading requires both, physical information on material and energy flows and in particular monetary information. This applies to all corporate functions that collect, process, use and communicate information – from procurement/sales market research through controlling/accounting to investment/financing and production planning as well as stakeholder dialogue. Details of material and energy flows are needed on the GHG involvement of a company's own production and processing activities and of the possible alternative investments of the strategy mix (e.g. the potential of JI/CDM to reduce CO_2). Moreover, maintaining a CO_2/GHG inventory is advisable merely due to the various re-

[19] The minimum expenditure refers to the legally regulated, compulsorily specified area of responsibility; cf. Antes (1996), pp. 228-30.

[20] See footnote 10.

porting obligations (Art. 7, 8 Kyoto Protocol; Art. 14 EU Directive) and interests (cf. Table 1) (WBCSD/WIR 2001, pp. 10-12; CCAP 2001; Scharte 2002, pp. 227-31; Betz et al. 2002). The methodological basics are already familiar from eco-balancing and life cycle assessment. Various initiatives currently are pressing for the international standardization of different emissions trading systems.[21] However, one generally new aspect for environmental management is the need for monetary information stemming from market coordination, especially in connection with trading certificates and the effects on accounting (for more on this see Scharte 2002, pp. 231-36). If emissions trading (i.e. market coordination) is also introduced internally, this will place additional demands on information management that can scarcely be met by the expertise of conventional environmental protection officers and departments.

3. *Coordination.* Internal coordination in environmental management has hitherto been hierarchical, including in hybrid forms (cooperation along energy and material flows). The arrival of emissions trading (external market coordination) now impacts on certain areas, namely the company's decisions and activities which cause GHG involvement. Two questions arise, only the first of which can be tackled here: the isolated view (how can this externally standardized task be transposed to internal environmental management?) and the overall view (how can the existing and new forms of coordination be integrated into a system of environmental management without reducing their effectiveness and while minimizing transaction costs (institutional interplay vs. overlapping))?[22]

Transposition could take place analogously by reproducing the GHG emissions trading system within a company to coordinate the various business units (cf. 4.1). However, so far the examples of BP Amoco and Shell have been exceptions; more attention is paid to more or less strong hierarchical adaptations. But whatever course of action is chosen, the importance of certain administrative functions – especially financial aspects which previously played only a marginal role in environmental management – will grow. This must be taken into account in the development of corporate structures. Moreover, as far as structural development is concerned, another issue which needs to be taken into account is that owing to the type of institutional regime of emissions trading, certain tasks (including in a decentralized corporate structure such as the organization of divisions) can only be carried out in a concentrated or centralized manner, such as by a 'Trading Team' (BP Amoco) or 'Trading Manager' (Shell) (Knoedel 2000, p. 44; Royal Dutch/Shell Group 2003a, pp. 8 f.) This includes all the responsibilities to the national regulatory authority (e.g. reporting obligations) and, in connection with the

[21] E.g. the "Greenhouse Gas Protocol Initiative" of the World Business Council on Sustainable Development and the World Resources Institute as well as efforts to put forward a draft ISO standard.

[22] See Young 2002, pp. 22-26; Kim 2003. The problem is particularly apparent from Sorrell's description of the UK emissions trading scheme as "institutional confusion" (Sorrell 2002).

institutionalization of an internal market, the functions of an equivalent internal authority which cannot be delegated (allocation, monitoring/controlling, maintenance of the trading platform, and if necessary imposing sanctions).[23] An (almost) exclusive concentration of GHG management tasks in special units can hence be ruled out for ecological and economic reasons. Firstly, an exclusively additive-functional structure formation would lead to problems in the accomplishment of preventive environmental protection (Antes 1996, 1999). Secondly, the economic success of the division of labour is based on paring off homogenous subtasks from previous processes and bringing them together in their own unit(s) to enable specialization advantages. By contrast, GHG management (as shown above) encompasses very heterogeneous tasks. Building up the common institution in Germany of the environmental officer (in charge of air pollution) into a central position under the banner of GHG management (Scharte 2002, pp. 220-24) therefore only makes sense at first sight, i.e. if the current tasks merely undergo a quantitative (rather than a qualitative) increase. After all, these statutory officers are mainly assigned to technical departments and generally have a scientific and engineering background (Schreiner 1996, pp. 93 f.), while the qualifications needed for market coordination are usually significantly underrepresented or even completely absent in most existing environmental protection units. During own interviews conducted with 'direct participants' of the UK emissions trading scheme, it became apparent that in most cases the inducement to (voluntarily) participate in emissions trading did not come from the environmental departments but instead from elsewhere in particular from the financial department and compliance officers. However, since engineering and scientific expertise are still needed (e.g. for measuring GHG emissions and evaluating technological alternatives), one alternative could therefore be to develop augmentative secondary organizational structures (i.e. lateral groups).

References

Antes R (2004) Nachhaltigkeit und Betriebswirtschaftslehre - Eine wissenschaftstheoretische und institutionelle Perspektive (Sustainability and business administration science: a methodological and institutional perspective), forthcoming

Antes R (1999) Die Aufbauorganisation des Umweltschutzes im Entwurf des Umweltgesetzbuches - Ein Beitrag zur nachhaltigen Unternehmung (The organizational structure of environmental management in the draft for a environmental civil code - a contribution for a sustainable company?). In: Seidel E (ed), Betriebliches Umweltmanagement im 21. Jahrhundert, Berlin et al., pp. 269-85

Antes R (1996) Präventiver Umweltschutz und seine Organisation in Unternehmen (Ecological prenvention and its organization in companies), Wiesbaden

Antes R (1992) Die Organisation des betrieblichen Umweltschutzes (The organization of corporate environmental management). In: Steger U (ed), Handbuch des Umweltmanagements. München, pp. 487-509

[23] In general Kreikebaum (1992), p. 2603; for emissions trading Trautwein (2002), pp. 77-83.

Antes R, Prätorius G, Steger U (1992, 1993) Umweltschutz und Transportmittelwahl - Ergebnisse einer empirischen Untersuchung im Güterverkehr (Environmental protection and transport decisions - empirical findings in the freight traffic). In: Die Betriebswirtschaft, no 6: pp. 735-59; errata, no 4, p. 576

Baumast, A (2002) Environmental management systems and cultural differences - An explorative study of Germany, Great Britain, and Sweden, St. Gallen 2002, Elektronische Dissertation der Universität St. Gallen, URL: http://verdi.unisg.ch/www/edis.nsf/www DisplayIdentifier/2659/$FILE/dis2659.pdf.

Betz R et al. (2002) Flexible Instrumente im Klimaschutz (Flexible Mechanisms of Climate Protection), (ed) by Ministerium für Umwelt und Verkehr Baden-Württemberg, revised edition, Stuttgart

Cames M, Deuber O, Rath U (2004) Emissionshandel im internationalen zivilen Luftverkehr (emissions trading and civilian aviation), (ed) by Institute for Applied Ecology, Freiburg/Br. 2004, URL: http://www.oeko.de/oekodoc/185/2004-001-de.pdf? PHPSESSID=57ee702b07e84c6a0ec70357f8acca1b.

CCAP Center for Clean Air Policy (2001) Study on the monitoring and measurement of greenhouse gas emissions at the plant level in the context of the Kyoto mechanisms, final report, Washington/DC

Deuber O (2002) Einbeziehung des motorisierten Individualverkehrs in ein deutsches CO_2-Emissionshandelssystem (Integration of the motorized private traffic with the german CO_2 emissions trading scheme), Freiburg/Br., URL: http://www.oeko.de/dokum.php? setlan=&vers=&id=185&PHPSESSID=57ee702b07e84c6a0ec70357f8acca1b.

DiMaggio PJ, Powell WW (1991) on cage revisited: institutional isomorphism and collective rationality in organizational fields, In: Powell WW, DiMaggio PJ (eds), The new institutionalism in organizational analysis, Chicago/London pp. 63-82, revised version of: American Sociological Review, vol 48, 1983, pp. 147-60.

EEA European Environment Agency (2001) Environmental signals 2001: European Environment Agency regular indicator report, Copenhagen

EP European Parliament (2003) Greenhouse gas emission allowance trading ***II, In: Texts adopted at the sitting of Wednesday 2 July 2003, provisional edition

Fichtner W, Enzensberger, Rentz O (2003) CO_2-Emissionsrechtehandels-Regime und die Bedeutung des Zertifikatpreises (CO_2 emissions trading regimes and the price of the certificates). In: UmweltWirtschaftsForum, vol 11, no 3: pp. 48-51

Grassl H (1993) Umwelt- und Klimaforschung (Environmental and climate research). In: Held M, Geißler KA (ed), Ökologie der Zeit, Stuttgart pp. 75-84

Hinding B (2002) Muster der psychischen Verarbeitung des globalen Klimawandels und Energiesparen (Patterns of mental processing global climate change and energy saving). In: Umweltpsychologie, Heft 2 p. 26-44

Hove S, van den Menestrel M, de Le/Bettignies HC (2002) The oil industry and climate change: strategies and ethical dilemmas. In: Climate Policy, vol 2, issue 1/May 2002, pp. 3-18

Innovest (2002) Climate change and the financial services industry, Module 1 & 2, (ed) by UNEP Finance Initiatives, URL: http://unepfi.net/

Kim Joy A (2003) Regime interplay: a case study of the climate change and trade regimes, submitted thesis to the School of Environmental Sciences of the University of East Anglia, Norwich

Knoedel P (2000) Der Emissionshandel bei BP Amoco (Emissions Trading at BP). In: UmweltWirtschaftsForum vol 1: pp. 41-45

Kreikebaum H (1992) Zentralbereiche (Central departments). In: Frese E (ed), Handwörterbuch der Organisation, 3. edn, Stuttgart , pp. 2603-2610

Meyer JW, Rowan B (1991, 1977): Institutionalized organizations: formal structure as myth and ceremony. In: Powell WW, DiMaggio PJ (eds), The new institutionalism in organizational analysis, Chicago/London, pp. 41-62, reprinted: American Journal of Sociology, vol 83, no 2:, pp. 340-63

Nicholls M (2003) Oil firms learn trading lessons. In: Environmental Finance, February 2003, pp. 13-15

North D C (1990) Institutions, institutional change and economic performance, reprint of the 1st edn. (1990), Cambridge et al.

OECD Organisation for Economic Co-Operation and Development (2001) OECD Environmental Outlook, Paris Cedex

Ortmann G, Zimmer M (1998) Strategisches Management, Recht und Politik (Strategic management, law and policy). In: Die Betriebswirtschaft, no 6: pp. 747-69

Pawlik K (1991) The psychology of global environmental change: some basic data and an agenda for cooperative international research, In: International Journal of Psychology, vol 26: pp. 547-63

PCF Prototype Carbon Fund: Annual Report (2001) URL: http://prototypecarbonfund.org/docs/AR_download.htm.

PCF Prototype Carbon Fund: Annual Report (2002) URL: http://prototypecarbonfund.org/docs/2002AnnualReport.htm.

Pfriem R (2000) Jenseits von Böse und Gut - Ansätze zu einer kulturwissenschaftlichen Theorie der Unternehmung (Beyond bad and good - cultural approaches to the theory of the firm). In: Beschorner T, Pfriem R (eds), Evolutorische Ökonomik und Theorie der Unternehmung, Marburg, pp. 437-476

Picot A, Dietl H, Franck E (1999) Organisation - Eine ökonomische Perspektive (Organization - an economic perspective), 2. edn, Stuttgart

Porter ME (1998a) Competitive strategy, New York

Porter, ME (1998b) Competitive advantage, New York

Preuss S (1991) Umweltkatastrophe Mensch – Über unsere Grenzen und Möglichkeiten, ökologisch bewusst zu handeln (Ecogical catastrophe man – On our barriers and potentials to behave ecologically conscious), Heidelberg

Richter R, Furubotn E (1996) Neue Institutionenökonomik (Institutions and economic theory), Tübingen

Rosenzweig R et al. (2002) The emerging international greenhouse gas market, ed by PEW Center on Global Climate Change, Arlington, URL: http://www.pewclimate.org/pro jects/trading.pdf.

Royal Dutch/Shell Group (2003a) The Shell tradeable emission permit system, o.J., URL: http://www.shell.com/static/royalen/downloads/steps.pdf.

Royal Dutch/Shell Group (2003b) Shell tradable emissions permit system (STEPS) - Learning and experience, URL: http://www.shell.com/static/royalen/downloads/steps _learning.pdf.

RSU, Rat von Sachverständigen für Umweltfragen (2002) Umweltgutachten (Environmental report), Stuttgart

Sandhövel A (2003) Emissionshandel aus Bankensicht (Banks and emissions trading). In: UmweltWirtschaftsForum vol 11, no 3: pp. 39-43

Santarius T, Ott H E (2002) Attitudes of german companies regarding the implementation of an emissions trading scheme, Wuppertal Papers no 122e, Wuppertal

Scharte M (2002) Klimapolitik und Treibhausgas-Management (Climate policy and greenhouse gas management), St. Gallen, electronical dissertation at ther University St. Gallen, URL: http://verdi.unisg.ch/www/edis.nsf/wwwDisplayIdentifier/2599/$FILE/dis2599.pdf.

Schneidewind U (1998) Die Unternehmung als strukturpolitischer Akteur (The company as a political actor of structuration), Marburg
Schreiner M (1996) Umweltmanagement in 22 Lektionen (Corporate environmental management in 22 lessons), 4. edn, Wiesbaden
Seo MG, Creed WED (2002) Institutional contradictions, praxis, and institutional change: a dialectical perspective. In: Academy of Management Review vol 27, no 2: pp. 222-247
Sorrell S (2002), The climate confusion: implications of the EU emissions trading directive for the UK climate change levy and climate change agreements, SPRU (Science and Technology Policy Research), University of Sussex, Brighton, URL: http://www.susex.ac.uk/spru/environment/research/ccfr.pdf
Trautwein S (2002) Chancen und Probleme des betriebsinternen CO_2-Zertifikatehandels - am Beispiel des Otto Versand (Chances and risks of a company internal CO_2 emissions trading - the example of the mail order business Otto-Versand), Hamburg/Lüneburg 2002
UNEP Finance Initiatives (2002) Climate change and the financial services industry - Ececutive briefing paper, CEObriefing
Villiger A, Wüstenhagen R, Meyer A (2000) Jenseits der Öko-Nische (Beyond the eco niche), Basel et al.
Vorholz F (2003) Was kostet die Luft? Der Markt hält Einzug im Umweltschutz - und sorgt in der Industrie für den großen Verteilungskampf (What are the costs of clean air? The market keeps move in environmental protection and provides the great distribution fight in the industry. In: Die ZEIT, no 29, 10.07.2003, p. 17
WBCSD World Business Council for Sustainable Development, WIR World Resources Institute (2001) The greenhouse gas protocol: a corporate accounting and reporting standard, URL: http://www.wbcsd.org/DocRoot/NlH8ZRdvbUZhkGFQaOuj/ghg-protocol.pdf.
Whittaker M, Kiernan M, Dickinson P (2003) Carbon finance and the global equity markets, (ed) by Carbon Disclosure Project, Februar 2003, URL: http://www.cdproject.net/
Wüstenhagen R (2000) Ökostrom - von der Nische zum Massenmarkt (Green electricity - from the niche to the mass market), Zürich
Wurzel R et al. (2003) Das britische Emissionshandelssystem (The UK emissions trading scheme). In: UmweltWirtschaftsForum vol. 11, Nor. 3: pp. 9-14
Young O R (2002) The instituional dimensions of environmental change: fit, interplay, and scale, Cambridge Ma., London
Zabel HU (2001) Ökologische Unternehmenspolitik im Verhaltenskontext (Ecological sound business policy - a behavioural approach), Berlin

Accounting for emission rights

Edeltraud Günther

Dresden University of Technology
Professorship for Environmental Management
Münchner Platz 1/3, 01187 Dresden, Germany
bu@mailbox.tu-dresden.de

Abstract

From 2005 companies have to account for emission rights. This article presents an overview on the international, European and national climate policy and shows alternatives for greenhouse gases to be reported. The focus of the article is based on the influence of climate policy on accounting. Moreover, the interdependencies of the development of accounting and the treatment of ecological resources are shown.

Keywords: Reporting of GHG, accounting for emission rights, climate policy

1 Introduction

From January 2005, greenhouse gas emissions trading within the European Union will be regulated by an allowance process, which has been designed following the Directive of the European Parliament and of the Council establishing a scheme for greenhouse gas emission allowance trading within the Community[1]. Each member state will implement the directive on a domestic level by drawing up a national allocation plan according to Article 9 of the directive.

The programme starts with an assessment of CO_2 emitted by those power and heat fossil plants, which generate 20 MW or more; mineral oil refineries; coking plants; iron production and processing plants; pig iron or steel production plants; cement clinker, lime, glass including glass fibre, ceramic products, cellulose, paper and cardboard production plants.[2] 46% of the emissions of CO_2 and 38% of the expected emissions of greenhouse gases (GHG) listed in the Kyoto Protocol (Carbon dioxide, methane, nitrous oxide, hydrofluorocarbons, perfluorocarbons and sulphur hexafluoride) should hereby be reported.[3] This should also help to minimize monitoring costs.[4]

By 2005, all companies whose installations are covered by the directive will need to decide on how the greenhouse gas emission allowances should be integrated into the balance sheet.

The objective of this paper is to answer the following questions:

1. How had the international, European and national climate policy evolved up until the greenhouse gas emission allowance programme was introduced?
2. How will GHG emissions be reported?
3. How does the evolution of the climate policy influence accounting?
4. Which developments of the accounting are of relevance to the treatment of ecological resources, particularly to the GHG allowances?
5. How else can ecological issues be integrated into annual accounts and into the annual report?

---

[1] See European Parliament and Council (ed): Directive 2003/87/EC of the European Parliament and of the Council establishing a scheme for greenhouse gas emission allowance trading within the Community and amending Council Directive 96/61/EC. Brussels 13th of October 2003.

[2] See European Parliament, ibid. Annex 1.

[3] See Commission of the European Communities (ed) (2001a): Proposal for a Directive of the European Parliament and of the Council establishing a scheme for greenhouse gas emission allowance trading within the Community and amending Council Directive 96/61/EC, Explanatory Memorandum, COM (2001) 581, Brussels 2001, section 11, p. 11.

[4] See Rheinisch-Westfälisches Institut für Wirtschaftsforschung (RWI) (ed), Die Klimagasemissionen in Deutschland in den Jahren 2005/2007 und 2008/2012. Endbericht zum Forschungsvorhaben im Auftrag des Bundesverbandes der Deutschen Industrie (BDI), Essen 2003, p. 1.

2 Evolution of the climate policy up until the introduction of the scheme for greenhouse gas emission allowance trading

The Intergovernmental Panel on Climate Change (IPCC) since 1988

The IPCC was established in 1988 under the auspices of the United Nations Environment Programme (UNEP) and the World Meteorological Organization (WMO). To date, its purpose has been to relate the results of climate change studies and to establish a broad scientific consensus.[5]

The Framework Convention on climate change of the 1992 Rio Earth Summit

After having identified the climate change related issues, the United Nations called for a two week Convention on the Environment and Development (3-14 June 1992) in Rio de Janeiro. Major agreements and conventions were adopted: the Rio Declaration on Environment and Development, the Biological Diversity Accord, the Statement of Forest Principles, Agenda 21 and the United Nations Framework Convention on Climate Change (UNFCCC), which is here of particular relevance.[6]

The objective of the UNFCCC is to achieve the "stabilization of greenhouse gas concentrations in the atmosphere at a level that would prevent dangerous anthropogenic interference with the climate system"[7]. At that time, no concrete reduction rate was defined.

The Kyoto Protocol of 1997

Five years later, the signatory countries of the Kyoto protocol made the legally binding commitment to reduce their GHG emissions, thereby specifying the objectives of the UNFCCC. Industrialized countries and countries with economies in transition listed in Annex I of the UNFCCC committed themselves to reduce their GHG emissions set out under Annex A (mentioned as CO_2 equivalents) by at least 5% below the 1990 levels in the commitment period from 2008 to 2012.[8] In practical terms, the European Union needs to reduce the emissions of carbon dioxide equivalents to 8% below the 1990 levels in the period running from 2008 to 2012, according to Annex B of the Kyoto Protocol. To achieve this common objective, the EU Member States agreed in June 1998 on the creation of the EU burden sharing, which re-defines emission reduction rates. Germany thereby committed itself

---

[5] See Intergovernmental Panel on Climate Change (ed), About IPCC. 31 Aug 2003 http://www.ipcc.ch/about/about.htm.

[6] See United Nations (ed), United Nations Framework Convention on Climate Change, New York 1992. The Convention was adopted on 9 May 1992 in New York and signed a few weeks later in Rio.

[7] See United Nations, ibid. Article 2.

[8] See United Nations (ed), Kyoto Protocol, Article 3. Kyoto Protocol to the United Nations Framework Convention on Climate Change, Kyoto 1997.

to reduce its CO_2 equivalent emissions by 21%, 18.9% of which were already achieved between 1990 and 2000. During the same period, the European Union reduced its emissions by 1%.[9]

The European Climate Change Programme (ECCP) of 2000

In 2000, the European Union ratified the European Climate Change Programme, the purpose of which is to support the definition of basic conditions for the climate protection on a European level.[10] The introduction of an European emissions trading belonged to the proposals of the ECCP.

The directive establishing a scheme for greenhouse gas emission allowance trading of 2003

All six gases listed in the Kyoto Protocol are in principle going to be traded. During the first phase, which runs from 2005 to 2007, the trading will be restricted to carbon dioxide and to some industries. Installations bound to the commitment are listed in Annex I of the directive (see above). According to Article 27, some of the installations or industries may however be excepted if they manage to reduce their greenhouse gas emissions by other means than the trading scheme. 1.849 installations are affected in Germany.[11] The number and the allocation of permits remain at the Member States' discretion. These are reflected in national allocation plans established in accordance with Article 9 and Annex III of the directive, and are afterwards checked by the Commission for consistency with the Kyoto targets. The national allocation plans should also take into account early actions, i.e. reductions achieved from 1990.[12] Once they have been allocated, permits may be pooled, so that the firms can control them together. The allocation of permits for the period 2005-2007 is free of charge. The Member States may allow permit transfers to the second period. From 2008 on, all transfers will have to be allowed. The emission rates produced by the emitting installations must be reported to the legal entities named by the respective Member State, and analyzed according to criteria described in Annex V of the directive. For every tonne of CO_2 equivalents emitted that is not covered by an allowance, a company will have to pay a penalty of 40 € in the first period of the directive and 100 € thereafter. Moreover the have to hand in the certificates later on.

---

[9] See Zeiger, in: IHK Dresden: EU-Emissionshandel - Ein neues klimapolitisches Instrument, Dresden, 2003: p. 3.

[10] See Commission of the European Communities (ed), Communication from the Commission to the Council and the European Parliament on EU policies and measures to reduce greenhouse gas emissions: Towards a European Climate Change Programme (ECCP), COM (2000) 88, Brussels 2000, p. 4 f., as well as Annex 3.

[11] See Zeigler, Loc.cit. (Footnote 9), p. 10.

[12] See Zeigler, Loc.cit. (Footnote 9), p. 10.

The German National Allocation Plan

The National Allocation Plan is established on the basis of the emissions reported by involved companies in 2000. The German Federal Environment Ministry has been assessing data for the first National Allocation Plan, encompassing the installations committed to the scheme from 2000 to 2002 and their production during the same period, according to Annex I of the directive. A 2-step process has therefore been established[13]: during the first step, the 2000 emission reports are analyzed by the *Länder*; the second one includes a query of facility operators. The objective of this query is to double-check the 2000 emission reports, to update the data reported for 2001 and 2002 by the facility operators and to give information regarding the products for the period 2001-2002 as far as their type and amount are concerned. CO_2 emissions are to be calculated according to emission factors provided by the Federal Environment Ministry. The German National Allocation Plan should be a part of the Climate Change Plan and also has to take into account the firms' commitments. How to consider early actions and find legal successors for given companies in the former East Germany[14] are still open issues. A large part of emission cuts already achieved in Germany is due to firms' closure in former East Germany. Not considering these cuts as early actions because of missing legal successors was interpreted as discrimination by the affected *Länder*. One also has to take into account that international companies may be bound to several allocation plans and that their installations are handled differently depending on the country they are in.

3 Greenhouse gas emission reporting

The reporting of emissions on the level of the company is a critical condition for the introduction of a GHG allowance trading as well as its implementation into annual accounts. There is at the moment no standard for the reporting of GHG emissions, either worldwide or Europe-wide. In June 2002 in Johannesburg, the International Organization for Standardization decided to establish "Guidelines for Measuring, Reporting and Verifying Entity- and Project-Level Greenhouse Gas Emissions".[15] In August 2002, the German DIN NAGUS Working Committee 7 was created.[16]

---

13 See IHK Dresden (ed), Emissionshandel - BMU erstellt ersten Nationalen Allokationsplan. Ermittlung der Daten für den Allokationsplan erfolgt in zwei Stufen. 25 August 2003.

14 Saxony-Anhalt reduced its CO_2 emissions by 56% since 1990, only a fraction of which were taken into account, though, see Zeigler, Loc.cit. (footnote 9), p. 12.

15 See International Standards Organization (ISO) (ed), ISO/TC 207 - Environmental Management: Tenth Meeting 2002-06-10 & 2002-06-16, Johannesburg, South Africa: Resolutions. ISO/TC 207 N 567 (E), 2002, p. 3.

16 See Deutsches Institut für Normung e.V. (DIN) (ed), DIN Committee created for GHG emissions. 20 August 2002, http://www2.din.de.

Because there is no available norm for the reporting of GHG emissions, already existing emissions trading systems or organizations, e.g. the World Business Council for Sustainable Development (WBCSD), have established several guidelines. By analysing the content of 23 existing guidelines[17], as well as the theoretical possibilities of window dressing, the features mentioned below appeared to be critical. It is important to note that all these factors influence the valuation of emission rights to be accounted for. One is therefore dealing with an indirect arrangement of reporting relevant facts. Those possibilities have first been analysed on a general level, and the corresponding item of the EU Directive has then been defined.

- *Legal system boundary:* In case of proportionate ownership, it has to be determined which shares will be considered for the emission reporting. The handling of outsourcing and franchising will also have to be clarified. Management control and equity share may be distinguished for the allocation of emission rights to the participating companies, similarly to full and proportional consolidations in consolidated accounts. The reporting by management control includes participations that are under the company's dominant influence. The WBCSD and the World Resources Institute (WRI) assume that this is the case when an entity owns 50% or more of the voting interests.[18] In case of dominant influence, 100% of the GHG emissions will be reported. In regard to the reporting by equity share, the emissions reported are based on the ownership interests. According to the WBCSD and the WRI, the reporting is optional for less than 20% ownership.[19] No concrete mention related to this has been made in the EU Directive as this rather falls instead under the responsibility of the authors of the national allocation plans.
- *Temporal system boundary:* The determination of a base year can serve as a basis for the objective description and help the trend definition, e.g. as far as meeting the commitments made in Kyoto is concerned. Acquisitions or spin-offs during the reporting period, which result in the change of the company structure will require adjustments of the emissions to the new structure. Also, the handling of emission cuts achieved in the past (early actions) has to be clarified. It is planned in Annex III (7) of the EU Directive that the national allocation plans will take early actions into consideration. The *Länder* of the former East Germany demand that 1990, being the reference year of the Kyoto protocol, be accepted as base year[20], whereas the current version

[17] See Öko-Institut, TU Dresden (ed), Eine Analyse von Leitfäden zur Erfassung von Treibhausgasemissionen, Berlin/Dresden 2003, forthcoming.

[18] See World Business Council for Sustainable Development and World Resources Institute (ed), The Greenhouse Gas Protocol - A Corporate Accounting and Reporting Standard. Washington D.C. September 2001, p. 15 ff.

[19] See World Business Council for Sustainable Development and World Resources Institute, Ibid., p. 15 ff.

[20] See Sommer, Umsetzung der EU-Emissionshandels-Richtlinie in Sachsen. In: IHK Dresden (ed): EU-Emissionshandel - Ein neues klimapolitisches Instrument, Dresden 2003, p.

of the National Allocation Plan intends to report emissions for the period 2000-2002.

- *Spatial system boundary*: GHG certificates should be allocated so as to avoid the creation of hot spots and to enable the formation of individual market zones with specific pollution limits in a given area. Emission certificates should be allocated so as to guarantee that the standards can be met in the respective zones. However, the arising of many different license markets remains hereby a major drawback. The Kyoto protocol proposes three flexible mechanisms to give the states the capability of meeting their commitments at minimum costs: Joint Implementation allows for the consideration of emission cuts achieved by projects undertaken in other industrialized countries and countries with economies in transition. The Clean Development Mechanism extends Joint Implementation to developing countries, i.e. countries, for which no quantitative reduction targets have been defined. The objective here is to transfer more modern and cleaner technologies to these countries. Emissions trading allows for emission certificate trading between Annex B countries. If such projects are led in different trading areas or national allocation regimes, the emission assessment may yield different results. The same applies within the EU, due to the national allocation plans to be drawn up. Article 9, 12 and 28 provide respectively for the establishment by the countries of domestic allocation plans, for the trading of rights within the Community and for the so called pooling between different operators. The latter may prove interesting, for example, within a branch of business.
- *Nature and source of emissions:* The emissions involved can be distinguished on the one hand by their nature and on the other one by their sources. As far as the emissions´ nature is concerned, 76% of all analyzed guidelines[21] recommend reporting all Kyoto gases. Reporting only carbon dioxide emissions may also be an alternative. As regards emission sources, those of a direct and indirect nature may be distinguished. Direct emissions include emissions from a source owned by the company or under its dominant influence, e.g. its own power generation or processes. Indirect emissions are a consequence of the company's activity, whereas their sources are owned by another company or under its dominant influence, e.g. as a result of the use of its products. In the particular case of a car manufacturer, the question of reducing direct emissions from the production or indirect emissions from the use of its cars may arise. The reporting of gases or sources partly requires the definition of a minimum emission rate..The EU Directive refers to direct emissions exclusively. The Annex I of the EU Directive lists sectors (already mentioned in the introduction), installations of which can

1 on 31 March 2003, Mr. Platzeck, Minister-President of Brandenburg wrote to Chancellor Gerhard Schröder to demand the adapted recognition of emission cuts achieved in installations of the former East Germany since 1990 by the allocation of emission rights specific to each installation. Furthermore, the German federal reserve for new entrants should not come from emission cuts achieved since 1990 by installations in the former East Germany.

21 See Öko-Institut, TU Dresden, Op.cit. (footnote 17).

only be operated if covered by GHG emission allowances. The trading will first cover carbon dioxide only, however, the directive plans to broaden its scope to methane, nitrous dioxide, hydrofluorocarbon, perfluorocarbon and sulphur hexafluoride[22].

- *First allocation:* For the first allocation, i.e. when emission rights are first put into operation, rights may be auctioned, sold or allocated without financial compensation. In the case of auctioning, facility operators already in possession of a permit to emit given polluting substances will also have to acquire emission rights. This almost equates to the suspension of all granted operating licenses, emission rights and missing provisions made to safeguard existing standards in older installations. These provisions will also be cancelled for older installations in case these are sold. Furthermore, the state or licensing authorities must define the market clearing price when selling emission rights, this would require, however, that they know the emitters' marginal avoidance costs, i.e. costs required to avoid the emission of one additional CO_2 equivalent tonne. In the case of free allocation, emitters are allocated certificates allowing them to emit the amount of gases corresponding to their actual emission. These certificates may be traded. Planning uncertainty and financial burden can hereby be avoided. Free allocation may nevertheless favour emitters that have been responsible for heavy emissions until now, or have only a little invested in clean technologies, contrary to market entrants. The EU Directive is planning to allocate free of charge at least 95% of the allowances between 2005 to 2007 and at least 90% of the allowances between 2008 and 2012. How the installations responsible for the emissions should be dealt with is to be defined in the national allocation plans. During the latest discussions, the following variables proved critical: proportional allocation to the sectors, allocation basis (absolute or specific emissions), the providing of legal evidence, the consideration of early actions, particularly for the former East Germany, the definition of a regulatory institutional structure (centralized or not, official or collateral organizations), the handling of new entrants and the definition of a facility (all installations of a given site or each facility must generate at least 20 MW)[23].
- *Target scope of rights:* Meeting the immission standards (first-best solution) and meeting the emission standards (second-best solution) are two different concepts[24]. The first one is theoretically not feasible, since each emitter would then have to be aware of the precise location and the effects of its emissions. For the second one, a certain immission load indicated by the state is to be converted into the correspondent total emission rate. This is ex-

[22] See European Parliament and Council, Loc.cit. (footnote 1), Annex II.

[23] See Zeigler, Loc.cit. (footnote 9), p. 12.

[24] Emissions take into account environmental factors (e.g. CO_2), which influence the source. These are transported and transformed through a multitude of processes (transmission), to have an effect on a system (plants, animals, humans and materials). Environmental influences are called immissions. See Etterlin, Hürsch, Topf, Ökobilanzen - ein Leitfaden für die Praxis, Mannheim 1992, p. 19.

actly the process the international climate policy has followed, by developing emission standards on the basis of scientific knowledge regarding the effects of greenhouse gases. The EU Directive takes over the emission reduction objectives of the Kyoto Protocol as well as those defined by distributing the total amount within the EU (see above). Its purpose is also to introduce the EU greenhouse gas emission allowance trading.

- *Limitation of rights:* Certificates that are limited in their duration allow each license holder to emit a corresponding amount of gases during the given period. By the end of this period, the certificates are no longer valid and new licenses (also of finite duration) are to be allocated by the environment authorities. The rights granted by unlimited allowances are permanent. The directive runs first for a three-year period from 2005 to 2007 and afterwards for successive five-year ones. According to Article 13, the rights are no longer valid four months after the determined deadline.
- *Collection of emission data:* Emission can in principle be determined according to three methods: they may be measured, calculated or evaluated[25]. The emissions can be calculated by multiplying the energy consumption of a company or a site by an emission factor, which represents the conversion factor for the emitted greenhouse gases. Emission factors may refer to input values, e.g. the raw material consumption, or output values, e.g. the product. A comparison of these factors partly resulted in major deviations. For example, the German emission factors for 1990 from the guidelines "The GHG Indicator" of the United Nations Environmental Programme and "the Greenhouse Gas Protocol" of the WBCSD and the WRI showed a 0.105 kg CO_2/kWh difference.[26] Using the higher value as reference, this amounts to a 19% deviation. The choice of the emission factor is a leverage for the valuation in the balance sheet, particularly for companies with a high energy consumption level. According to the directive, the emission information can be calculated either by using the following formula, referred to in Annex IV:

---

[25] The Directorate-General for Environment of the European Commission gave the following definitions in its guideline for the implementation of a European Pollutant Emission Register (EPER): Emission data are measured by standardized or accepted methods; additional calculations are often required to convert the measurement results into annual emission data. Emission data are calculated using methods nationally or internationally agreed methods and emission factors that are representative for the respective industry sector. Emission data is estimated on the basis of non-standardized estimations derived from industrial best assumptions or expert guesses. See Directorate-General for Environment of the European Commission (ed), Guidance Document for EPER implementation according to Article 3 of the Commission Decision of 17 July 2000 (2000/479/EC) on the implementation of a European Pollutant Emission Register (EPER) according to Article 15 of Council Directive 96/61/EC concerning Integrated Pollution Prevention and Control (IPPC), Brussels 2000.

[26] See Thomas/Tennant/Rolls, The GHG Indicator: UNEP Guidelines for Calculating GHG Emissions for Business and Non-Commercial Organisation, Geneva 2000, p. 47 and World Business Council for Sustainable Development and World Resources Institute (ed), Calculating CO_2 emissions from the combustion of standards fuels and from electricity/steam purchase: Calculation worksheets, Washington 2001.

Activity data * Emission factor * Oxidation factor[27]

or by measurement using accepted methods. As no specific value has to be evaluated, the scope of discretion is not more broadened by this step.

- *Conversion to CO_2 equivalents*: In 1996, the IPCC worked out conversion factors for the greenhouse effect of other gases, based on existing values for CO_2[28]. All information reported in the EU Directive is based on CO_2 equivalents. The emission rate for other greenhouse gases can be obtained by conversion.
- *Price per tonne of CO_2 equivalents:* If the allowances are auctioned, their price will depend on the demand. In the case that they are sold, the government or its representative will have to assess a market clearing price. In the case of first free allocation, allowances will gain value only after a period of time. It is stated under Article 16 of the EU Directive that a company should pay a 40 € penalty for every excess emitted tonne of CO_2 equivalents emitted in the first commitment period, if the company does not produce the allowances by the end of April of the next year, the penalty for the second period will be 100 €. Moreover the emission rights have to be handed in later on. If allocated free of charge, the gas tonne will go up in value only upon first allocation, the range 20 to 33 €[29] is expected, depending on the circumstances.
- *Verification by an independent expert:* This step is compulsory or recommended in more than 70% of all guidelines concerned. The proposed examinations affect, for example, the emission data, the calculation methods or self-developed emission factors. It is important to note that a control is required also for this market economy tool of license trading, a drawback, which is often associated to regulatory bodies. A market for the verification of emission data must first be developed.

[27] If the emission factor does not take into account the fact that some of the carbon is not oxidised, then an additional oxidation factor shall be used. See European Parliament and Council, Loc.cit. (footnote 1), Annex IV.

[28] "Carbon equivalents of non-CO_2 GHGs are calculated from the CO_2-equivalents, using the 100-year global warming potentials". 21 for methane and 310 for nitrous dioxide were calculated. See Intergovernmental Panel for Climate Change (ed), Technologies, Policies and Measures for Mitigating Climate Change, Technical Paper I, Cambridge 1996, p. 9.

[29] The allowance price of 20 € results from the study "Economic Evaluation of Sectoral Emission Reduction Objectives for Climate Change" of May 2001. This study ascertained that the Kyoto commitments made by the Community - taking into account all 6 GHG listed in the Kyoto Protocol - could be fulfilled with a cost of up to 20 € per tonne of CO2 equivalent. The study "The economic effects of EU-wide Industry-Level Emissions trading to Reduce Greenhouse Gases - Results from PRIMES model" of February 2003 came to a price of 33 €. This study was the basis for the Green book and took only carbon dioxide emissions into account (contrary to the above-mentioned first study). See Commission of the European Communities, Loc.cit. (footnote 3), p. 50.

The following spreads summarize the current status of the investigation and also reflect the scope of the valuation to be accounted for:

Feature	Spread
Legal system boundary	50%, depending on whether management control or equity approach is used.
Temporal system boundary	18.9% emission rate difference between 1990 2000, depending on the German emission basis year.
Spatial system boundary	Depending on the scenario.
Type of emission	21.1% depending on whether CO_2 only or all Kyoto gases are affected.
Source of emission	If only the installations listed are included, 62% of all Kyoto gases and 54% of all CO_2 emissions will be disregarded.
First allocation method	Relevant only in case of allocation with financial compensation of 5 or 10% of the rights.
Target scope of the rights	Set out in the Kyoto Protocol.
Allowance limitation	Definition of periods to determine supply and demand, i.e. the price.
Reporting of emission data	Emission factors vary of up to 19% with the higher value as reference.
Conversion into CO_2 equivalents	Determined with factors from the IPCC based on the greenhouse gas potential over 100 years.
Price of the tonne of CO_2 equivalent	39.4% based on the highest expected market price and 60% on the highest penalty.
Examination by an expert	At most at the discretion of the expert.

Fig. 1. Scope of the window dressing

The trading system plays, therefore, an essential role in the valuation, i.e. the value of rights to be accounted for depends on the specific features of the trading system. In the particular case of the EU Directive, the valuation depends on the arrangement of the national allocation plans. The entity and the accountant preparing the balance sheet have, therefore, to consider the system arrangements.

4 Treatment of emission rights in the annual accounts

Once the value of emission rights is known, the method by which they will be recognized in the annual accounts has to be defined. The International Financial Reporting Interpretations Committee (IFRIC) of the International Accounting Standards has addressed this question by writing a statement dealing with how to account for emission rights.[30] As there are neither previous guidelines addressing

[30] See International Financial Reporting Interpretations Committee (ed), Draft Interpretation D1 Emission Rights, London 2003.

this question, nor practical definitions, the following sections will focus exclusively on the IFRIC statement as well as the comments available to the author[31].

The objective of the IFRIC is to take up the current evolution of the climate policy. The IFRIC based its work on the smallest common denominator of the trading systems currently being designed, one of the aims of which is the fulfilment of the Kyoto protocol's objectives. The IFRIC is thereby aiming at supporting a broad majority of companies. All systems, which the IFRIC referred to are already designed, i.e. already regulating one trade. In the European Union the system will be launched on 1 January 2005.

The IFRIC worked on the following **assumptions** to establish its draft[32]:

- Rights to emit at a specific level are allocated free of charge or sold by a government or a government agency.
- At the beginning of the given compliance period, allowances are allocated, and at the end of this period, actual emissions are verified.
- Participants are free to buy and sell allowances. They may: a) emit pollutants up to the level of allowances they were allocated, b) emit lower than this level and sell or carry forward the excess allowances or c) emit a higher level, and either buy additional allowances or pay a penalty. Participants may also sell their allowances and later buy some equal to their actual emissions.
- At the end of each compliance period, participants are required to deliver allowances to their actual emissions. If a participant cannot deliver sufficient allowances, it will incur a penalty, e.g. a cash payment, a reduction of allowances allocated for a subsequent period, or a restriction on its operation.
- The scheme allows brokers and position-taking institutions to have access to the trading system.

From these assumptions that are consistent with the EU Directive arrangement, the IFRIC proposes the following accounting method for emission rights[33]:

An allowance is considered as an intangible asset and accounted for according to IAS 38. Allowances which are allocated for less than their fair value, e.g. free of charge, shall be measured initially at their fair value. Allowances shall not be amortized, but may be impaired.

- If allowances are allocated for less than their fair value, the difference is considered as a government grant according to IAS 20. The grant is recog-

[31] The comments from Deloitte Touche Tohmatsu, from the Deutschen Rechnungslegungs Standards Committee e.V. (DSR), from the European Financial Reporting Advisory Group (EFRAG), from the Hong Kong Society of Accountants, from the Institut der Wirtschaftsprüfer (IdW), from the International Emissions Trading Association and from the Japanese Institute of Certified Public Accountants were at the author's disposal.

[32] See International Financial Reporting Interpretations Committee, Loc.cit. (footnote 30), p. 1 f.

[33] See International Financial Reporting Interpretations Committee, Loc.cit. (footnote 30), p. 3 ff.

nized initially as deferred income and subsequently as income over the compliance period.

- As emissions are made, a liability is recognised for the obligation to deliver allowances equal to emissions that have been made. This liability is a provision that falls under the scope of IAS 37. The liability is settled by delivering allowances, incurring a penalty or a combination of both. This liability shall be measured at the best estimate of the expenditure required to settle the obligation on the balance sheet data. The market price is then accordingly assessed, or, if a cash penalty is incurred, it will correspond to the cost of the penalty.
- Certain assets may become impaired if the cash flows (e.g. of a particular facility) which are expected to be generated by those assets are reduced as a result of the scheme.

The following questions, which the IFRIC has addressed[34], are the background to this proposition:

1. Does an emission right scheme give rise to a) a net asset or liability or b) an asset (for allowances held) and a liability, a deferred income and/or an income?
2. If a separate asset is recognised, what is the nature of that asset?
3. If a separate liability, deferred income and/or income is recognised, what is the nature of that item and how is it measured?

ad 1.: The IFRIC chose a gross approach, i.e. the assessment as an asset **and** liability. An allowance meets the definition of an asset in the Framework for the Preparation and Presentation of Financial Statements, namely it is a resource that can be controlled by the company. Once emissions have occurred, the entity has a liability, i.e. the obligation to deliver allowances or to pay a penalty. These positions are independent of each other. Depending on the market situation, a company may decide to pay a penalty, even though it holds rights, or to use rights allocated under the scope of different trading systems. Moreover, the gross approach also has the advantage of enabling the same handling of rights, whether these are sold or allocated free of charge. Finally, the IAS 20 requirements are not met.

ad 2.: The asset is described as intangible, since it meets the definition of intangible assets in IAS 38.7, namely it is a non-monetary property item without physical substance and held for production purposes. IAS 38.8, listing examples e.g. fishing licenses, also allows this classification. Under IAS 38.93, the measurement of an intangible asset is estimated using market prices prevailing at the time of initial recognition (benchmark treatment) or at the time of a subsequent revaluation if the entity adopts the alternative treatment under IAS 38. The value of an allowance should not be amortized. In accordance to IAS 36 however, an impairment loss should be planned. Allowances are not accounted for as financial assets under IAS 39, since they are neither equity instruments nor contracts for other financial assets. Allowances may be best recognized at fair value, since they can be traded

[34] See International Financial Reporting Interpretations Committee, Loc.cit. (footnote 30), p. 3.

in a ready market, as opposed to other intangible assets. The income statement is to be amended accordingly. This is not possible under IAS 38 which only permits fair value as an allowed alternative, and changes in fair value are to be reported in equity. The IFRIC has omitted to ask the Board about this matter since the arrangement is still under discussion at the moment.

ad 3.: The IFRIC agreed that a liability, i.e. an obligating event under IAS 37, arises only once emissions are made. At the start of the compliance period when allowances are allocated, there is no liability yet in agreement with IAS 37.19, since obligations can only arise from past events. Once emissions are made, a liability will arise and will be measured according to IAS 37.36 at the best estimate of the expenditure required to settle the obligation at the balance sheet date. This amount corresponds either to the market price or the penalty. As there is no liability at the beginning of the compliance period, the IFRIC considered that the allocation of allowances for less than their fair value (e.g. free of charge) gives rise to a government grant, as mentioned in IAS 20. In the case of emission rights, the government may grant resources to the company relating to its operations. A non-monetary transfer comes within IAS 20.23, which states that these grants be estimated at fair or nominal value. The IFRIC agreed that the estimation at nominal value should not be permitted for emission rights, since this would not enable the equal treatment of allowances allocated free of charge and sold allowances. Government grants should initially be recognised as deferred income, which will subsequently be amortized depending on how allowances are used.

This example of alternative treatment[35] illustrates the proposal. A short summary will follow:

Starting point: Year 01 of the compliance period	
01-01-01:	Company A is allocated, free of charge, allowances for the year to emit 12,000 tonnes of carbon dioxide equivalents. Market price: 10 € per tonne.
06-30-01:	Company A has emitted 5.500 tonnes, 12.000 tonnes emissions are expected for t_0. Market price: 12 € per tonne.
12-31-01:	Company A has emitted 12.500 tonnes. At the end of the year, it buys allowances for 500 tonnes emissions, market price: 9 € per tonne.
No allowances are sold during the year.	
Company A decides whether it will use the alternative treatment under IAS 38. Allowances will be measured at fair value with changes in value above cost reported in equity and changes in value below cost reported in the income statement.	

B/S 01-01-01 (in 000s)

allowances	120	government grants	120

[35] See International Financial Reporting Interpretations Committee, Loc.cit. (footnote 30), p. 5 ff.

B/S 06-30-01 (in 000s)

allowances[1]	144	Δequity[1]	24
		government grants[2]	65
loss	11	liability[3]	66

1 Market price raise 12.000t * (12/t – 10/t) = 24000

2 120.000 – retransfer 5.500t * 10/t = 65.000

3 Actual emission cost first half-year 5.500t * 12/t = 66.000

Income statement first half year 01 (in 000s)

expense for liability for emissions[3]	66	income for amortization of grant[2]	55
		loss	11

B/S 12-31-01 (in 000s)

allowances[4]	112.5	Liability	112.5
Δcash	-4.5		
loss[7]	4.5		

Income statement second half-year 01 (in 000s)

decrease in value of allowances[5]	12	income for amortization of grant	65
expense for liability for emissions[6]	46.5		
profit	6.5		

4 Original allowances at allocated price and purchase of new allowances: 12.000t * 9/t + 500t * 9/t = 112.500

5 Market price decrease 12.000t * (12/t – 9/t) – retransfer Δequity = 12.000

6 Actual emission cost second half year (12.500t * 9/t) – (5.500t * 12/t) = 46.500

7 Net loss for the first half-year – annual net profit second half year 11.000 – 6.500 = 4.500

The example describes the above-mentioned major items characterising the accounting for emission rights in annual accounts.
Gross approach of assets and government grants;

- Treatment as intangible asset;
- In the case of alternative treatment, market price increases and market price decreases find expression in equity capital and income statement respectively;
- Expenditure distributed according to the use of allowances;
- Impairment of the allowance value.

Comments about the IFRIC's proposition could be submitted until 14 July 2003. Suggestions from Deloitte Touche Tohmatsu, Deutscher Rechnungslegungs Standards Committee e.V. (DSRC), the European Financial Reporting Advisory Group (EFRAG), the Hong Kong Society of Accountants, the Institut der Wirtschaftsprüfer (IdW), the International Emissions Trading Association and the Japanese Institute of Certified Public Accountants are already available to the author in addition to her own ones. Essential remarks will then be summarized so that the reader can assess the possible development.

The recognition as intangible asset,[36] government grant and liability are generally well accepted. The chosen gross approach[37] and the use for systems that are already being used, met with general approval.

---

[36] The International Emissions Trading Association agrees with the statement as financial asset and a treatment in accordance with IAS 39, the main objective being hedging. To the author's mind, information concerning the considered allowances are hereby disregarded.

The major concern voiced in the comments is the valuation of asset and liability:

Intangible assets should initially be measured at fair value. To date there is no active market and hence a market price does not exist. It has been suggested that the penalty be used as surrogate for fair value. The IFRIC is required to establish rules concerning this matter. On a general level, the assessment of a fair value should be better defined. At least during the first period of its existence, the required market is limited by regional characteristics or pooling-opportunities. In subsequent periods, increases in value are either not recognized (benchmark treatment) or recognized and cause increases in equity (allowed alternative treatment).

The liability, on the other hand, is measured at fair value on each balance sheet day with changes recognized in profit or loss. This is why revenues and expenditures can be treated in a different way without an economic justification. IAS 37.36 and IAS 37.37 require that the expenditure be measured at the closest estimate required to settle the obligation on the balance sheet date, i.e. the amount that an entity would actually pay to settle the obligation. If the benchmark treatment is applied, the best estimate value matches the current value of the emission rights, as far as the ownership of rights is concerned. The best estimate value of the required rights is either the market price or the penalty. In the particular case of hedging however, IAS 39 applies.

The IFRIC requires that government grants are accounted for at fair value, although IAS 20.23 also permits nominal value. Rights allocated free of charge are not included in the balance sheet since their nominal value is zero, contrary to purchased rights. This is why most of the comments agree with the use of the fair value. The question may be raised of whether such a decision is not outside the IFRIC's remit. This issue is also of particular importance since this statement may be of concern to other cases.

Beside these essential comments, the following points still need to be clarified:

- The use of IAS 2 for brokers should be defined.
- Two questions have arisen as far as the expenditures for emissions are concerned: Shall emissions be capitalized as inventory costs according to IAS 2.14? Shall they be recognized as costs of manufacturing or as penalties for polluting?
- The IFRIC requires the valuation of the emission right expenditure on the balance sheet date. Pollution, hence the use of allowances, is an ongoing process as is the expenditure of materials and personnel. As a consequence, it is recommended that a methodical valuation be performed at average costs over the concerned period.
- Considering that emission rights may constantly be purchased and sold, and that they are recognized as similar assets, the question may arise of whether a group valuation procedure is required at all.

[37] In its comment, the International Emissions Trading Association for example has pleaded for a netto approach.

After these questions were clarified, and the IFRIC proceeded accordingly, there are still open items within the individual trading systems, which the IFRIC has not addressed. Specific rules are required regarding these items.

5 Evolution of accounting at the international level

The way emission rights are reported in the annual accounts is considered as a special case of the above-mentioned IAS 20, 36, 37 and 38. This means that the evolution of accounting in general may be of relevance, also in particular, for the treatment of emission rights.

Three essential features of the treatment of emission rights in accounting, as demonstrated by the IFRIC, will have to be discussed, since they will prove of importance for the future accounting of ecological resources:

- Intangible assets
- Gross approach of intangible assets and liabilities
- Character of the liability

ad a) To the author's mind, particularly the recognition of emission rights as an intangible resource is a revolutionary decision for the treatment of ecological resources. A comparison of the characteristics of ecological resources and of "traditional" intangible resources shows points of similarity that may cause an equivalent treatment in the annual accounts – as in the case of emission rights. These encompass the following features[38]:

- Both ecological and intangible resources are requirements for the establishment of total performance and for the company's success. A capitalization is consistent.
- Both resource packages arise outside the company's system. One of the consequences is that companies use pollution possibilities or human capital, although they do not own the resources. Allowances for these resources are the property of a third party or of the general public, i.e. these have not been explicitly allocated. This hinders the capitalization of these resources in the balance sheet, although it amounts to a resource "leasing".
- Prices for both resources can only partially be determined by the markets, or with restriction, as shown in the previous sections regarding emission rights.
- It appears impossible to establish a complete recording of all factors influenced by and influencing ecological i.e. intangible resources. It seems very unlikely that a complete (material) stocktaking will be possible.
- There are no definitions which allow a clear distinction between the two types of resources. Generally only a qualitative measurement can be performed, therefore limiting the objectivity of the process.

[38] Günther/Günther, Controlling 2003, p. 191-199.

- The responsibility for both resource packages goes to not directly value-adding cross-sectional departments of a company, (e.g. R&D, environment protection or human resource department).
- Due to the yet novel nature of intangible assets, principles of proper accounting for their recording, valuation and recognition have not yet been established, in contrast to the centuries-long precedence of double entry book-keeping for financial and tangible resources.

These characteristics, as well as the indication regarding the treatment of Renewable Energy Permits in the IFRIC-Statement[39] show that other ecological resources are expected to be recognised as intangible assets, provided they are referred to in annual accounts.

ad b) The IFRIC proposes a gross approach of intangible assets on the one hand, and of liabilities on the other. When reading the balance sheet, one can visualize the global amount of GHG emission allowances and at the same time their actual use. The assignment remains a simple process as long as only one type of allowance is considered. However, the balance sheet may lose in clarity despite the use of the gross approach if more than one type of right for the use of ecological resources is summarized. The use of the gross approach is also of importance for other ecological resources. This alternative indeed also allows the gross reporting of accruals for inherited burdens and the corresponding plot of land. The plot of land may even have a negative market price, i.e. a "purchaser" has in fact to be paid to acquire it. A negative market price may also arise for stocks of materials, the use of which is forbidden by law and the disposal of which gives rise to liabilities. In order to provide the reader with a clear picture of the financial situation and the profitability of the company, it is recommended to detail the respective positions in an annex of the annual accounts[40].

ad c) The treatment of ecological resources should make a clear distinction between liability, accrued liability and contingent liability, as they are defined in the international accounting standards. The European Communities' recommendation of the recognition of environmental issues in the annual accounts and annual reports of companies[41] distinguishes between two types of obligations, which may give rise to a liability: a legal or contractual obligation and a constructive obligation, the latter must be so clearly defined, that the company has no discretion to avoid the action any longer. Cases such as this will arise frequently, since companies often commit themselves to protect the environment, they should therefore consider the consequences before they undertake such an action. Accrued liabili-

[39] See International Financial Reporting Interpretations Committee, Loc.cit. (footnote 30), p. 10.

[40] See Commission of the European Communities (ed), Recommendation of 30 May 2001 on the recognition, measurement and disclosure of environmental issues in the annual accounts and annual reports of companies. In: Official Journal of the European Communities, N L 156/33-42, 2001, Annex 3.7.

[41] See Commission of the European Communities, Ibid. (footnote 40) Annex 3.1.

ties should not be used for the value adjustment of assets.[42] As a consequence, the cases outlined under b) may either lead to the adjustment of the valuation of assets or give rise to an accrued liability.

To summarize, the implementation of ecological issues in annual accounts is a special case of several regulations. Recognizing them as intangible resources may, however, suggest how they should be dealt with.

6 Possible development of the recognition of environmental issues in annual accounts

Companies are used to writing up an annual report for the shareholders and a separate environmental report for so-called interested parties. In regard to this, the Commission of the European Union has worked out a recommendation on the recognition, measurement and disclosure of environmental issues in the annual accounts and annual reports of companies, the purpose of which is to better relate environmental reports with annual reports. The working committee for environmental management of the Schmalenbach-Gesellschaft für Betriebswirtschaft e.V. (AKUM) worked out a position paper for environmental reporting by companies. This paper focuses on the role played by environmental reporting. In the following, three of the possible developments are expanded: (1) a more clearly defined relationship between management control and provision of information, (2) an accurate preparation of the balance sheet items and (3) the choice of the discount rate.

ad (1) The theses of the AKUM for environmental reporting in companies[43] aim at connecting management control to provision of information. According to thesis #1 "*External transparency follows internal transparency*", which follows the management approach which states that an environment oriented decision-making system is the sine qua non for environmental reporting. Thesis #2 concerns only internal transparency and will be disregarded here. Thesis #3 entitled "*The value of external transparency consists in nurturing trust and safeguarding reputation*", bases itself on the public mistrust phenomenon arising when a single company refuses environmental disclosure. External transparency is therefore absolutely essential. Thesis #4 "*External expertise can help to install a climate of trust*" ex-

[42] See Article 20 paragraph 1 and Article 31 paragraph 1 bb) of the Fourth European Directive.

[43] See Arbeitskreis Umweltmanagement (AKUM) der Schmalenbach-Gesellschaft für Betriebswirtschaft e.V. (ed), Umweltwirtschaftsforum, Issue 4, 2003. The following members worked out the position paper: Dieter Arnold, Fraport AG, Prof. Dr. Günter Beuermann, University of Cologne; Joachim Ganse, Gerling AG; Prof. Dr. Edeltraud Günther, Dresden University of Technology; Prof. Dr. Wolf-Dieter Hartmann, Klaus Steilmann, Institut f. Innovation u. Umwelt GmbH; Prof. Dr. Manfred Kirchgeorg, Leipzig School of Management; Dr. Jochen Rudolph, Degussa AG; Prof. Dr. Marion Steven, Ruhr-Universität Bochum; Dr. Udo Weis, ABB AG; Klaus Wilmsen, Karstadt AG; Dr. Claudia Wöhler, BDI.

plores the trust arising with the involvement of a third party. The above-mentioned theses demonstrate on the one hand the essential role of external transparency and on the other hand its necessary connection with the instruments of decision in the company. In regard to this, the author also recommends interpreting the proposals of the European Commission, i.e. additional information will be available to the reader only if annual accounts and the annual report reflect decisions having taken into account environmental aspects (e.g. purchase of an end-of-pipe technology or of an integrated technology).

ad (2) As a consequence, a differentiated identification of expenditures aimed at the protection of the environment, for example, can give information about the company's priorities as far as process, product and organizational decisions are concerned. A proposal for the distinction of environmental expenditures has been worked out in the context of the VDI Directive, and it expands the definition given by the German Federal statistic law. (See fig. 2)

All expenditures stated individually in the proposal aiming at the protection of the environment already influence the value of the annual account items. However, a differentiated statement is usually missing, or only appears in the environmental report, with no clear reference to the environmental report in the annual report. Therefore, annual accounts and annual reports do not provide the reader with any information about the company's operations for the protection of the environment, and hence about possible starting points for control. Establishing a better connection between the environmental report and the annual report should not require a high expenditure. The arrangement must always be considered taking into account the groups the reports are aimed at. The AKUM suggests selecting the available information, which can be assessed by external stakeholders.

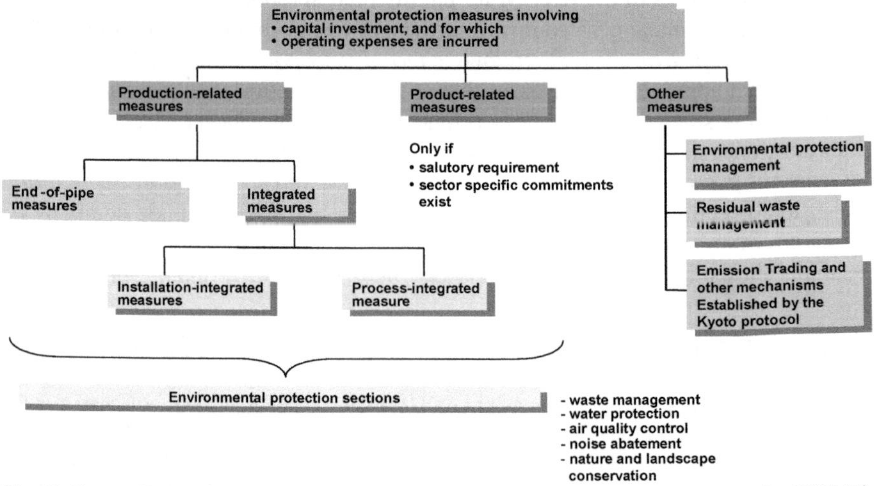

Fig. 2. Expenditures for the protection of the environment in accordance with the VDI Directive 3800[44]

[44] See Verein Deutscher Ingenieure (ed), Guideline VDI 3800: Determination of costs for environmental protection measures. Berlin 2001, p. 8.

ad (3) It is interesting to see that the European recommendation has taken up the question of the choice of the discount rate, which has been discussed for decades. The discount rate reflects the decision-maker's time preferences and allows the assessment of cash inflows and outflows that arise at different times. Several characteristics can be observed by choosing the discount rate when considering ecological issues. Discounting is for example not appropriate if environmental damages accumulate in such a way that they become irreversible and an immediate action is required. In this case, a time preference being a basis for future damages goes against sustainable management. Future technical developments may however enable the repairing of damages considered as irreversible up to now, and again give rise to discounting. An accumulation may even be required if there is a hardening of the regulatory framework due to a change in the behaviour of stakeholders, in order to assign a higher time preference to the immediate action. Finally, a long term calculation based on discounting cannot be considered as an exclusive decision criteria, since the interests of future generations would thereby be neglected.

7 Summary

The setup of markets for emission rights challenges external accounting, since technical, political, economic, ecological and commercial points of view have to be interconnected. On 1 January 2005, the European Union introduced the emission allowance trading. Already in summer 2003, the IFRIC proposed to recognize these emission allowances as intangible assets, and the obligation to deliver allowances corresponding to the actual emissions as a liability (gross approach). Moreover, the allocation free of charge of these pollution rights gives rise to accrued items that will have to be retransferred during the allowance validity period.

In addition to details of the accounting procedure for emission allowances, analysis of the current legal situation has resulted in the following three items:

- Because they can be considered as usufructs, ecological resources will be accounted for as intangible assets.
- The accounting for emission rights is the first broad integration of financial and environmental facts and figures.
- In the future the entity preparing the balance sheet and qualified auditors will have to assess the facts behind usufructs.

The role of stakeholder driven corporate governance – the example of BP's climate change strategy

Thomas Langrock

Trelleborger Straße 9, 13189 Berlin, Germany
thomas_langrock@yahoo.de

Abstract

BP has adopted a commitment to reduce the GHG emissions from its operations. This commitment has been and will be implemented using a variety of measures, including emissions trading and emission credits. This article analyses why BP has adopted this GHG commitment. The research approach rests on using policy network analysis and evaluation.

Keywords: Evaluation, emissions trading, policy networks, corporate governance

Acknowledgement: The author wishes to thank the Economic and Social Research Institute (ESRI, Cabinet Office Japan) for the opportunity to contribute to their Collaborations Projects Programme. The presented work is one outcome of the research efforts undertaken under the umbrella of the Wuppertal Institute's Collaboration Project "Governance of Sustainability". The results of this study were published during 2004 (Bleischwitz et al. 2004).

1 Introduction

BP plc. is one of the largest companies in the oil and gas sector: In December 2002 around 115,250 employees generated 179 Billion US Dollar revenue. In total the company consists of 38 chemical sites and 24 refineries worldwide (BP 2004). This truly trans-national company has become one of the corporate forerunners with regard to climate change mitigation. The company acknowledges climate change is a serious issue that the oil and gas sector faces, and it has come up with a wide-ranging climate change strategy. According to the website "Our position", "an integral and key element within our {BP's} strategy to address climate change" is the commitment to control the greenhouse gas emissions (GHG emissions) of BP's operations (herein abbreviated GHG commitment) (BP 2003b).

BP's climate change strategy fits into a larger class of voluntary initiatives by trans-national companies with the aim of contributing to sustainable development. From studying these initiatives and the discourse on corporate social responsibility, the Working Group on Eco-efficiency and Sustainable Enterprises of the Wuppertal Institute extracted a concept called "responsible corporate governance" (RCG). It is defined as "a stakeholder-oriented policy approach allocating responsibilities to societal actors, who will drive corporate accountability" (Kuhndt et al. 2004, p. 15). This article is part of a series of projects in the Wuppertal Institute that seek to understand how RCG can contribute to sustainable development. From these projects the following interpretation of RCG emerged: *A company that adopts Responsible Corporate Governance establishes a policy network (consisting of its stakeholders), which can influence its policy-making and can contribute to the policy implementation.* Research questions that naturally arise with such an interpretation are: What is the steering potential of these policy networks? Can network governance of companies substitute hierarchical steering by the nation state (and to which extent)? Can network governance of trans-national companies (partially) fill the governance gap that exists on the international level?

It is the purpose of this article to understand why BP has adopted its climate change strategy and how this can be interpreted in light of the last question. More specifically the author will test the following research hypothesis:

1. The policy making process that led to the adoption of the GHG commitment has been influenced by a policy network of actors / stakeholders outside BP.

Various stakeholders/actors outside BP have contributed to the implementation of the GHG commitment through the mobilisation of their resources.

2. This will be done through an analysis of the interactions between BP and its stakeholders that actually took place and through an analysis of the policy outcomes that can be interpreted as a coordinated mobilisation of resources that contribute to implementation of the GHG commitment. Before this is done an evaluation of the GHG commitment will serve as a means of presenting the substance of the GHG commitment as well as a means of testing whether the adopted climate policy really is a significant step towards climate change mitigation. The article rests almost entirely on the information that BP plc. provides on its internet site (http://www.bp.com). In addition to

this the author has approached individuals inside and outside BP in order to better understand some of the issues presented here. However, the author refrains from quoting these individuals in this article.

2 Background: evaluation research and policy network analysis

In this paper the author will make use of a research methodology and a concept from the social sciences: evaluation and policy network analysis. The following two sections seek to introduce the reader to the two concepts. Naturally, the scope of this article only allows a very crude and simplified introduction of these two branches from the social sciences.

2.1 Evaluation

Evaluation is a technique that is widely applied in public administration and the social sciences. Very different objects can be evaluated, e.g. research programmes, the performance of professors at university, large-scale government subsidy programmes and projects that are financed by official development assistance (Stockmann 2000). In the last 15 years the evaluation of (environmental) policy instruments has emerged as a new subject of policy analysis. It is this branch of evaluation that shall be used here.

From his survey of literature (Rossi, Freeman, Lipsey 1999; Vedung 1999; Huppjes 2000; Stockmann 2000; Schmitt von Sydow and Summa 2001) the author deduced the following components that are usually part of an evaluation. The first comprises the analysis of the objectives of a policy instrument. Sometimes these policy objectives are spelled out explicitly, EU Directives in the environmental field almost always contain such a description. Very often the policy objectives can only be deduced indirectly. The second component is an assessment of the attainment of the objectives (Effectiveness). Some evaluations also include an assessment of the objective itself (Relevance), i.e. they judge the contribution of the policy instrument to the solution (mitigation) of the overall problem. The third and most typical component of evaluations is an assessment of the (cost) efficiency.

In this paper the author transfers the evaluation of policy instruments to the company level. That is, he will understand BP's climate strategy as a set of policy instruments that BP has implemented. He will then assess the relevance, effectiveness and costs of this set of policy instruments.

2.2 Policy networks

Policy Networks are extensively used as a means for studying the policy formulation and implementation process (Börzel 1998; Kenis and Schneider 1991; Marsh and Smith 2000; Carlsson 2000). The underlying model of policy networks is relatively simple: There is a relatively small number of actors, which have a mutual interest in their resources (resource interdependency). Examples of such actors and resources are industry associations that dispose of steering capacity with respect to their members, local authorities that dispose of knowledge about the local situation, non-governmental organisations that have the capacity to set standards etc.

As a consequence of the resource dependencies, relationships between actors in a policy network are not hierarchical in nature. Policy networks are generally characterised by rather informal structures, which nevertheless exist relatively stable in time. Due to these characteristics the typical mode of interaction within policy networks is negotiation in order to achieve palpable policy outcomes. It is important to note that these interactions can assume two completely different shapes. There is first the mobilisation of disperse resources such that collective action becomes possible (Carlsson 2000). Then the negotiations are about who contributes which resource and how much of it. The second mode of interaction, sometimes described as negative coordination, is detectable whenever one actor threatens to use his resource in case a policy outcome is agreed upon that he does not regard acceptable (Scharpf 1993, 1997). It is for this reason that such an actor can participate in the negotiations.

This paper rests on the premise that the general idea of policy networks can be transferred to the corporate level. Thus, the relationship between BP's stakeholders and the usual governing bodies of BP is assumed to be similar to the relationship between the policy making process of Government and Parliament on the one hand and the interactions with interest groups, associations and corporate actors on the other hand.

3 Evaluation and assessment of the GHG commitment

3.1 The policy objectives

John Browne announced the commitment to monitor and control the greenhouse gas emissions of BP in a landmark speech delivered at Stanford University in May 1997, a few months before the adoption of the Kyoto Protocol (BP 1997). This commitment was quantified in 1998: BP pledges to reduce its "emissions of greenhouse gases by 10 percent from a 1990 baseline over the period to 2010." (BP 1998). In March 2002 John Brown publicly stated, "Now, five years on, I'm delighted to announce that we've delivered on that target. That means that our emissions of carbon dioxide have fallen to almost 80 Million tonnes, 10 Million tonnes below the level in 1990 and 14 Million tonnes below the level that they had reached in 1998." (BP 2002).

In the same speech Brown announced the new target: BP will "hold the emissions from our operations at 10 percent below 1990 levels, through 2012, with approximately half of that coming from improvements in internal energy efficiency, and half from the use of market mechanisms, generating carbon credits." (BP 2002). This new target is thus significantly different from the first target as it introduces the principle of neutralisation: Emissions reductions that are achieved at installations outside BP's operations result in emission credits (or emission certificates) and these emission credits neutralise emission from BP's operations. This means that the actual emissions from BP's operations can still rise as long as enough emission credits are generated somewhere else.

Of course, the two targets need further qualification as regards the system boundary and other technical details. BP publishes various documents that provide clarification: among them the environmental reporting guidelines (BP 2000), and various audit notes (KPMG, DNV, ICF 2000, 2001, 2003). All these documents primarily focus on monitoring of BP's emissions; however, almost all include references to BP's GHG commitment. From these documents it can be deduced that the GHG commitment consists of controlling the direct equity share emissions of BP. Thus, the system boundary is drawn such that both are included: all emissions from sites that BP owns completely, as well as an appropriate share of the emissions of those sites that BP partly owns.

Table 1. Elements of BP's Climate Change Strategy that accompany the commitment to reduce greenhouse gas emissions

Promoting flexible market instruments	"We (BP) will continue to promote the flexible market instruments, including emissions trading, Joint Implementation (JI) and the Clean Development Mechanism (CDM)." (BP 2003b)
Participating in the policy processes	"We (BP) are committed to take an active part in the climate change policy debate and to investigate innovative ways of reducing GHG emissions."(BP 2003b)
Co-operation in order to accelerate new energy technologies	"We (BP) intend to work with other industries to advance the development of energy efficient technologies and to make these available to our costumers." (BP 2003b)
Investing in research	"BP plans to continue to invest in and support science, technology and policy research." (BP 2003b)

As table 1 shows the GHG commitment is only one part, although the most significant part, of BP's climate strategy.

3.2 An assessment of the relevance

An evaluation of the relevance of a policy outcome must include an assessment of the contribution to the overall problem, which is climate change. The GHG commitment clearly is a contribution to the mitigation of climate change. Its relevance can be assessed by putting the commitment in a context with the emissions BP can theoretically control as well as with the emissions of various nation states.

In the year 2002 BP emitted 80.5 million tonnes of carbon dioxide equivalent into the atmosphere. To put this in perspective, this is less than 10 percent of the emissions of Germany in 2002. Hence, a reduction of 10 percent of BP's emissions can only contribute little to climate change mitigation. But of course it is millions of different actors that must contribute to the mitigation of climate change. Therefore, an assessment of the relevance of the objective should not stop at an assessment of the absolute emission reductions. In relative terms, BP's contribution is, so to speak, above average, i.e. far above the minus eight percent that the Kyoto Protocol demands from the European Union or the minus seven percent that it stipulates from the United States of America.

3.3 Effectiveness ex post: looking into the past

It has been described above that BP announced a first target in 1998. BP perceives this target as already achieved: According to the website "Our Performance" (BP 2003c), BP has reduced its direct equity share emissions in 2001 to 10 percent below 1990. This fact was verified by KPMG and Det Norske Veritas (DNV), two respected accountancy companies. BP has put the audit notes on its website (KPMG, LLP, DNV 2003; KPMG, DNV, ICF 2000, 2001). Although the audit paints a comprehensive picture of the emission trends and thus indirectly the achievement of the GHG commitment (for example a lot of effort has been invested into properly treating acquisitions and divestments), it must be noted that as of yet, there is no formal process that assesses compliance with the GHG commitment. In other words, a statement is missing which clearly states that the commitment has been achieved. Instead, John Browne announced the compliance with the first target.

BP has implemented the GHG commitment through a series of activities both inside and outside the system boundary. The most important instrument inside the system boundary has been the setting up of an internal BP wide emissions trading scheme. In 1999 a pilot scheme was introduced and during 2000 and 2001 BP wide emissions trading took place (BP 2003d). In 2002 the trading scheme was abandoned in order to give room to the public emissions trading schemes that are being set up. By nature of its construction emissions trading only makes sure that the lowest cost potential for emission reductions is tapped. The actual implementation of the GHG commitment comprises a variety of technical measures on the business sites. BP itself mentions investment into energy efficiency, reduced leaks of methane in the natural gas business and reduced flaring and venting. With such activities inside the system boundary, BP achieved significant emission reductions.

3.4 Effectiveness ex ante: looking into the future

BP announced a second or revised GHG target in 2002, which will last until 2012 and thus its effectiveness cannot be evaluated in the strict sense until after 2012. However, what can be done is a first ex ante look at the measures that BP intends to implement in order to achieve this second GHG target.

The speech, in which Lord Browne announced the new target (BP 1998), contains a few hints how BP intends to achieve the commitment. It wants to expand its business and, therefore, include further emission sources inside the system boundary. This will clearly lead to an increase in GHG emissions inside the system boundary, if no action is taken. Lord Browne estimated that emissions might rise as high as 130 million. Reconciliation of this 'business as usual' scenario with the GHG commitment will be executed through substantial investments into emission reductions both inside and outside the system boundary. More concretely, BP intends to achieve emission reductions of around 25 million tonnes of carbon dioxide equivalent inside and 25 million tonnes outside the system boundary. The emission reductions outside the system boundary shall then neutralise the same amount of emissions within the system boundary. In doing so BP would actually emit 105 million tonnes carbon dioxide equivalent inside its system boundary, of which 25 million is regarded as neutralised due to the realisation of emission reductions outside the system boundary. Thus, BP achieves the required stabilisation at a level of 80 million tonnes of carbon dioxide equivalent.

Neutralisation of emissions through emission reductions outside one's own sphere of control should be regarded as an accepted principle. The two Kyoto-Mechanisms Joint Implementation (JI) and Clean Development Mechanism (CDM) rest on the same principle. Nevertheless, neutralisation poses some questions with regard to effectiveness:

Firstly, the quality of the emission reductions must be very carefully controlled. The Kyoto-Mechanisms – JI and CDM – offer a platform for producing high quality emission reductions; yet it is questionable whether BP can and is willing to realise all the announced emission reductions within this legal framework. Already BP has announced that it wishes to work with car manufacturers in order to reduce the emissions of the cars they sell. Such activities, however, may not be eligible under the mechanisms CDM and JI. Additional measures to guarantee the quality of these emission reductions will, therefore, be necessary. Secondly, the Kyoto Protocol stipulates so-called supplementarity, i.e. the usage of neutralisation should only be a supplement for domestic emissions reductions. If BP follows the announcements of John Browne, then supplementarity would hold.

3.5 The efficiency: cost reduction

BP itself states that the benefits associated with the emission reductions have so far exceeded the costs for implementation. BP states, "The outcome was that our first target was delivered at no net cost to our business – we estimated that our reduction projects delivered over $600 million of net present value from the fuel saved." (BP 2004h). Unfortunately, BP does not provide further details regarding this calculation.

Every evaluation of BP's GHG commitment must remain preliminary as it is running until 2012. The simplified ex-ante evaluation above has shown that BP's commitment clearly is an above average contribution to climate change mitigation. Summarising, BP's climate strategy clearly is a significant step towards meeting the challenges of climate change. With this evaluation the ground is prepared for a closer look at the policy making process that may have led to the adoption of the GHG commitment. This should be done bearing in mind the overall question: Did a policy network of stakeholders have a significant influence on the formulation of this GHG commitment?

4 The policy network approach applied to the BP plc. policy making process

On its website "Working in Partnership" (BP 2004) BP enumerates a lot of partners that it works with in order to tackle its greenhouse gas emissions. However, these partners are not the only actors involved, there are of course public actors who need to be considered as well as actors from the financial community and from the critical NGO sector. The following chapter describes the identified interactions with these actors and attempts to derive the resource dependency relationships that might exist.

4.1 BP and the non-governmental organisations

The term NGO is used for a wide range of organisations. Many are not-for-profit entities that fulfil economic functions or provide public goods, some work as policy think tanks and the most well known are the critical NGOs that engage in the policy making process through campaigning. BP maintains relationships to NGOs of all three types.

BP acknowledges the work with Nature Conservancy (BP 2004d), a US-American non-profit NGO that has the mission "to preserve the plants, animals and natural communities that represent the diversity of life on Earth by protecting the lands and waters they need to survive." (Nature Conservancy 2004). This NGO is very active in setting up and managing natural parks, thus providing a public good. Together with BP it has set up carbon sequestration projects. In addition to this, BP sponsors the exploratory work of Nature Conservancy, research institutes and other organisations that seek to find out how biological carbon sequestration projects can be designed such that they contribute both to the conservation of biodiversity and the mitigation of climate change. On its website BP refers to its participation in the climate and biodiversity alliance, which among many activities aims to develop an independent standard for biodiversity and restoration projects that mitigate climate change (sometimes referred to as the Blue Chip Standard).

BP further maintains very close relationships to two US American policy think tanks: Environmental Defence and the Pew Centre on Climate Change. The cooperation with Environmental Defence started in 1997 with the objective of building an internal emissions trading scheme (Environmental Defence 1997). BP publicly acknowledges the important work that Environmental Defence has undertaken in formulating and implementing the commitment to control BP's greenhouse gas emissions (BP 2002). The cooperation with Environmental Defence led to further outcomes: In October 2000 Environmental Defence together with seven big greenhouse gas emitting companies – among them BP –launched the "Partnership for Climate Action" (Environmental Defence 2000). The partnership was designed as a platform that every company can join as long as it agrees to a fixed set of objectives and common principles. Salient among these principles and objectives is the willingness of each member to "publicly declare a global greenhouse gas emission limitation commitment backed by management actions, policies and incentives to achieve them." (PCA 2000).

The working relationship with the Pew Centre on Climate Change is, to some extent, similar to the one with Environmental Defence. BP has joined the Pew Centre's Business Environmental Leadership Council (BELC), which is a group of "leading companies worldwide that are responding to the challenges posed by climate change" (Pew 2004). The website at the Pew Centre states several core beliefs with regard to climate change that all members of the BELC share. The website furthermore lists the company profiles and the individual GHG reduction targets for each member of the group. The website does not contain references to concrete action by the Business Environmental Leadership Council.

BP is also a member of the International Emissions Trading Association (IETA) - an NGO, which strives a) "to promote an integrated view of the emissions trading system as a solution to Climate Change, b) to participate in the design and implementation of national and international rules and guidelines and c) to provide the most up-to-date and credible source of information on emissions trading and greenhouse gas market activity" (IETA 2004). The actual interaction between this NGO and BP that could be detected through internet research is that a senior group advisor of BP is at the same time director of IETA (IETA 2004).

As has been mentioned, a variety of NGOs pursue their objectives through campaigning. The oil and gas sector has repeatedly been exposed to such campaigns. Most prominent has been Greenpeace's campaign against the dumping of the Shell oil platform Brent Spar in 1995 (Greenpeace 1995). More recently Greenpeace demanded - based on a study commissioned from the Wuppertal Institute (Luhmann et al. 2002) – that the oil and gas producing companies invest more heavily into renewable energies. At the moment Greenpeace, Friends of the Earth and Planet & People are carrying out a campaign in the UK that focuses on ExxonMobils branch Esso, which operates petrol stations worldwide. It asks its UK customers to boycott Esso until ExxonMobil stops pushing its sceptical position on climate in various political arenas. The campaign further targets the political influence that ExxonMobil is supposed to exert on US foreign policy (Greenpeace, FoE, Plant and People 2003). As of yet, none of the campaigns against companies in the oil and gas sector have focused on greenhouse gas emissions from the operations. And none of the campaigns have targeted BP.

The capacity to carry out such campaigns can be interpreted as a resource that an actor uses in order to enforce negative coordination, as has been explained in the section on policy networks. BP maintains a variety of relationships with NGOs on a direct contact basis and there is certainly a number of critical NGOs among these. On its website "NGOs", BP states that "early identification of key emerging issues" is one example how BP benefits from discussions with NGOs (BP 2004f). According to this website "discussing issues with people who often have very different views to your own is difficult, but often raises ideas and concerns that would not normally be considered in day-to-day business operations" (BP 2004f).

4.2 BP and international law-making bodies

The Conference of the Parties to the UN Framework Convention on Climate Change (CoP) is the law-making actor regulating greenhouse gas emissions on a global level. The Kyoto-Protocol is one of the major policy outcomes of this negotiation body. The Kyoto-Protocol compels the parties who are listed in its Annex B (mainly the OECD countries and the Eastern European countries with economies in transition) to limit the amount of greenhouse gases that they emit during the years 2008 to 2012.

The Kyoto-Protocol as a legal document only contains provisions for subjects of international law (nation states, the European Union, some international governmental organisations). Thus, it compels nation states to introduce policies and measures that are suitable for reducing the amount of greenhouse gas emissions from their territory. As regards BP, the CoP obviously has no mandate to establish regulation that has a direct impact on BP's operations. The only way in which BP will be affected is when governments of parties to the Kyoto-Protocol implement the provisions. Due to the architecture of the Kyoto-Protocol, only the countries listed in Annex B of the Kyoto-Protocol are likely to do so.

As a novelty within international law, the Kyoto-Protocol contains the so-called flexible mechanisms International Emissions Trading, Joint Implementation (JI) and Clean Development Mechanism (CDM). JI and CDM are instruments for cooperation on a project-by-project basis (the so-called CDM and JI projects) between parties to the Kyoto-Protocol. These CDM and JI projects are not developed by the parties themselves but by private entities, notably private companies. The basic idea of CDM and JI is that these entities invest in climate friendly technologies and receive a certain amount of internationally accepted emission certificates in return. BP announced that it wants to use CDM and JI.

However, the actual interactions between BP and international law making bodies have, thus far, only been informal or indirect. Neither the websites of BP, of the UNFCCC Secretariat or of the NGO CDM-Watch mention that BP has formally submitted a project design document of a CDM project. Thus, there are no formal interactions between BP and the international law-making bodies. Yet, BP participates in the Prototype Carbon Fund of the World Bank and therewith it is involved in a variety of CDM and JI projects. Thus, there are indirect interactions between BP and the international law-making body.

4.3 BP and the European Union

The European Union has adopted the Emissions Trading Directive 2003/87/EC (EU 2003). Participation in EU emission allowance trading, which commenced in 2005, is mandatory for certain types of industrial sites. It is important to note that the design of the EU-wide emissions trading scheme significantly deviates from that of the UK so that it remains to be seen, whether the UK will be forced to revise its own scheme (Sorrell 2003). As regards BP, the differences may be significant.

In the beginning, methane as well as other greenhouse gases will not be included in the EU-wide scheme and as the directive stands now, oil and gas production sites will not be part of the scheme. Thus, it is very likely that a significant share of BP's emissions in the EU member states will not be part of the EU-wide emissions trading scheme. It is not possible to predict so far whether the EU will incorporate other measures that target the operations of BP.

4.4 BP and the UK Government

As a multinational company, BP must naturally accept the legal and economic framework that national governments set under their jurisdiction. However, in the various nation states there is only very little implemented policy that directly targets greenhouse gas emissions from industrial processes. In the United Kingdom, where the headquarters of BP are based, numerous policy instruments are in place which directly target the operations of BP (Sorrell 2003, Langrock 2001).

The business units of BP UK that belong to the chemicals operations pledged to cut emissions in a negotiated agreement with the UK Government. In exchange for delivery of the corresponding emission reductions, the involved business units will receive an 80 % reimbursement of the UK climate change levy (an energy tax). The negotiated agreement with the UK Government is not publicly available, so that it is impossible to assess the proportion of BP UK total emissions that are subject to this agreement. Furthermore, the research methodology applied here does not allow for assessing the magnitude of the Climate Change Levy reimbursement that BP receives in exchange.

Further parts of BP UK participate in the UK emissions trading scheme as so-called direct participant, from 2002 onwards. BP UK entered an agreement with the UK Government, wherein it accepted to report its emissions from a defined set of sources of GHG during a specified commitment period. More importantly, BP UK accepted to hand in an amount of emission allowances that is equivalent to the amount of emissions by the participating business units at the end of the commitment period. As participating business units, BP mentions its refineries and the various oil and gas production sites in the North Sea as well as onshore in the UK. BP expects to realise a large magnitude of the emission reductions through reduced gas flaring at the oil and gas production sites as well as through increased energy efficiency. In exchange for a pledged reduction of 350.000 t CO_2 eq. BP UK received around £ 20 million as incentive money from the government (BP 2002; DEFRA 2002).

5 Conclusions from the case study

The GHG commitment of BP clearly mirrors the structure of the Kyoto Protocol. The many references to this treaty that BP has made indicate that BP clearly anticipated legislation regulating greenhouse gas emissions. Yet, the analysis has also shown that BP has taken on a commitment to reduce greenhouse gases that is significantly broader in scope than any of the coming legislation. This includes legislation on the International, European and national policy level. Thus, interactions with International, European or national lawmakers do not suffice as an explanation why BP has taken on its GHG commitment.

The screening of BP's interactions with other stakeholders (particularly NGOs) does not indicate that there have been strong impulses that ultimately lead to the adoption of the GHG commitment. Due to limits of capacity the role of stakeholders from the financial markets could not be analysed. However, first screenings suggest that none of these players has pushed BP into the adoption of the GHG commitment.

With a look back to the introduction, it can be said that the first hypothesis cannot be verified. There have been no indications that a policy network of stakeholders has significantly influenced the policy making process of BP with regard to the GHG commitment. From the author's point of view, this statement is robust with regard to refinement of the research methodology: None of the experts inside and outside of BP ever pointed to impulses from outside BP.

The analysis of BP's interactions with its stakeholders has shown that BP intensively cooperated and cooperates with NGOs in order to implement the measures that are necessary for achieving the GHG commitment. Particularly, knowledge with regard to emissions trading, monitoring and emission offset projects (JI, CDM) could be mobilised through the policy network of stakeholders.

Additionally, BP will make use of the steering and law-making capacity of the International, European and National lawmakers. BP already announced its willingness to use CDM and JI as a means of acquiring emissions certificates that shall offset emissions inside BP's system boundary.

A very interesting feature of the case study is BP's interaction with Environmental Defense, The Pew Center on Climate Change and the International Emissions Trading Association. BP cooperates with these NGOs in order to influence the policy making process in the USA and on the International policy level. In the language of the policy network approach, BP benefits from mobilising the resource "authoritative steering capacity" of these stakeholders. Concluding from these three paragraphs the second hypothesis can be verified.

As the research approach is based on the transfer of concepts from policy analysis at the company level, a few remarks concerning it may be appropriate. From the author's point of view, the policy network approach has provided terms and concepts (resources, policy formulation, policy implementation) that proved helpful for describing the phenomena. The article has shown that a pure Internet-based research design can bring about significant results with reasonable effort. What is particularly striking is the huge amount of primary sources (audit notes, speeches, reports) that BP makes available on its Internet site. The limits of this approach are also quite evident: Many interactions between BP and stakeholders are informal and can, therefore, not be detected through pure Internet research. And the enormous complexity of a trans-national company poses a high hurdle to sensibly applying this approach.

Apart from testing the two hypotheses, the analysis also results in a new interpretation of the GHG commitment: By adopting the GHG commitment BP initiated an internal and external learning process. The first target helped BP learn about the reduction of GHG emissions and the use of emissions trading. The new target will focus on emission credits. With this knowledge, BP has been and will be a key driver in the debate on emissions trading and emission credits. The knowledge that BP has generated is incorporated into the design of UK emissions trading and EU emissions trading.

References

Bleischwitz R, Acosta J, Budzinski O, Hennicke P, Karius O, Kuhndt M, Langrock T, Ramesohl S, Schubert UM, Supersberger N, Tuncer B, Wallbaum H (2004) Governance of Sustainability: Market Creation via Synergies Between Corporate and Political Governance., Tokyo: Economic & Social Research Institute., URL: http://www.esri.go.jp/en/prj-rc/kankyou/kankyou16/syousai-e.html

BP (1997) Addressing Climate Change: A speech by John Browne, Chief Executive, BP, delivered at Stanford University, California, May 19, 1997, London: BP plc., URL: http://www.bp.com/centres/press/s_detail.asp?id=29, (05.11.2003)

BP (1998) Leading a Global Company: The case of BP: Sir John Browne, Group Chief Executive, The British Petroleum Company plc, Yale School of Management - "Perspectives of Leadership", 18 th September 1998, London: BP plc., URL: http:// www.bp.com/centres/press/s_detail.asp?id=50, (30.10.2003)

BP (2002) Beyond Petroleum. Business and the Environment in the 21st Century: Speech by John Browne, Group Chief Executive, BP: hosted by Stanford Graduate School of Business, March 11 2002, London: BP plc., URL: http://www.bp.com/centres/press/stanford/highlights/index.asp, (30.10.2003)

BP (2003a) Climate Change: Our Position. London: BP plc., URL: http://www.bp.com/environ_social/environment/clim_change/position.asp, (05.11.2003)

BP (2003b) Climate Change: Emissions Trading. London: BP plc., URL: http://www.bp.com/environ_social/environment/clim_change/emissions.asp, (05.11.2003)

BP (2003c) Climate Change: Greenhouse Gas Audit. London: BP plc., URL: http://www. bp.com/ environ_social/environment/clim_change/greenhouse.asp, (05.11.2003)

BP (2003d) Climate Change: Our Performance London: BP plc., URL: http://www. bp.com/environ_social/environment/clim_change/perform.asp#f, (05.11.2003)

BP (2004) BP at a glance. London: BP Plc., URL: http://www.bp.com/ sectiongenericarticle.do?categoryId=3&contentId=2006926, (11.1.2004)

BP (2004a) An Overview of BP. London: BP plc., URL: http:// www. bp.com/sectiongene ricarticle.do?categoryId=14&contentId=2002063, (13.1.2004)

BP (2004b) BP. London: BP plc., URL: http://www.bp.com/generiarticle.do?categoryId=9 &contentId=2002350, (13.1.2004)

BP (2004c) HSE. London: BP plc., URL: http://www.bp.com/genericarticle.do?catego ryId=27&contentId=2000432, (13.1.2004)

BP (2004d) Working in Partnership. London BP plc, URL: http://www.bp.com/genericarti cle.do?categoryId=27&contentId=2000432, (13.1.2004)

BP (2004e) Our Performance (Biodiversity). London: BP plc., URL http://www.bp.com/ extendedgenericarticle.do?categoryId=56&contentId=2006886, (15.1.2004)

BP (2004f) NGOs. London: BP plc., URL: http://www.bp.com/sectiongenericarticle.do?ca tegoryId=802&contentId=2001461, (16.1.2004)

BP (2004g) Products and the Environment. London: BP plc., URL: http://www.bp.com/sec tiongenericarticle.do?categoryId=57&contentId=2000094, (16.1.2004)

BP (2004h) Climate Change: Overview. London: BP plc., URL: http://www.bp.com/gene ricarticle.do?categoryId=55&contentId=2004291, (15.1.2004)

Börzel T (1998) Organizing Babylon - On the Different Conceptions of Policy Networks. Public Administration 76: 253-273.

Carboncredits.nl (2003) Get a higher return on your investments in sustainable energy and energy efficiency, The Hague: Senter International, URL: http://www.senter.nl/asp/ page. asp?id=i000000&alias=erupt, (05.11.2003)

Carlsson L (2000) Policy Networks as Collective Action. Policy Studies Journal 28(3): 502-520

DEFRA (2002) UK Emissions Trading Scheme: Frequently Asked Questions, London: Department for the Environment and Rural Affairs (DEFRA), URL: www.defra.gov.uk/ environment/ climate change/trading/faq.htm, (05.11.2003)

Environmental Defence (1997) British Petroleum announces Plan to Measure and Report Greenhouse Gas Emissions; To Set Targets; And to Practice Emissions Trading. Washington: Environmental Defence (News Release; 9/30/1997), URL: http://www. environmentaldefense.org/pressrelease.cfm?ContentID=1931, (11.1.2004)

Environmental Defence (2000) Global Corporations and Environmental Defence partner to reduce Greenhouse Gas Emissions: Partnership Pioneers Real-World Solutions. Washington: Environmental Defence (News Release; 10/17/2000), URL: http://www.envi ronmentaldefense.org/article.cfm?contentid=503, (11.1.2004)

European Commission (EU) (2002) Amended Proposal for a Directive of the European Parliament and the Council establishing a scheme for greenhouse gas emission allowance trading within the community and amending Council directive 96/61/EC. Doc. 14935/02, Brussels: Council of the European Union

Greenpeace (2003) Greenpeace Brent Spar protest in the North Sea. Amsterdam: Greenpeace International, URL: http://archive.greenpeace.org/~comms/brent/brent.html, (05.11.2003)

Greenpeace; Friends of the Earth (FoE), Planet & People (2003) Don't buy Esso., URL: http://www.stopesso.com, (05.11.2003)

Huppjes G (2000): Environmental Policy Instruments in a New Era. Berlin: Wissenschaftszentrum Berlin, FS II 01-404

International Emissions Trading Association (IETA) (2004) About IETA. Geneva: International Emissions Trading Association, http://www.ieta.org

Kenis P, Schneider V (1991) Policy Networks and Policy Analysis : Scrutinizing a New Analytical Toolbox. In: Marin B, Mayntz R (1991): Policy Networks - Empirical Evidence and Theoretical Considerations. Campus Verlag, Westview Press

KPMG, DNV, ICF (2001) BP's Greenhouse Gas Emissions : Audited 2000 Data, 28 February 2001, URL: http://www.bp.com/environ_social/downloads.asp (31.10.2003)

KPMG, Det Norske Veritas, ICF Consulting, BP Amoco (2000) Development and implementation of a process to audit BP Amoco's greenhouse gas emissions: Audit Overview, February 2000, URL: http://www.bp.com/environ_social/downloads.asp, (31.10.2003)

KPMG LLP, Det Norske Veritas (2003) BP's 2002 Greenhouse Gas Emissions : Independent Greenhouse Gas Audit Report, 3 March 2003, URL: http://www.bp.com/environ_social/downloads.asp, (31.10.2003)

Kuhndt M, Tuncer B, Anderson KS, Liedtke C (2004) Responsible Corporate Governance : An Overview of Trends, Initiatives and State-of-the-art Elements. Wuppertal: Wuppertal Institute for Climate, Energy and Environment, Wuppertal Paper No. 139

Langrock Thomas (ed) (2001) Emissions Trading and Joint Implementation as a Chance for CEECs, Berlin: Federal Ministry of the Environment, Nature Protection and Nuclear Safety. URL: http://www.wupperinst.org/Publikationen/TerraTec2001-Dokumentation/index.html, (05.11.2003)

Luhmann HJ, Müller E, Nitsch J, Ziesing HJ (2002) Oil Corporations and the Destruction of the Climate. How investment can be diverted into renewable energies to protect the climate, Hamburg: Greenpeace Germany

Marsh D, Smith M (2000) Understanding Policy Networks : towards a Dialectical Approach. Political Studies 48: 4-21

Pew Center (2004) Working Together ... Because Climate Change is Serious Business. URL: http://www.pewclimate.org/companies_leading_the_way_belc/, (11.1.2004)

Rossi PH, Freeman HE, Lipsey MW (1999) Evaluation : A systematic Approach. 6th Edition, Sage Publications, Thousand Oaks, London, New Delhi

Sorrell S (2003) Turning an Early Start into a False Start: Implications of the EU emissions trading Directive for the UK Climate Change Levy and the Climate Change Agreements. Paris: Organisation for Economic Cooperation and Development (OECD), CCNM/GF/SD/ENV(2003)7/Final, 28 Pages

Schmitt von Sydow H, Summa H (2001) Report of the Working Group "Evaluation and Transparency" (Group 2b). Brussels: European Commission

Scharpf F (1993) Positive und negative Koordination in Verhandlungssystemen. In: Heritier, A (ed) (1993): Policy-Analyse : Kritik und Neuorientierung. Politische Vierteljahresschrift 34(34): 57-83

Scharpf F (1997) Games real Actors Play. Westview Press, Boulder

The Nature Conservancy (2004) How we work., URL: http://nature.org/aboutus/howwework/, (15.1.2004)

The Partnership for Climate Action (2004) About PCA. Washington: Environmental Defence, URL: http://www.pca-online.org/about/, (15.1.2004)

Van den Hove S, Le Menestrel M, de Bettignies HC (2002) The oil industry and climate change. Strategies and ethical dilemmas, Climate Policy 2(1): 3-18

Vedung E (1999): Evaluation im öffentlichen Sektor. Böhlau Verlag, Wien, Köln, Graz

Emissions trading and effects on financial markets

Timo Busch

Research Fellow
Wuppertal Institute for Climate, Environment and Energy
Sustainable Production and Consumption Department

Swiss Federal Institute of Technology Zurich (ETH Zürich)
Department for Management, Technology, and Economics
Group for Sustainability and Technology
Kreuzplatz 5, KPL J 16, 8032 Zurich, Switzerland
tobusch@ethz.ch

Abstract

Climate change and its effects on business has become a focal discussion point in relation to corporate financial performance. As emissions trading is one of the closest and most self-evident influences on climate change, many companies have to face new financial constraints, especially in emissions-intensive sectors. However, these direct and indirect effects of emissions trading are not only affecting single companies and entire industry sectors. Due to the financial links to companies, there is also a strong linkage to financial markets; new business opportunities and challenges emerge. Furthermore, financial institutions can contribute to establishing and fostering emissions trading as a business case in general by a proactive involvement. Thus, all actors in financial markets should anticipate the business opportunities and assume a proactive role in supporting ET to on its path to success.

Keywords: Emissions trading, business opportunities in financial markets, role of financial institutions

1 Introduction

Climate change (CC) and its effects on business has become a focal discussion point in relation to corporate financial performance. In the international context, political institutions (e.g. UNEP, EU Commission) and scientific organisations (such as IPCC, WRI, GRI) as well as business associations (for example the "GHG Protocol Initiative" of WBCSD or "Development and Climate Change Project" of OECD) are coming to realise that CC is a vital topic for future economic development.

CC Risks caused by the impact of companies' activities (in a widespread sense) can be mitigated by an early assessment of climate related issues and the development of appropriate adaptation strategies. Furthermore, by presently engaging in this, companies are enabling a better position for generating new business opportunities and for outperforming their competitors in future markets. These influences comprise a wide range of different aspects, from those that will definitely affect the business (highly systematic risk) to others that could instead affect single sectors or companies (non-systematic risk with potential for diversification).

In order to fulfil the obligation to contribute to the Kyoto reduction aims, the European Union (EU) laid down the regulations for the European emissions trading (ET) system in 2002. This system is one of the closest and most self-evident influences of CC, which many companies have to face, especially in emissions-intensive sectors such as the oil/gas or automobile industry and the electronic sector.[1]

The focus of this paper is on ET and its effects on financial markets. Consequently, the first step illustrates the general interfaces of financial markets and sustainability. Secondly, the impact of ET on companies is discussed. Based on this, it is then shown to what extent current developments result in new business opportunities in financial markets. Finally, conclusions are drawn.

2 Interactions of financial markets and sustainability

Sustainability is considered more and more as a topic that influences companies' competitiveness and, consequently, their financial performance. Many different surveys and studies have been conducted regarding this issue. A detailed discussion of the results is not the subject of this paper; nevertheless, the prevalent point can be described as:

Sustainable development is becoming a decisive theme, notably in a long-term perspective. Considering the correlation of sustainability issues and financial exposure in ex-post analysis, sustainability assets do not automatically result in a better financial performance, but they have no clear disadvantage compared to conventional assets.[2]

---

[1] See Whittaker M, Kiernan M, Dickinson P (2003).

[2] Compare Schröder M (2003); Plinke E (2002).

This development is not only affecting single companies or entire industry sectors. Due to financial links to companies, financial markets generally interact with sustainability issues in a number of ways:[3]

- As investors – supplying the investment needed to achieve sustainable development;
- As innovators – developing new products and services in financial markets to encourage sustainable development, e.g. carbon funds;
- As valuers – pricing risks and estimating returns, from a sustainable perspective for companies and projects;
- As powerful stakeholders – influencing the management of companies as lenders, shareholders or members of supervisory boards of directors;
- And as victims of environmental change – e.g. huge weather related financial burdens for reinsurance companies.

Thus, sustainability related risks are also affecting banks and insurance companies as well as institutional and private investors. For example, risks related to CC are already affecting financial markets due to a negative effect on companies' performance, as is illustrated by Xstrata (FTSE 100 listed company):[4] In 2002 the Japanese government announced that they were considering a coal levy, which would be put into effect in Oct 2003. The shares of Xstrata – one large coal exporter to Japan – fell almost 10% due to the likely impact of such a move.

3 Emissions trading and effects at a company level

In April 2002 the European Council agreed to ratify the Kyoto Protocol (Resolution 2002/358/EG). One result is a community-wide emissions trading scheme which was established by 2005.[5] In 2004 Member States were required to submit a national allocation plan for two trading periods (First period: 2005-2007, second period: 2008-2012). These plans had to include detailed information regarding to what extent the industries, companies and facilities would be affected by Kyoto restrictions and mechanisms. The European ET directive covers four industries at the current stage: Energy, iron and steel, paper and pulp, and minerals. Plants above a defined size receive a fixed amount of allowances, whereby the amount will decrease due to tightening emission targets over time. In the case that the allocated amount is not sufficient, it is up to them to decide whether to improve efficiency by modernizing their production facilities or to purchase further allowances to allow excess emissions through permits.

As a matter of fact, this procedure influences companies' financial performance. It is an individual issue whether a single company or facility is affected in a

[3] Derived from Delphi/Ecological (edn) (1997) The Role of Financial Institutions in Achieving Sustainable Development, Report to the European Commission.
[4] Compare UNEP FI (edn) (2003), p. 12, Kiernan M (2002).
[5] See for further details of this chapter European Parliament, The Council (edn) (2003).

positive or negative way by greenhouse gas emission permits and emission-allowance trading. However, the results are monitored by a company's profit and loss account. To date many companies are not well prepared for the upcoming ET scheme and struggle to adapt their production processes due to the new constrains. In the end many fail to reduce their emission output in order to comply with limits imposed and they exceed the limit of their permits. For these companies it is necessary to acquire new allowances by purchase. Otherwise, they will be fined € 40 per ton of CO_2 equivalent during 2005-2007 and € 100 per ton during 2008-2012. If companies are unable to transfer the costs to the end-users, these costs will be as tangible as traditional costs and hence influence the economic performance, e.g., the free cash flow and the shareholder value.

In this context it also has to be mentioned that the Kyoto based flexible mechanisms are supposed to be set into action in addition to the already settled EU trading scheme.[6] The two project-based mechanisms, Joint Implementation (JI) and the Clean Development Mechanism (CDM), enable parties to meet part of their emission limits by reducing greenhouse gas emissions in other countries (emission credits).[7] Even though the detailed procedure is not yet clear, this will provide opportunities of financial advantage for companies when these reductions are possible at a lower cost than at home.

As a result, the understanding of ET as an impact factor on financial performance is gaining more of a mainstream perspective. In this context, for example, Standard & Poor's (S&P) stated in August 2003 that

- Implementation of the EU ET directive "will impose an additional level of debt and/or result in a reduction in the operating cash flows of many companies";
- The measures are "likely to have a significant impact on costs of production in the European power sector;"
- "Virtually all sectors [will] be affected by increased power costs" and
- Eventually, "these costs could have an impact on companies' business and financial profiles and, consequently on the ratings assigned to them."[8]

It is a crucial fact for capital markets that even organisations like S&P have begun to realise potential impacts of ET on companies, as it is demonstrated by the following example: In March 2003 Munich RE was rated down from AA+ to AA- by S&P due to a lack of capitalization.[9] As a result, the company decided on cash-raising exercises, which were followed by a drop of up to five percent in the stock market.

The ET scheme has drafted a new element of climate policy; the discussion on internalisation of external effects at a company level has shifted to a new dimension: A good management of external effects (focused on emissions) results for

---

[6] Compare Commission of the European Communities (edn) (2003) and Langrock T, Sterk W, Wiehler HA (2003).

[7] See for a detailed explanation of CDM and JI Bundesministerium für Umwelt, Naturschutz und Reaktorsicherheit/Umweltbundesamt (edn) (2003).

[8] Standard & Poor's (edn) (2003).

[9] See Reim M (2003).

the first time directly in extra profit/reduced costs and vice versa.[10] The ET scheme can be seen as one of the first tangible indicators that reveals certain industries, sectors, and single companies as having environmentally related risks. Companies have to realize that dealing with emissions in an irresponsible way is no longer considered only a problem of society in general, but serves as a reflection of their own business and management strategy.

As a conclusion, in terms of biasing business in a negative monetary way, the potential cost of ET/penalties might just be a minor consideration. From a financial perspective, it seems to be more crucial to scrutinize stakeholders' reactions to the expected announcements by national governments or the EU. Referring to the commission's directive "in order to ensure transparency, the public should have access to information relating to the allocation of allowances and to the results of monitoring of emissions" and further "member states shall ensure publications of the names of operators who are in breach of requirements to surrender sufficient allowances."[11]

This entails stakeholders' receiving decisive information about a company's emission strategies – at least when the company needs further emission allowances or there is a breach of rules. Thus, based on this information stakeholders can decide whether to support best practice or to reject certain company´s respective production methods. Furthermore, stakeholders could perceive this as an indication of a company's attempt to transfer the costs to the end-users.

However, the impact of ET-related issues on companies due to stakeholder reactions has not yet been evaluated. Nevertheless, it seems to be obvious that when crucial stakeholders[12] take into account environmental aspects (such as commitment and policy towards climate change), companies will have to tackle these issues. If companies do not develop a strategy for meeting the expectations and needs of stakeholders, the resulting indirect effects on their reputation and economic performance could be much more intense than the direct costs due to ET or penalties.

To sum up, ET has serious effects at a company level.[13] Beyond the obvious emissions-intensive sectors[14], the financial impact also is tangible for companies in the "financial services, transportation, semi-conductor, telecom, electronic equipment, food, agriculture, and tourism sectors."[15] Referring to CERES, an international coalition of investors, environmental and public interest groups, it is said that "companies with significant presence in Europe will have to address the issue earlier than others, but ultimately everyone will be affected."[16] However, the

---

[10] Compare Sandhövel A (2003), p. 40 f. for an estimation of impacts of CO_2 emissions trading on corporate values.

[11] European Parliament/The Council (edn) (2003), p. 5 and p. 20.

[12] In the context of ET these crucial stakeholders can be defined as consumers, suppliers and customers as well as investors, banks and insurance companies.

[13] See also Henderson Global Investors (edn) (2002).

[14] See Whittaker M, Kiernan M, Dickinson P (2003) and chapter one

[15] Whittaker M, Kiernan M, Dickinson P (2003).

[16] Investor Responsibility Research Center (edn) (2003).

divergences in corporate strategies[17] illustrate that many companies have not yet realized this trend and have minimum or no knowledge regarding upcoming opportunities.[18] It is in the company's best interest to acknowledge this and start to anticipate and mitigate the related risks through the following steps:[19]

- Collaborate in experimental programmes and gain experience in the field of ET, JI and CDM;
- Establish baselines for emissions and set targets for reductions;
- Increase energy and material efficiency, strengthen the use of clean and renewable energy, and minimise emissions of climate pollutants;
- Report progress so that investors and stakeholders can assess the amendments and consequently adjust their strategies and decision-making processes.

4 New business opportunities in financial markets

Both the direct as well as the indirect effects of ET, are impacting companies' financial performance. As previously discussed, there is always a strong link to financial markets: "[...] substantial emission reductions – like any other strategic global business challenge – ultimately become a financial issue."[20]

One main objective of participants in financial markets is to analyze and assess opportunities and risks of companies or corporate clients. Generally speaking, realising any new dimension of opportunities or risk factors as early as possible enables them to generate a better earnings/risk ratio for their portfolios. This provides a competitive advantage regardless of whether the focus is on loans, project finance, investment banking, insurance business, or asset management. And ET determines companies performance and, therefore, affects these areas.

This provides several options for new products and services within the financial service sector. Through the development of special ET-products, investment opportunities or new methods of quantifying and assessing ET-risks in various portfolios, banks, insurance companies, and investors can harness the positive and eliminate the negative effects of ET on companies. These new business opportunities in financial markets can be separated into different types of management functions, products, and services.

[17] Compare the findings of Cogan DG (2003).

[18] See Kreditanstalt für Wiederaufbau (edn) (2003).

[19] Compare Bergius S (2003) The steps are derived form Investor Responsibility Research Center (edn) (2003) and Henderson Global Investors (edn) (2002).

[20] United Nations Environmental Programme (edn) (2002), p. 2.

Table 1. New business opportunities in financial markets

Business opportunities of ET in financial markets		
management	products	services
impact assessment	fund/ asset management	Project certification and validity
risk control surveying	project finance	clearing function
brokerage	corporate finance/ lending	consulting/ advisory boards
research	insurance/ reinsurance	network building
	derivatives	

4.1 Management functions

Impact assessment. As previously discussed, the implications of ET affects companies as well as financial markets. The basic management function for participants in financial markets is to scrutinise and evaluate the different impacts. Financial institutions have to derive adequate strategies through a concise impact assessment for all relevant business areas. It is a challenge for management to extend and improve the existing methods of quantifying and assessing risks and opportunities in this dimension. Notably, an adapted impact assessment has to anticipate two factors: A precautionary action is required and the resulting yields are going to increase or decrease business benefits.[21]

Risk control surveying. The second step after assessing the impact is a steady survey of risk control. This should be applicable to all products, both in terms of an initial analysis of the underlying risk as well as a continuous evaluation thereof, until the expiration of the contracts. Financial institutions should incorporate this into their regular systems and client policies, thereby fully pricing in risks from ET (e.g. loan agreements or insurance premiums).

Brokerage. There is not a huge difference between ET and regular stock market trading.[22] Started in 2005, allowances are traded and consequently, there are a new market and the need for further brokers and traders.

Research. In addition to the previous point, it is urgent both that brokers and traders are familiar with the latest trends and developments, and that analysts and investment bankers are capable of anticipating market risks and opportunities. As a condition, financial institutions have to enable them to make rational decisions

[21] Compare Henderson Global Investors (edn) (2002), p. 5.

[22] See Sandhövel A (2002), p. 46.

and provide the needed research information. This can be considered as a new business field insofar as there is currently no concise information tool/system available and, therefore, a huge market potential for specialized research organisations exists.

4.2 New products

Fund/asset management. In terms of new products for fund and asset management, two different types of funds can be distinguished. On the one hand, carbon funds are investment funds that are created by JI and CDM projects.[23] The aim of this investment strategy is to reduce the amount of allowances needed and to sell certificates on the market. The return of these sales can be seen as an additional cash flow into the regular earnings (of the project's revenue). On the other hand, investment companies can set up special mutual funds that are focused on shares of companies running their business in ET/carbon markets. These companies could be, for example, corporations in the field of low emission technologies[24] or renewable energy generation (e.g. wind or tidal), firms located in Non-Annex-I-Countries that are cooperating in light of the flexible mechanisms, or specialised trading agencies. It is also possible to initiate specialized funds that pursue the best-in-class approach due to ET related opportunities. In all of these cases the investment strategy is based on the assumption that the ET-market is going to be a lucrative business sector and, therefore, companies in this field are going to outperform the market.

Project finance. In this business area, single JI/CDM projects or emission reducing projects can be initiated and/or financed by a financial institution. These projects are mainly in the field of adaptation and mitigation. Also, already-existing projects have to be considered, as to whether they can be modified or extended to JI or CDM projects.[25] In both cases, the investment strategy is again to generate additional profits.

Corporate finance/lending. In corporate finance and lending activities, the focus is not on original ET investments or projects, but rather on the impact on companies that are affected by related threats and opportunities. For example, this impact could involve costs for necessary investments in reduction technologies or unplanned costs due to non-fulfilling of ET regulations. For banks, two business opportunities rise in this context for extending their product portfolios: By refinancing these expenditures, by granting loans for energy efficient investments and by developing special contract conditions for clients with low ET related risk.

Insurance/reinsurance. For insurers and re-insurers new risk factors are, of course, combined with new insurance options. In light of the ET scheme, these are mainly in the field of insurances for JI/CDM projects and increased demand for risk transfer. The latter is particularly visible in terms of insurances for production

[23] See Sandhövel A (2002), p. 46.
[24] See Sustainable Asset Management (edn) (2002), p. 10.
[25] See Sandhövel A (2002), p. 46.

capacities. In such cases, companies expect that they might not have sufficient allowances or intent to extend their production. The insurance would cover the costs of new allowances. However, there is also another area that might become relevant in the future; companies could face lawsuits when they are not able to comply with the ET regulations over a certain time-period, or are fined due to violations. Thus, insurance companies could also develop special insurances for liability risks.

Derivatives. As well as futures, options and swaps are quite common in the different areas of capital markets; there is also going to be a demand for hedging due to the uncertainties of emissions, ET and related projects.[26] In these instances, financial institutions can provide hedging products (derivatives).

4.3 New services

Project certification and validity. As a first step, companies involved with the ET scheme have to evaluate their emissions exactly.[27] The result has to be validated. In addition, for determination of future related reductions or all realized reductions since 1990 this procedure is going to be applicable. Furthermore, the JI/CDM projects have to be certified and the results have to be monitored. When the so-called "Gold Standard" has become reality, there will be the necessity for experts to assess different types of projects. In all these areas, new business options arise for financial institutions.

Clearing function. In regular clearing-houses, checks and bills are exchanged among member banks so that only the net balances need to be settled in cash. This clearing function can be extended to transactions with emission allowances. Furthermore, new "emission clearing houses" can bring together seekers and providers of allowances and thus, match demand with supply.

Consulting/advisory boards. Beyond consulting and advising companies regarding risks and opportunities of ET in general, two major business opportunities can be identified: On the one hand, ET results in different competitive conditions within the EU due to the national allocation plans. Therefore, companies need support in terms of defining the most effective location where to invest. On the other hand, companies have to face new variables in the process of planning long-term investments. These variables will alter, for example, either regulations/conditions of ET over time or opportunity costs of potential alternative (more efficient) technologies. In both cases, a new dimension of services is rising for financial institutions: First of all, excellent consulting and advertising give top-quality support to existing clients. Furthermore, it enables them to set up a completely new branch of investment banking and consulting, thereby attracting new clients.

---

[26] See Lafeld S, Sandhövel A (2003).

[27] Compare for detailed information on this - Bundesministerium für Umwelt, Naturschutz und Reaktorsicherheit/Umweltbundesamt (edn) (2003).

Network building. The new ET market with all its possibilities and huge variety of different investment strategies is still immature and cannot be opened up to single players in an effective and efficient manner. The expertise in managing technology and investments varies enormously. Thus, it is in participants' best interests to build national and global alliances and networks. These alliances and networks provide essential information about the market and, therefore, contribute to generating the basis for best solutions, for their own, and for their clients' strategies.

5 Current developments and a framework for pro-active involvement

According to a report of Innovest and UNEP-FI the annual market volume of ET could be worth as much as US$ 2 trillion by 2012.[28] This demonstrates that there is a surge of new business opportunities, but even so, up to now the entire business concerning "ET and Effects on Financial Markets" is almost unknown. There is now the challenge for financial institutions to develop new products and services, which meet these future demands of the market: "Yet opportunities are emerging that should allow the financial services industry to reduce or hedge against the risks and even help to curb emissions of the greenhouse gases […]."[29]

Even before the ET start in 2005, initial steps were conducted by the launch of the "Hesse Tender"[30] project in Germany. A group of various participants (including financial institutions such as Dresdner Bank AG, Deutsche Ausgleichsbank (DtA) and xlaunch/Gruppe Deutsche Börse) set up a CO_2-trading simulation program, which tried to create a realistic market-frame for an emissions trading concept. The project's results prove that ET is actually working and that there is a new market with new business opportunities, for both companies and financial institutions:[31] Six companies committed to reducing their emissions by about 1.3 mill. tonnes from 2005 to 2009. Combined with a price of € 6.58 per tonne at the latest auction in December 2002, this comes to a potential trading volume of € 8.6 mill.

Another best practice example is the project "Implementing the Kyoto-Mechanisms: Contributions by Banks and Insurance Companies."[32] The project focused on the development of innovative financial products. The objective was to meet the needs of prospective users like financial institutions, investors, companies and other participants in the ET system. In the project two financial institutions were also involved, Gerling (insurance company, Germany) and San Paolo

[28] See Innovest Strategic Value Advisors/UNEP-FI (edn) (2002).
[29] United Nations Environmental Programme (edn) (2002).
[30] See http://www.dresdner-bank.com/meta/kontakt/03_dresdner_bank/06_nachhaltigkeits bericht/HesseTender_ExecutiveSummary.pdf.
[31] See Kreditanstalt für Wiederaufbau (edn) (2003).
[32] See http://www.iwoe.unisg.ch/IMKYM-COFIN.

Imi (universal bank, Italy). One result of the investigation is the fact that Gerling decided to withdraw from investments in carbon sensitive sectors and companies in 2003.

To continue this development and to broaden the perspective on ET as a business field, there is need for a framework that provides a proactive involvement for financial institutions. This framework can be described by internal and external conditions which determine the further integration of the topic ET in the financial community. Furthermore, it focuses on how financial institutions can contribute towards establishing and fostering ET as a business in general steered by a proactive engagement process.

Table 2. Framework for proactive involvement

financial institutions: framework for proactive involvement

internal conditions	engagement	external conditions
commitment/ policy	capacity building	policy support/ reliability
adapted risk approach	information tools	market transparency
education/ qualification	shareholder activism	global competitiveness
knowledge management	awareness building	non trading boundaries

5.1 Internal conditions

Commitment/policy. A financial institution's commitment and policy towards the issue of climate change and ET is the fundamental point for the overall framework and the further aspects. Where internal criteria allow, banks and insurers should only agree to engage (in terms of investments, loans etc.) in companies from carbon intensive sectors that can demonstrate leadership or best practice in their approach to climate change and ET. Otherwise, financial institutions should reject non-responsible practises and focus on non-carbon intensive business possibilities. This strategy should also be extended to suppliers and customers.

Adapted risk approach. As mentioned before, it is in the financial institutions' own interests to acknowledge the need to adapt and extend existing methods of risk determination and evaluation.[33] As one main internal condition, management

[33] Compare also Sandhövel A. (2003), p. 41.

has to anticipate the new circumstances and commit to a pro-active risk approach. Therefore, they have to be aware of two major facts that are apparent: On the one hand, science is clear in determining that climate change has become a reality.[34] One the other hand, the policy framework is in place and is definitely going to affect companies' financial performance.

Education/qualification. One important condition for financial institutions to utilize ET and generate business opportunities is the education and qualification of employees. Banks and insurers have to ensure that the relevant employers (such as brokers, asset managers, risk analysts) have sufficient experience and knowledge to make responsible and sensible decisions regarding ET related risks and opportunities.

Knowledge management. A functioning knowledge management is the baseline for the utilisation of required skills and experiences, especially when it comes to complex structured figures such as CC and ET.[35] This can be considered the internal counter part to the earlier discussed aspect of alliances and networks.

5.2 Engagement

Capacity building. The most effective means of supporting a wide spread involvement of market participants in this framework can be achieved through the creation and application of financial incentives. In this context, capacity building is meant to be an engagement process wherein financial institutions try to broaden, at the company level, the insights into ET related impact. Financial institutions should develop an engagement strategy that encourages companies to consider their individual position and manage derivable risks. This could be done, for example, by monetary incentives in loan conditions as a result of improved eco-efficiency or new ET-contracts. If financial institutions grant better contract conditions due to these aspects, more companies will see the benefit in committing to pro-active involvement.

Information tools. Corporate as well as private clients do not seem to encourage financial institutions to interact in the field of sustainability at all, nor do they require that information on this topic be submitted to them. At least this is the impression one gets in light of the current development of the social responsible investment (SRI) market: One reason that SRI is still a niche market is the fact that SRI opportunities are almost unknown. Thus, it is up to financial institutions to provide their clients with detailed information regarding the impact of ET. Different information tools and discussion methods/platforms are needed for this (e.g. the carbon disclosure project). These are special tools that deliver specific information for investors (e.g. about the risk/return ratio of carbon projects) and tools, which companies can use to build different scenarios in terms of their own ET

[34] See Intergovernmental Panel on Climate Change (edn) (2002) or United Nations Environmental Programme/Climate Change Secretariat (edn) (2002).

[35] See Busch T and Orbach T (2003), p. 32.

strategy (e.g. to decide, from a financial perspective, whether to invest in new production systems or to engage in a joint implementation project).

Shareholder activism. Asset managers of banks and insurance companies as well as private and institutional investors (e.g. pension funds) should request information about the ET strategy and attitude of companies they invest in. This should consist of detailed information about how the sectors in general and single factories in particular are affected by ET. Furthermore, they should demand that management declares how the company is going to tackle these issues and what the coping strategies towards the new demands and expectations look like.

Awareness building. This last aspect of how financial institutions can engage in the process of fostering an ET framework refers to their role in building public awareness. The majority of society is, to date, not definitely sure what ET actually means. Moreover, people are not aware that this topic is also referring to likely risks and opportunities for their own investments or, more generally speaking, what the impact of climate change is about. Financial institutions should help boost understanding of these impacts, risks and opportunities (e.g. by specific advertisements).

5.3 External conditions

Policy support/reliability. In terms of supporting the described framework, national governments are being urged to adopt a durable ET plan at an international level. At the EU level it is crucial that regulations on JI and CDM projects are decided and implemented as soon as possible. Reliability and certainty of the regulations and laws is a vital condition for markets functioning in an effective way. Without this, there is no secure foundation for investments and corporate decisions and the success of ET as a means of reducing emissions is endangered. Furthermore, policy can support fostering and extending the ET market through tax incentives. For example, one option could be to reduce income taxes for special carbon funds.

Market transparency. Transparency in capital markets concerning opportunities, risks, conditions, expected returns, etc. is essential for all participants. Only transparent markets can enable efficient and effective investment decisions, both for companies and for investors. It is in the best interest of all participants and policy makers to be eager in ensuring a transparent ET market in terms of fulfilling the overall EU/Kyoto goal. In this context, governments again play an important role: They should enforce regulations that require companies to disclose financial-impact related to ET and companies' strategies for addressing this issue.[36]

Global competitiveness. The ET market still has to mature and gain transparency. One central element in building up the discussed framework is competitiveness in national and global markets. Broad participation in the ET market and easy access to various sources of information and technology must be a top priority. For establishing an effective ET market, it has to be ensured that all participants

[36] See Cogan DG (2003), p. iii.

are sufficiently informed and are able to determine their decisions under competitive conditions.

Non-trading boundaries. In terms of a successful global realization of the flexible Kyoto mechanisms, it is also necessary that investments in JI and especially CDM projects can be made without different contract conditions or obligations. These could mainly result from trading boundaries such as different international taxes, local subsidiaries, or limitation of maximum investment per country etc.

6 Conclusions

The topic emissions trading is supposed to move from matters that are not subject to management decisions, to matters that need to be addressed both by companies and financial markets due to their inherent financial relevance.

To act and interact in these new markets and to develop new methods of quantifying and assessing risks is an important and challenging task. The two examples in chapter five, illustrate that some players on financial markets are aware of the positive and negative effects ET has on their business. Notably, it is now the proper time for action and for generation of future business opportunities. And by doing so, participants in financial markets will increase the contribution of financial markets to a sustainable development, and they will be able to utilize this development for optimising their own risk/earnings ratio.

Obviously, not all financial institutions are well prepared, and it is an incorrect assumption to remain complacent when time is a crucial component due to emerging risks. Thus, all financial institutions should anticipate the business opportunities and assume a pro-active role in supporting ET on its road to success by[37]

- Generating and spreading out an efficient and transparent market for ET products and services;
- Incorporating carbon risks and ET considerations into all business processes as well as risk assessments and the development of new financial tools;
- Encouraging best practice and supporting requests for greater disclosure;
- Broadening the discussion with industries, fund managers, the public, and policy-makers;
- Cooperating with other participants involved in ET markets and seeking expert science and policy advice;
- Identifying and assessing new paths for accelerated development of eco-efficient and low emission technologies.

[37] Derived from Dlugolecki A (2003) and Cogan DG (2003).

References

Bergius S (2003): Handeln für die Umwelt - Erwerb und Verkauf von Emissionsrechten für CO_2 startet in Europa schon 2005, in: Handelsblatt, 23.4.2003, p. B1

Bundesministerium für Umwelt, Naturschutz und Reaktorsicherheit/Umweltbundesamt (ed) (2003) Leitfaden für klimapolitische Bewertung von emissionsbezogenen JI- und CDM-Projekten, Berlin

Busch T, Orbach T (2003) Zukunftsfähiger Finanzsektor - Die Nachhaltigkeitsleistung von Banken und Versicherungen, Wuppertal Paper 129, http://www.wupperinst.org/Publikationen/WP/WP129.pdf

Cogan DG (2003) Corporate Governance and Climate Change - Making the Connection, CERES/IRRS-Report, Washington/Boston, http://www.ceres.org, http://www.irrc.org

Commission of the European Communities (ed) (2003) Proposal for a Directive of the European Parliament and the Council, amending the Directive establishing a scheme for greenhouse gas emission allowance trading within the Community, in respect of the Kyoto Protocol's project mechanisms, COM (2003) 403

Dlugolecki A (2003) Wenn nichts geschieht, droht der Kollaps - Der Klimawandel kann den finanziellen Ruin bedeuten, in: Versicherungswirtschaft, 58. Jahrgang, Januar 2003, Heft 1: 6-10

European Parliament/The Council (ed) (2003) Directive of the European Parliament and the Council establishing a scheme for greenhouse gas emissions allowance trading within the Community and amending Council Directive 96/61/EC, PE-CONS 3659/03

Henderson Global Investors (ed) (2002) Socially Responsible Investment - Climate Change Position Paper, Internet: http://www2.henderson.com/sri/index.asp

Innovest Strategic Value Advisors/UNEP-FI (ed) (2002) Climate Change and the Financial Services Industry,http://www.unepfi.net

Intergovernmental Panel on Climate Change (ed) (2002) Climate Change 2001 - The Scientific Basis, Geneva

Investor Responsibility Research Center (ed) (2003) Top Greenhouse Gas Emitters in Oil, Utility and Auto Industries Not Disclosing, Action on Financial Risks of Climate Change, press release 9-Jul-03, http://www.irrc.org

Kiernan M (2002) Taking control of climate, in Financial Times, 25/11/2002

Kreditanstalt für Wiederaufbau (ed) (2003) Emissionshandel - ein neuer Markt entsteht, Presseerklärung, http://www.kfw.de

Lafeld S, Sandhövel A (2003) Kyoto's Impact on Risk Management, http://www. dresdner-bank.de/meta/kontakt/03_dresdner_bank/06_nachhaltigkeitsbericht/Allianz_Risk_Management_kurz.pdf

Langrock T, Sterk W, Wiehler HA (2003) Akteursorientierter Diskussionsprozess "Senken und CDM/JI", Endbericht, Wuppertal Spezial 29, Wuppertal

Plinke E (2002) Aktienperformance und Nachhaltigkeit - Hat die Umwelt- und Sozialperformance einen Einfluss auf die Aktienperformance? Studie der Bank Sarasin & Cie AG, Basel

Reim M (2003) Gerüchte um Kapitalerhöhung der Münchener Rück - Rating-Agentur setzt weltgrößten Rückversicherer unter Druck. In: Süddeutsche Zeitung vom 27.08.2003, Internet: http://www.sueddeutsche.de

Sandhövel A (2002) Geld verdienen mit CO_2-Zertifikaten? Emissionshandel aus Finanzdienstleistersicht. In: Elektrizitätswirtschaft, Jahrgang 101, Heft 14: 44-47

Sandhövel A (2003) Emissionshandel aus Bankensicht - Strategien, Chancen und Risiken für Unternehmen. In: UmweltWirtschaftsForum, 11. Jahrgang, Heft 3: 39-43

Schröder M (2003) Socially Responsible Investments in Germany, Switzerland and the United Staates - An Analysis of Investment Funds and Indices, ZEW Discussion Paper, Mannheim

Standard & Poor's (ed) (2003) Emissions Trading: Carbon will become a taxing Issue for European utilities, publication date: 21-Aug-03, http://www.standardandpoors.com

Sustainable Asset Management (ed) (2002) Changing Climate in the Energy Sector - a new Wave of Sustainable Investment Opportunities Emerges, http://www.sam-group.com

United Nations Environmental Programme (ed) (2002): Financial Sector, Governments and Business must act on Climate Change or face the consequences, UNEP News Release, http://www.innovestgroup/pdf/UNEP_10_08_02.pdf

United Nations Environmental Programme/Climate Change Secretariat (ed) (2002): Climate Change - Information kit, Zimbabwe

United Nations Environmental Programme Finance Initiative (ed) (2003) 0.618 ... The golden ratio - building quantity without sacrificing quality, issue 3 Jan. 2003, http://www.unepfi.net

Whittaker M, Kiernan M, Dickinson P (2003) Carbon Disclosure Project - Carbon Finance and the Global Equity Markets, Innovest (ed), http://www.cdproject.net

Part D

Effects of emissions trading schemes existing and being implemented

The EU emissions trading scheme and its competitiveness effects upon European business – results from the CGE model DART

Sonja Peterson

Kiel Institute for World Economics
Department of Environmental and Resource Economics
Duesternbrooker Weg 120, 24105 Kiel, Germany
speterson@ifw-kiel.de

Abstract

The EU emissions trading scheme will not only affect the cost structures and competitiveness of the sectors covered directly by the trading scheme, but – through changes in energy demand and thus energy prices – will also have repercussions on the entire EU market. As Europe is closely tied to the rest of the world by international trade, it will also change the position of European business on the world markets and is likely to influence international energy prices as well.

So far, research has been mainly confined to a qualitative analysis of the EU. Few studies – mainly focussing on Kyoto trading and not the EU scheme – have tried to give quantitative results. Other EU specific studies estimate abatement costs of emission reductions and prices, but ignore international effects. Against this background, the aim of this study is to assess the range of possible implications of the trading scheme using the computable general equilibrium model DART that accounts for European and international linkages. As a result, it is possible to estimate the direct economic costs for the European business as well as the trade and competitiveness effects in a globalizing world.

Keywords: Emissions trading, EU, competitiveness, computable general equilibrium model

1 Introduction

The EU-wide trading scheme for CO_2 emissions (further on denoted as ETS) started in January 2005. The EU directive requires each member state to impose binding, absolute caps on emissions of facilities in energy activities, the production and processing of ferrous and non-ferrous metals, the mineral industry and the pulp, paper and board production. Altogether, around 10,000 industrial sites, responsible for about 45% of the EU CO_2 emissions, will be allocated emission allowances and be able to trade emissions from 2005 onwards. Countries are free to include other sectors. The first trading period until 2007 is seen as a test for the second trading period from 2008-12 that is also the first commitment period of the Kyoto Protocol. For a summary of the EU directive, see Gagelmann and Hansjürgens (2002).

It is clear that this single largest market for emissions will affect the cost structures and competitiveness not only of the sectors covered directly by the ETS but – through changes in energy demand and thus energy prices – will have repercussions on the whole EU market. As Europe is closely tied to the rest of the world by international trade and as the ETS covers sectors where international competition is strong, it is especially important to account for international effects: The ETS will change the position of European business, whose main players are in fear of losing competitiveness in the world markets due to higher energy prices. In addition, international linkages imply that the effects of the EU ETS for European business depend on climate policies in other major economies, e.g. the USA, which may or may not undertake emissions reductions and participate in international Kyoto emissions trading.

Many existing studies only estimate abatement costs of emission reductions and permit prices for different emissions trading regimes, ignoring international effects (e.g. Capros and Mantzos 2000a,b; Capros et al. 2002). General equilibrium studies on emissions trading in the EU either focus on Kyoto trading (Viguier et al. 2003; Zhang 2002) or analyse scenarios that are not very close to the trading regime as it is outlined in the EU directive (e.g. Böhringer 2002). Even though the details of the ETS and especially the amount of emissions allocated to it are still not entirely clear, it is now possible to simulate the ETS more realistically. The aim of this paper is to assess the range of possible competitiveness effects of the ETS for European business accounting for European and international linkages. Section 2 defines competitiveness and analyses the different international effects. Section 3 and 4 then attempt to assess the implications of the ETS quantitatively, using the computable general equilibrium (CGE) model DART. DART is designed for analysis of international climate policies and is especially able to capture general equilibrium and international effects on energy prices as well as on the different sectors in the different regions. Section 5 concludes.

2 Emissions trading and competitiveness in a globalizing world

The impact of environmental regulation (including emission targets) on competitiveness has received considerable attention in the last few years (Jenkins 1998). In the context of emissions trading and environmental policy in general, the discussion centres on so-called international industry level competitiveness, which is best defined in relation to the respective industry in other countries. This competitiveness consists of the ability of specific industries to compete for market share with businesses located in other countries, which affects the location of production across countries (Sinner 2002). Thus, it is usually related to performance in international trade and in trade theory; this kind of competitiveness is denoted comparative advantage. The main concern with environmental policies is that production may be shifted to countries with less strict environmental standards (Sinner 2002). The effect that energy-intensive industries, due to emission restrictions within a certain region, move to non-abating regions with increasing emissions is denoted 'leakage effect'.

Measures of industry level competitiveness are usually based on total or net exports (= exports minus imports) and then adequately normalized (Jenkins 2002). They are generally referred to as Revealed Comparative Advantage (RCA) and there are different RCA indices (see for example Balance 1988). For this paper the RCA of commodity X is defined as the logarithm of the ratio of the value of exports to the value of imports of commodity X normalized/divided by the ratio of the value of all exports in one country to the value of all its imports. Other relevant variables for industry level competitiveness effects are sectoral output, sectoral exports, market shares and gross and net energy prices (Klepper 2003).

Turning to the effects of the European ETS on industry level competitiveness, the driving forces are the energy prices. Compared to the business as usual (BAU) scenario without any European emission restrictions, the ETS increases gross energy prices in the sectors covered by the ETS and implies a change in relative prices of the goods produced. The substitution effect decreases demand for energy and for energy intensive goods. This lowers net energy prices, not only in Europe, but since Europe is responsible for a considerable share of world energy demand, in the rest of the world as well. On the other hand, sectors not covered by the ETS face higher prices for intermediate inputs from sectors subject to emission constraints. In addition, emission restrictions decrease European income (income effect) leading to lower domestic and import demand. Via the same mechanisms, the policy of the remaining countries can influence the European economies. In fact, the competitiveness effects of emissions trading schemes in general depend largely on the coverage of the scheme and the extent to which foreign governments follow a similar policy for regulating emissions (Klepper 2003). As a result, the magnitude and – in sectors not covered by the ETS and in the non-EU countries – also the overall direction of the effects depends to a large degree on the magnitude of reductions in regions outside the European ETS. Summarized, a multilateral emissions trading regime, like the EU scheme, leads to the following:

a restructuring of demand and production towards less energy-intensive goods in the participating countries, an overall reduction in production and a shift of energy-intensive production to the non-participating countries. At the same time, even though all firms face the same permit price in the trading scheme, the effect on industries in different countries differs since the CO_2-intensity of economic sectors may vary widely between economies (Klepper 2003).

Naturally, these effects become stronger with increasing emission restrictions. For this reason the number of permits allocated to the ETS in each European country makes a big difference. In addition, the ETS is only one part of the European strategy towards reaching its Kyoto target. If and how this overall target is reached – in the sectors covered by the ETS as well as in the sectors outside – has thus a large impact as well. Yet, what will happen, particularly in the sectors not covered by the ETS is quite unclear.

Another important issue is the choice of the counterfactual (Klepper 2003). Evaluating the EU ETS always implies a comparison to some other scenario. This counterfactual scenario is usually a BAU scenario in which there are no additional climate policy measures that go beyond the current state of regulation. Against a BAU scenario the European ETS induces economic costs. An alternative approach is to assess the efficiency gains of the ETS by comparing it to a policy alternative that achieves the same emission target but with a different policy instrument. Compared to unilateral action, for example, a European ETS is less costly.

Existing empirical studies are mostly concerned with environmental stringency of taxes but not explicitly with emissions trading. Most studies cannot find significant competitiveness effects of conventional environmental expenditures of firms, and the competitiveness of an industry rather depends on general conditions such as the trade intensity, labour costs, exchange rates and innovation (Shin 2003; Barker and Johnstone 1998). On the other hand, competitiveness effects may become more important as it is not only likely that future climate polices induce higher economic cost than past measures but also they are pervasive throughout the economy and require a fundamental restructuring of the economy (Barker and Johnstone 1998).

In this paper, the computable general equilibrium model DART will be used to quantify the different effects of the European ETS under different reduction scenarios.

3 Simulation of competitiveness effects of the EU emissions trading scheme

3.1 The DART model

The DART (**D**ynamic **A**pplied **R**egional **T**rade) Model is a multi-region, multi-sector recursive dynamic CGE model of the world economy. Here, it is used with an aggregation covering 17 regions and 12 sectors (which are summarized in Table 1) and the production factors labour, capital, land and natural resources. The economic structure is fully specified for each region and covers production and final consumption. Each market is perfectly competitive. Output and factor prices are fully flexible. For each region, the model incorporates two types of agents: producers distinguished by production sector and the final consumer, which comprises a representative, household and the government.

The DART model is recursive-dynamic, and thus solves a sequence of static one-period equilibria for future time periods. The major driving exogenous factors of the model dynamics are population change, the rate of labour productivity growth, the change in human capital, the savings rate, the gross rate of return on capital, and thus the endogenous rate of capital accumulation. The savings behaviour of regional households is characterized by a constant savings rate over time. The static part of the DART-Model is currently calibrated to the GTAP5 database that represents global production and trade data for the year 1997. In addition, the elasticities of substitution for the energy intermediate goods coal, gas and crude oil are chosen to reproduce the emission projections of the EIA (EIA 2002). For a more detailed description of the DART-Model, see Springer (2002), Klepper et al. (2003).

Table 1. Dimensions of the DART-Model

Country/Region		Production Sectors	
GBR	United Kingdom	Energy	
DEU	Germany	COL	Coal
FRA	France	GAS	Natural Gas
ITA	Italy	CRU	Crude Oil
SCA	EU-Scandinavia	OIL	Refined Oil Products
BEN	Benelux	EGW	Electricity
SEU	Spain, Portugal, Greece		
EEU	Eastern Europe	Non Energy	
REU	Rest of EU	IMS	Iron Metal Steal
EFT	Norway, Iceland, Switzerland	PPP	Pulp & Paper Products
USA	USA	CEP	Chemicals Products
FSU	Former Soviet Union	AGR	Agricultural Production
PAO	Rest Annex B	TRN	Transport Industries
MEA	Middle East, North Africa	MOB	Transportation
CPA	China, Hong-Kong	Y	Other manufactures & services
IND	India		
ROW	Rest of the World	CDG	Savings Good

3.2 Formulation of policy scenarios

As the sectoral definitions in GTAP are hard to translate to those of the EU directive, and as it is not possible to separate installations that fall below the capacity thresholds, the European ETS is approximated by including the sectors IMS, PPP, OIL, CRU and EGW (see Table 1). They cover in the model about 45% of total

EU emissions, which is close to the estimated 46% of the actual trading regime. As discussed above, there are a number of conceivable scenarios that differ with respect to the countries that undertake emission reductions, the emission targets and how the targets are reached. In this paper, the analysis has to be confined to the comparison of only a few scenarios that highlight possible outcomes.

Besides the BAU scenario with no emission reductions, there are three scenarios of EU emissions trading. In the first, most lax scenario (ET-0), the European countries only restrict their emissions in the sectors covered by the ETS. In a second scenario (ET-K), all European regions also control their emissions in the sectors not covered by the directive to reach their overall Kyoto targets. This is intended to highlight the role of emission reductions in other sectors. While there are no emission reductions in the remaining countries in these first two scenarios, in the third scenario (AXB) the other Annex B countries also reach their Kyoto target. As it is not yet clear how Kyoto trading can be linked to the EU trading regime, it is assumed that the targets are reached unilaterally with unilateral CO_2-taxes.

The CO_2 targets in all scenarios are chosen in line with the Kyoto targets as specified in the EU Burden Sharing Agreement, the Kyoto Protocol itself. Sinks are not accounted for. For scenario ET-0, the target for the sectors included in the ETS is calculated by multiplying the sectoral emission share for DART's calibration year 1997 by the Kyoto target. This so-called historical approach is advocated by the European Commission (2003) as one possibility for defining targets. For the Eastern European countries, it is assumed that none of their excess emissions ("hot-air") is available in the ETS and the target is equal to the benchmark emissions. The question for scenario ET-K is how to define the sectoral targets outside the ETS. Some studies assume equal percentage reductions in all sectors. This leads to unrealistic high abatement costs in some sectors including the mobility sector and the gas sector. For this reason a least cost approach – also advocated by the EU Commission (2003) – is assumed for all European regions. To generate the least-cost targets, DART is first run under the EU Kyoto constraints with internal emissions trading in every European region. The resulting sectoral emissions are then used to set the targets in and outside the ETS. Figure 1 shows the emissions targets in the different scenarios compared to the BAU emissions. As the sectors outside the ETS have comparable high marginal abatement costs relative to those in the ETS, the least-cost emission allocation to the EU ETS used in the scenarios ET-K and AXB implies stricter emission targets for the sectors inside the ETS than the historical approach used in scenario ET-0. Altogether, 5% fewer emission rights are allocated to the EU ETS under the least-cost allocation than under the historical approach. Figure 1 also shows that, because of the Burden Sharing Agreement, the ETS sectors in Germany, Great Britain and the Benelux countries face the largest absolute emission reductions.

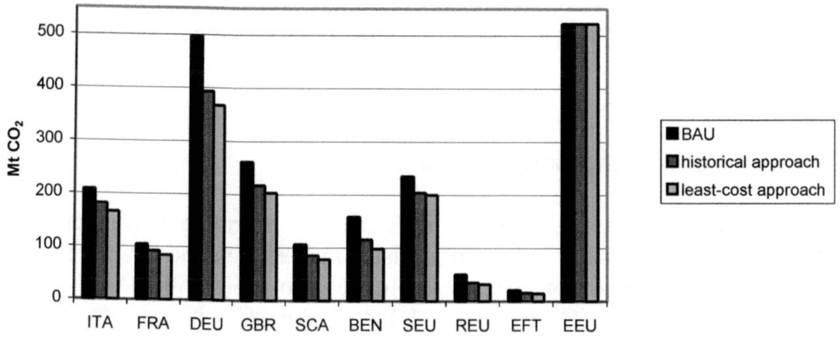

Fig. 1. BAU emissions and emission targets of the ETS sectors

In the sectors outside the ETS the targets are reached by a uniform regionally differentiated CO_2-tax[1]. The same assumptions are used for scenario AXB. Emission reductions always start in 2005 and continue linearly until the target is reached in 2012. Table 2 summarizes all scenarios[2].

Table 2. Policy scenarios analysed with DART

BAU	Business as usual; no emission reductions
ET-0	EU emissions trading in the sectors that are covered by the EU directive. The sectors not covered by the directive do not face any emission targets. The targets are calculated using the historical approach. No hot-air is included in the ETS. There are no emission restrictions in the remaining, non-European countries.
ET-K	The same as ET-0, but the sectors not covered by the EU directive also face emission targets that are achieved by uniform regional CO_2-taxes. The targets are calculated using the least cost approach.
AXB	The same as ET-K, but now the other Annex B regions restrict their emissions to reach their Kyoto target by means of a uniform (regional) CO_2-tax.

[1] This means that each region in the DART model individually sets a tax that is the same for all sectors outside the ETS in this DART region.

[2] Note that the BAU scenario does not account for measures to reduce CO_2 emissions that were implemented after the calibration year 1997, such as the German Ecotax. Including these measures would reduce the difference between the BAU scenario and any of the three ETS scenarios.

4 Simulation results

All results that are reported are for the year 2012, the end of the second trading period.

4.1 Permit prices and permit trade

One of the major outcomes of the EU ETS that will determine its competitiveness effects and that is of major interest to decision makers is the permit price. Current estimates vary between 5 and 30 € /tCO_2. In the simulations with DART[3] the price turns out to be 8.5 € t/CO_2 in scenario ET-0, 12.5 € t/CO_2 in scenario ET-K and 14.5 € t/CO_2 in scenario AXB. The first reason for this variation is that the emission target in scenarios ET-K and AXB is stricter than in scenario ET-0 (see Figure 1). The second reason originates from the differences in energy prices in the scenarios. The higher the emission restrictions in and outside the EU, the lower the world energy demand and the lower the energy prices in and outside the EU (see Appendix). Other things equal, lower energy prices imply an increased energy demand and increased emissions in the ETS sectors. Thus, the same target can only be achieved with increased shadow prices. As overall emission restrictions increase by moving from one scenario ET-0 to ET-K to AXB, so do permit prices.

Turning to permit trades, the simulation results show that the discussion concerning who is selling and who is buying permits in Western Europe becomes entirely redundant when the Eastern European countries join the ETS. Even though there is no hot-air allowed in the scenarios, Eastern Europe is basically the only seller of permits. Assuming that each sector receives permits in accordance with its historical emission shares, that is, the least-cost emission shares, Figure 2 displays the resulting net permit purchases in the different regions and sectors across the three scenarios. Eastern Europe (EEU), which is not included in the figure, sells the remaining emission rights.

What happens in Eastern Europe is that the highly emission-intensive electricity sector cuts down its production by 15 to 20% relative to BAU, depending on the scenario. Instead, electricity imports increase by 30 to 55% relative to BAU. Some of the electricity emissions are thus shifted to other countries. The resulting disposable permits of the electricity sector amount to almost 90% of total permit sales of Eastern Europe. To a smaller degree, the same happens in the other energy intensive sectors inside the ETS: they cut down production in order to reduce their emissions and sell permits, and domestic supplies are substituted by imports. Because of hot-air, the sectors outside the ETS do not face any emission restrictions in Eastern Europe, the factors previously employed in the ETS now move to the sectors outside the ETS where output increases.

[3] Assuming an exchange rate of 1.0971 US$ in 1997 = 1 € in 2000.

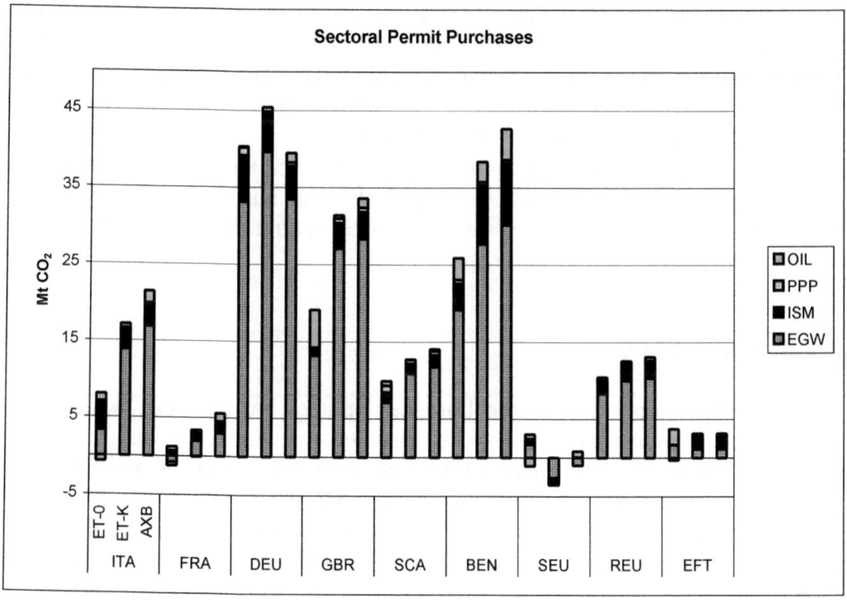

Fig. 2. Sectoral permit purchases in $MtCO_2$ in 2012

4.2 Scenario ET-0

The aim of the rather unrealistic scenario ET-0 with no emission reductions outside the ETS is to separate the "partial effects" of the ETS per se from those that include the reduction policies in the non-ETS sectors. The simulation results[4] show that the ETS as a stand-alone measure only has small macroeconomic effects. Welfare and GDP remain practically unaffected. Total output of the energy and energy intensive sectors taken together falls by less than 1% in all European regions except Eastern Europe. International competitiveness of EU business, as measured by the RCA (see Appendix), is hardly affected. As the European regions are mainly trading among themselves, the RCA in the sectors that are responsible for over 95% of exports (the non-energy sectors and in EFT also the crude oil sector) remains almost constant.

What is more interesting is that the indirect effects of the ETS that come through changes in gross energy prices and demand or prices of intermediate inputs tend to dominate the direct effects that are determined by whether a sector is covered by the ETS or not. In other words, the sectors covered by the ETS are not necessarily affected more considerably than the sectors not covered by the ETS.

[4] Results regarding sectoral output changes, the price of the energy aggregate and the RCA are reported in the Appendix. More detailed results e.g. regarding exports are available from the author upon request.

As coal is the most emission-intensive fossil fuel and is, moreover, mostly used for generating electricity, the coal sector exhibits the largest reductions in output, even though the sector is not covered directly by the ETS. As a result, coal output in the EU decreases by 6.5%. In the electricity sector, whose emissions are restricted, EU output decreases only by around 3%.

Another examples are the pulp and paper sector and the chemical sector. Both are energy intensive sectors, but the chemical sector outside the ETS is affected more considerably by the ETS (on average –0.3% output compared to BAU) than the pulp and paper sector (on average –0.2%).

4.3 Scenario ET-K

The EU ETS is only part of the European strategy for reaching the European Kyoto target. In scenario ET-K, it is assumed that the Kyoto targets will be reached efficiently by a regional CO_2-tax and that the reductions in and outside the ETS are determined in a cost-effective manner (least-cost allocation). The results in Table 3 show that the necessary CO_2-taxes are much higher than the permit price in all regions except Southern Europe. The reasons are that the sectors covered by the ETS are, in general, sectors with comparatively low abatement costs and that the ETS leads to an equalization of marginal abatement costs in the ETS sectors throughout Europe.

Table 3. Permit Prices and CO_2-tax in €/tCO_2 in scenario ET-K in 2012

Permits	Regional CO_2-tax								
in ETS	ITA	FRA	DEU	GBR	SCA	BEN	SEU	REU	EFT
12.5	22.8	16.6	21.6	26.4	25.6	46.8	11.2	48.9	46.9

Clearly, the sectors outside the EU ETS are thus affected more considerably than the sectors inside, which profit from the cheap abatement possibilities- especially in Eastern and Southern Europe. While the energy intensive sectors inside the ETS, for example, lose on average less than 1% in output compared to the BAU scenario, the chemical and mobility sector outside the ETS lose 2 to 2.5%.

4.4 Scenario AXB

Scenario AXB is again a stylised and rather unrealistic scenario (the USA, for example, will most likely not reach their Kyoto target in 2012), but it is intended to shed some light on the role of the non-European countries. The main effect of emission reductions in the other Annex B countries is that European energy prices decrease relative to the energy prices in the scenario with unilateral emission reductions. As a result, some production of energy intensive goods is shifted from the other Annex B countries to the EU, and European energy intensive sectors in scenario AXB gain in competitiveness compared to scenario ET-K. Looking at the RCA's (see Appendix) of the chemical sectors and the iron, metal, steel sectors, which are energy intensive, reveals that the major exporting regions such as Germany, Great Britain, the Benelux countries and the EFTA countries gain in competitiveness compared to scenario ET-K. For example, in the chemical sector in Great Britain, the RCA in scenario AXB is even higher than in the BAU scenario. The negative effects of the European emission restrictions are thus offset by the gains in competitiveness. The effects of scenario AXB can also be seen in Figure 3, which shows the relative changes in the combined output of the energy sectors and energy intensive sectors (IMS, PPP, CEP and MOB) in all three ETS scenarios. The decrease in output in scenario AXB is much less than in scenario ET-K. Some countries like France and the EFTA countries can even increase their output relative to the BAU scenario. Altogether, the EU is better off in scenario AXB than in scenario ET-K and profits from the emission reductions outside Europe.

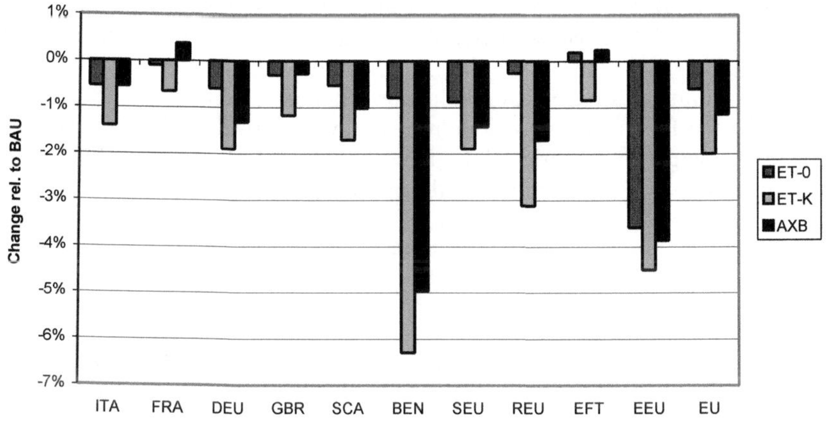

Fig. 3. Changes in total output of the energy and energy intensive sectors in 2012

4.5 Regional differences – the electricity and chemical sectors

What clearly matters for the strength of the effects in the different EU regions, are the emission intensities. Here, we take the electricity sector and the chemical sector as an example. Figure 4 shows the output changes in the three scenarios of these two sectors.

In the case of the electricity sector, the two most exceptional cases are those of France and the EFTA countries that include Norway and Iceland. In both regions electricity generation does not mainly depend on CO_2-intensive coal as in the rest of Europe, but in the case of Norway and Iceland it is generated by hydropower and in the case of France by nuclear power. As a result, the overall emission intensity of electricity is almost zero in EFT and still exceptionally low in France. As a result, the electricity sectors in both regions are not seriously affected by the EU ETS. They can even increase their output of low CO_2-intensive electricity within the more restricted overall world emissions. In scenario AXB, French electricity output is 1.7% higher than in the BAU scenario and electricity output in EFT even 7.3% higher. The electricity sectors in both regions are only buying low quantities of permits. Another special case is Eastern Europe. Here, emission intensity is still quite considerable in all sectors so that it is cheaper to reduce production and to sell permits (see also paragraph 4.1). It becomes clear though, that sectoral output is not always a good measure of competitiveness, as it ignores the fact that the sectors gain from selling their permits.

In the chemical sector, the large output loss in the Benelux countries is striking. The reason for this is the large CO_2-tax that is necessary to reach the overall Kyoto target in the Benelux countries combined with the comparatively high emission intensity which is, for example, about twice the emission intensity of the Scandinavian chemical sectors. Note also, that these results are based on the assumption of a regional uniform CO_2-tax in the sectors outside the ETS. Moving away from the least-cost scenario, the result would become even more pronounced.

Figure 4 illustrates once again the major differences between the three scenarios. While the ETS as a stand-alone measure has relatively small effects, output changes are larger in scenario ET-K. In scenario AXB, the energy intensive sectors might even gain from an improved competitiveness.

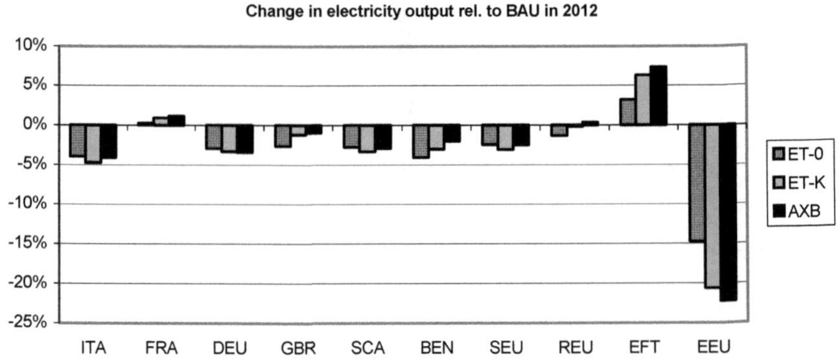

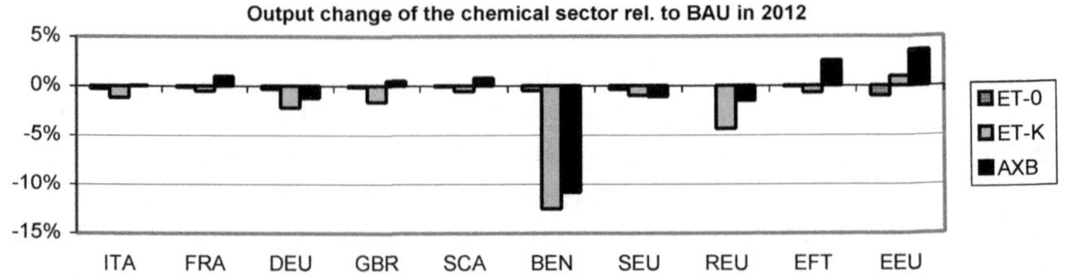

Fig. 4. Changes in the output of the electricity and chemical sectors relative to BAU

5 Summary and conclusions

The EU emissions trading scheme (ETS) has – directly or indirectly – repercussions on the whole EU market. This article discusses the likely competitiveness effects of the EU ETS for European Business taking into account the effects that occur due to European and international linkages. Major determinates of the strength and sometimes even the direction of the effects are (a) the amount of emissions that are allocated to the sectors inside the ETS, (b) the treatment, i.e. the magnitude of reductions in the sectors not covered by the ETS and (c) the magnitude of emissions reductions in the rest of the world. For a specific EU country, the strength of the effects is also determined by the sectoral emission intensities and, especially, the emission intensity of the electricity sector.

To quantify the effects, the computable general equilibrium model DART was used. DART covers the major European regions and 10 other major world regions. The 11 energy and production sectors include the sectors covered by the EU ETS. Besides the business as usual scenario (BAU), three emissions trading scenarios were analysed that differ in the dimensions listed above. In a first scenario, there are no emission reductions in sectors or regions outside the EU ETS. The amount of emission rights allocated to the ETS is calculated by multiplying each region's Kyoto target with the historical (year 1997) emission share of the sectors included in the ETS. In a second scenario, the sectors outside the ETS are taxed with a regionally uniform CO_2-tax to reach the overall Kyoto target in each country. The allocation to the ETS follows a least cost approach. In the final scenario, the Annex B regions PAO and USA also reach their Kyoto target by means of a unilateral CO_2-tax.

The most striking result is that the only region to be selling permits is Eastern Europe, even though no hot-air is included in the simulations. Further, the simulation results show that the overall macroeconomic effects of the EU ETS alone are almost negligible. The only sectors that suffer from a loss of output and exports are the energy sectors, here primarily the coal and electricity sector. In addition, the sectors covered by the ETS are not necessarily affected more than the sectors outside the ETS, as indirect effects via changes in the prices of intermediate inputs dominate. The strength of the effects also depends on the sectoral emission intensity in each country. France and Norway/Iceland, which produce low CO_2 intensive electricity by using nuclear power, i.e. hydropower, can even profit from the ETS. The energy-intensive production sectors are not seriously affected in the first trading scenario. This is different if emissions are restricted in the sectors outside the ETS. As these are sectors with relatively high abatement costs; their gross energy prices rise considerably. As a result, these sectors suffer more than the sectors inside the ETS that profit from an equalization of marginal abatement cost across Europe. Nevertheless, as the least-cost approach allocates fewer permits to the ETS than the historical approach, and because the emissions restrictions in the sectors now subject to a CO_2-tax have repercussions on the sectors inside the ETS, the competitiveness effects in the ETS sectors increase as well.

If the other Annex B countries of the Kyoto Protocol also reduce their emissions, energy-intensive production is, to some degree, shifted from the other Annex B regions to Europe. Thus, Europe benefits from emission reductions in other regions and EU welfare increases compared to a scenario involving no foreign emission restrictions.

Summarized, the simulations show that in our scenarios, the EU emissions trading regime alone will have only small competitiveness effects on the European Business and only a negligible effect on the macroeconomic performance of the EU. Only if the ETS is part of the EU strategy to reach the Kyoto targets, do the effects increase. Nevertheless, if the emissions are also restricted outside the ETS, the sectors inside the ETS profit from the cheap abatement opportunities (especially in Southern Europe and Eastern Europe) and are affected less than the sectors outside the ETS. The strength of the effects depends to a large degree on the CO_2-intensity of sectoral production in each country. If other countries such as the USA and the rest of the Annex B countries decide to restrict their emissions as well, the non-energy sectors gain from a reduction in a shift of production. Overall, the EU is better off.

Nevertheless, there remain a large number of open questions. First of all, this study was based only on hypothetical allocation plans. By now the allocation plans for the first trading period are known. Still, the reductions outside the trading scheme, the role of hot-air in and outside Europe, the role of the non-European countries and the role of the flexible Kyoto mechanisms is not yet clear. The current analysis is only capable of showing a range of possible outcomes. Once further information is available, it can be updated.

References

Balance RH (1988) Trade Performance as an Indicator of Comparative Advantage. In: Greenway D (ed) Economic Development and International Trade. MacMillan Education Ltd., Houndmills, pp. 6-24

Barker T, Johnstone N (1998) International competitiveness and carbon taxation. In: Terry Barker T, Köhler J (eds) International Competitiveness and Environmental Policies. Edward Elgar, Cheltham, pp. 71-139

Boemare C, Quirion P (2002) Implementing Greenhouse Gas Trading in Europe: Lessons from Economic Literature and International Experience. Ecological Economics 43: 213-230

Böhringer C (2002) Industry-level Emission Trading between Power Producers in the EU. Applied Economics 34: 523-533

Capros P, Mantzos L (2000a) The Economic Effects of Industry-Level Emission Trading to Reduce Greenhouse Gases. Report to DG Environment

Capros P, Mantzos L (2000b) The Economic Effects of EU-Wide Industry-Level Emission Trading to Reduce Greenhouse Gases. Institute of Communication and Computer Systems of National Technical University of Athens

Capros P, Mantzos L, Vaino M, Zapfel P (2002) Economic Efficiency of Cross-sectoral Emissions Trading in CO_2 in the European Union. In: Albrecht J (ed) Instruments for Climate Policy. Edward Elgar Cheltham, pp. 25-62

EIA (Energy Information Administration) (2002) International Energy Outlook 2002

European Commission (2003) The EU Emissions Trading Scheme: How to develop a National Allocation Plan. (Non-paper of the 2nd meeting of Working Group 3, Monitoring Mechanism Committee, Directorate General Environment, 1st April 2003)

Eyckman J, Cornillie J (2002) Supplementarity in the European Carbon Emission Market. In: Albrecht J (ed) Instruments for Climate Policy. Edward Elgar Cheltham, pp. 96-127

Gagelmann F, Hansjürgens B (2002) Climate Protection through Tradable Permits: The EU Proposal for an Emissions Trading System in Europe. European Environment 12: 185-202

Jenkins R (1998) Environmental Regulation And International Competitiveness: A review of Literature and Some European Evidence. Discussion Paper No 9801, United Nations University INTECH, Maastricht

Klepper G (2003) International Trade and Competitiveness Effects of Emissions Trading. CATEP Policy Brief

Klepper G, Peterson S, Springer K (2003) DART97: A Description of the Multi-regional, Multi-sectoral Trade Model for the Analysis of Climate Policies. Kiel Working Papers no 1149 Kiel Institute for World Economics

Shin S (2003) Kyoto-Protokoll, Wettbewerb und WTO-Handelssystem. HWWA Discussion Paper No 215, Hamburg

Siebert H (2002) Außenwirtschaft, Gustav Fischer Verlag, Stuttgart

Springer K (2002) Climate Policy in a Globalizing World: A CGE Model with Capital Mobility and Trade. Kieler Studien 320. Springer, Berlin

Viguier L, Babiker MH, Reilly JM (2003) The costs of the Kyoto Protocol in the European Union. Energy Policy 32: 459-481

Zhang ZX (2002) The Economic Effects of an Alternative EU Emissions Policy. Journal of Policy Modeling 24: 667-677

Appendix – detailed simulation results for selected regions

More detailed results can be obtained from the author upon request.

Table A-1. Percentage change in the price of the energy aggregate PE in 2012 relative to the BAU scenario in 2012 in the non-energy sectors in selected regions

	ITA	FRA	DEU	GBR	SCA	BEN	SEU	REU	EFT	EEU	PAO	USA	FSU
							ET-0						
EGW	11.1	4.3	12.0	10.7	12.5	12.9	11.4	9.9	2.6	20.5	-0.5	-0.3	-0.1
IMS	6.1	4.6	6.4	6.4	6.0	8.3	5.5	5.4	1.8	15.1	-0.3	-0.1	-0.1
PPP	5.7	3.6	5.1	5.2	4.3	5.6	5.1	3.8	1.7	13.8	-0.2	-0.1	-0.1
TRN	3.6	0.6	3.3	2.4	2.6	2.1	2.9	2.3	0.0	8.5	-0.1	-0.1	-0.1
MOB	0.5	0.6	0.8	0.6	0.3	0.3	0.6	0.6	0.6	1.2	-0.2	-0.2	0.0
CEP	1.4	0.6	1.3	0.9	1.0	0.9	1.1	0.9	0.4	2.7	-0.2	-0.1	0.0
AGR	2.2	0.6	2.2	2.2	1.4	1.8	1.7	1.4	0.4	3.4	-0.2	-0.1	-0.1
Y	3.1	0.6	2.0	1.8	2.0	1.2	2.2	1.7	0.3	6.9	-0.2	-0.1	0.0
							ET-K						
EGW	15.6	5.9	16.6	9.0	18.3	15.5	16.1	15.1	3.0	29.9	-0.9	-0.5	-0.7
IMS	8.3	6.1	8.5	4.9	8.4	9.7	7.5	7.8	1.8	21.6	-0.5	-0.3	-0.7
PPP	7.3	4.7	6.7	3.3	6.0	5.8	6.8	5.5	1.6	19.7	-0.4	-0.3	-0.7
TRN	9.4	5.6	7.7	7.6	8.6	17.0	6.5	11.2	5.3	11.9	-0.4	-0.2	-0.7
MOB	8.8	6.9	11.4	11.3	15.0	18.2	6.2	23.0	19.9	1.1	-0.7	-0.6	-0.6
CEP	9.5	7.1	11.2	10.8	12.5	19.7	6.3	23.7	16.9	3.2	-0.6	-0.4	-0.6
AGR	9.9	6.6	9.8	8.0	13.3	24.2	6.2	21.5	15.9	4.4	-0.5	-0.3	-0.6
Y	9.6	5.7	9.7	8.5	9.7	17.3	5.9	16.5	11.1	9.5	-0.4	-0.3	-0.6
							AXB						
EGW	14.1	5.1	17.9	8.2	18.0	14.0	14.8	15.9	2.7	32.3	60.2	53.9	-2.8
IMS	6.6	4.3	8.3	4.1	6.7	8.4	5.0	6.2	0.8	22.7	26.0	27.1	-2.7
PPP	6.7	3.7	6.5	2.6	5.5	5.2	4.7	4.8	0.2	20.5	21.6	23.9	-2.9
TRN	9.0	7.0	8.8	7.6	9.2	17.2	8.2	12.1	5.5	11.7	24.0	22.0	-3.4
MOB	6.0	4.6	10.7	9.5	16.4	16.9	7.4	23.1	20.6	-3.6	38.4	45.3	-3.8
CEP	7.6	5.7	11.6	9.5	13.5	19.4	7.6	24.8	17.5	0.0	32.6	35.1	-2.8
AGR	8.7	5.9	10.7	7.8	15.0	26.0	7.3	23.6	16.6	1.5	30.2	26.7	-3.2
Y	9.0	5.3	10.0	7.6	10.3	16.6	6.5	17.4	11.4	8.3	22.0	26.0	-2.7

Table A-2. Percentage change in total sectoral output in 2012 relative to the BAU scenario in 2012 in the non-energy sectors in selected regions[5]

	ITA	FRA	DEU	GBR	SCA	BEN	SEU	REU	EFT	EEU	PAO	USA	FSU	EU	NEU
ET-0															
COL	-28.4	-3.2	-8.3	-4.4	-3.3	-1.1	-4.9	-3.0	-3.0	-6.3	-0.7	-0.4	-0.6	-6.5	-0.4
OIL	-2.6	-1.1	-1.6	-0.6	-0.4	-1.9	-1.4	-1.7	-1.5	-1.6	0.1	0.1	0.5	-1.4	0.2
GAS	-1.7	-0.5	0.0	-0.1	-3.3	-0.6	-5.6	-3.1	-0.8	-0.6	-0.2	0.0	0.0	-0.8	-0.1
EGW	-3.9	0.3	-2.9	-2.6	-2.8	-4.1	-2.4	-1.3	3.2	-14.7	0.1	0.1	0.0	-3.0	0.3
IMS	-0.4	0.0	-0.4	-0.1	-0.8	-1.2	-1.3	-0.1	0.3	-5.6	0.2	0.2	0.8	-0.7	0.2
PPP	-0.2	0.0	-0.2	0.0	-0.5	0.0	-0.4	-0.3	0.0	-0.7	0.0	0.0	0.1	-0.2	0.0
CEP	-0.3	-0.1	-0.3	-0.1	-0.1	-0.5	-0.4	0.0	-0.1	-1.0	0.1	0.1	0.1	-0.3	0.1
*In the sectors CRU,Y,AGR,TRN and MOB the changes are below 0.4% in all regions															
ET-K															
COL	-39.5	-5.8	-14.2	-9.2	-5.2	-1.9	-7.5	-6.3	-5.3	-9.3	-1.1	-0.6	-0.8	-10.6	-0.7
CRU	-0.5	-0.3	-0.4	-0.5	-0.5	0.0	-0.1	-0.7	-0.3	-0.3	-0.2	-0.1	-0.2	-0.4	-0.2
OIL	-5.1	-3.2	-5.8	-4.1	-3.5	-6.9	-4.0	-8.1	-7.7	-1.8	0.5	0.6	0.8	-4.7	0.6
GAS	-16.6	-24.3	-0.9	-3.7	-22.0	-13.4	-52.4	-47.0	-18.7	-1.5	-0.8	0.0	-1.1	-12.7	-0.8
EGW	-4.7	0.9	-3.3	-1.2	-3.3	-3.0	-3.1	-0.2	6.3	-20.6	0.2	0.2	2.1	-3.1	0.4
Y	-0.1	-0.1	-0.1	-0.1	0.0	0.5	-0.1	0.1	0.0	0.1	0.0	0.0	-0.1	0.0	0.0
IMS	-0.9	-0.4	-0.8	0.1	-1.4	1.0	-2.0	0.2	1.6	-8.5	0.1	0.1	1.4	-0.9	0.2
PPP	-0.5	-0.3	-0.6	0.0	-0.9	1.1	-0.8	-0.5	0.4	-1.3	0.0	0.0	0.2	0.0	0.0
AGR	-0.4	-0.2	-0.1	-0.1	-0.5	-1.7	0.0	-0.9	-1.0	0.6	0.0	0.0	0.2	0.0	0.0
TRN	-0.7	-0.4	0.0	-0.1	0.2	1.5	-0.6	0.9	0.3	-0.6	0.0	-0.1	0.0	0.0	0.0
MOB	-0.9	-1.4	-2.1	-1.5	-2.2	-7.2	-1.9	-4.7	-5.0	0.9	0.3	0.5	0.5	0.2	0.4
CEP	-1.2	-0.6	-2.2	-1.6	-0.6	-12.6	-1.0	-4.3	-0.7	0.9	0.7	0.6	1.7	0.6	0.7
AXB															
COL	-68.8	-12.1	-19.0	-14.5	-9.6	-5.4	-13.0	-10.2	-11.4	-14.0	-9.7	-36.4	-2.4	-15.5	-16.7
CRU	-2.2	-1.7	-1.8	-2.2	-2.2	0.0	-0.4	-2.1	-2.1	-1.6	-2.6	-3.6	-1.4	-2.1	-2.0
OIL	-0.4	1.6	-2.2	-2.1	0.0	-1.2	-0.5	-3.1	-9.5	1.6	-14.1	-13.4	2.3	-1.1	-3.2
GAS	-26.4	-37.6	-2.4	-6.6	-38.0	-19.3	-76.7	-56.0	-26.4	-3.3	-31.9	-2.4	-2.6	-19.0	-6.9
EGW	-4.1	1.2	-3.4	-1.0	-2.9	-2.0	-2.5	0.3	7.3	-22.3	-6.6	-8.6	4.3	-2.9	-3.3
Y	-0.2	-0.2	-0.2	-0.2	-0.2	0.2	-0.1	-0.2	-0.5	-0.1	-0.3	-0.1	-0.3	-0.2	-0.2
IMS	-0.7	-0.4	-0.8	0.6	-0.5	1.6	-1.3	0.5	2.7	-9.2	-3.2	-1.7	4.2	-0.6	-0.1
PPP	-0.4	-0.3	-0.6	0.1	-0.7	1.2	-0.6	-0.4	0.8	-1.5	-1.1	-0.9	0.7	-0.3	-0.4
AGR	-0.3	-0.3	-0.1	-0.1	-0.6	-2.1	-0.1	-1.1	-1.1	0.8	-1.2	-1.4	0.9	-0.4	-0.4
TRN	-0.6	-0.5	0.1	0.2	0.7	1.7	-0.6	1.2	0.4	-0.4	-2.0	0.0	0.4	0.0	-0.4
MOB	0.5	0.4	-0.9	-0.4	-1.1	-4.4	-1.5	-3.2	-3.2	2.9	-2.6	-5.5	2.0	-0.6	-1.1
CEP	0.0	1.0	-1.2	0.5	0.8	-10.9	-1.1	-1.5	2.5	3.6	-5.8	-3.7	4.1	-0.9	-0.5

[5] The large changes in the GAS sectors in SEU and REU most likely stem from mistakes in the data base.

Table A-3. RCA in 2012 in selected regions (absolute values)

	ITA	FRA	DEU	GBR	SCA	BEN	SEU	REU	EFT	EEU	PAO	USA	FSU
							BAU						
COL	-5.21	-3.46	-2.25	-2.16	-5.66	-4.27	-4.33	-5.86	-1.87	1.23	0.39	3.06	0.88
CRU	-5.10	-4.09	-3.21	0.71	-1.47	-5.15	-5.78	-5.74	2.89	-4.83	-1.37	-3.46	1.82
OIL	-0.22	-0.07	-1.48	1.57	0.04	0.36	-0.42	-1.34	0.04	0.03	-0.52	0.30	1.64
GAS	-6.19	-3.45	-3.24	-0.14	0.70	0.88	-10.54		2.64	-6.57	0.32	-3.32	0.85
EGW	-3.48	2.73	0.07		-0.02	-1.17	0.41	0.68	1.21	-0.01	1.53	-1.46	-0.16
Y	0.35	-0.06	-0.04	-0.04	-0.07	-0.12	0.12	-0.08	-0.32	0.04	0.01	-0.08	-0.98
IMS	-0.15	0.06	0.19	0.20	0.01	0.25	0.02	-0.01	0.21	0.55	0.39	-0.26	1.11
PPP	-0.07	-0.16	0.10	-0.20	1.53	-0.12	-0.05	0.05	-0.36	-0.15	0.41	0.40	-0.41
AGR	-0.55	0.32	-0.42	-0.31	0.19	0.36	0.18	0.40	-0.77	-0.03	-0.66	0.74	-0.84
TRN	-0.22	0.39	0.63	-0.10	0.00	-0.10	0.12	-0.39	-1.56	-0.18	0.63	-0.04	-1.21
MOB	-0.32	-0.36	-0.93	-0.23	-0.35	-0.13	0.59	0.13	-0.58	0.77	-0.35	0.29	0.27
CEP	-0.13	0.13	0.36	0.31	-0.13	0.24	-0.47	0.45	0.61	-0.27	0.08	0.59	-0.51
							ET-0						
COL	-5.21	-3.26	-1.63	-1.89	-5.50	-4.12	-3.81	-5.70	-1.69	1.88	0.35	2.90	0.68
OIL	-0.23	-0.10	-1.51	1.55	0.03	0.34	-0.45	-1.36	0.02	0.02	-0.51	0.31	1.65
GAS	-6.20	-3.45	-3.24	-0.06	0.71	0.90	-10.56		2.67	-6.51	0.32	-3.32	0.83
EGW	-3.61	2.81	-0.02		-0.09	-1.29	0.34	0.70	1.36	-0.53	1.62	-1.38	-0.07
IMS	-0.16	0.06	0.18	0.20	0.00	0.24	-0.01	-0.02	0.22	0.46	0.39	-0.25	1.13
							ET-K						
COL	-5.21	-3.05	-1.27	-1.55	-5.45	-4.04	-3.57	-5.43	-1.36	1.36	0.32	2.82	0.61
CRU	-5.05	-4.06	-3.16	0.78	-1.44	-5.08	-5.73	-5.62	3.02	-4.80	-1.37	-3.46	1.78
OIL	-0.24	-0.11	-1.51	1.59	0.07	0.39	-0.46	-1.35	0.06	0.01	-0.52	0.30	1.62
GAS	-6.22	-3.49	-3.07	0.88	0.53	1.31	-11.17		2.74	-6.52	0.31	-3.35	0.77
EGW	-3.67	2.83	-0.06		-0.13	-1.28	0.30	0.69	1.42	-0.76	1.65	-1.36	-0.03
IMS	-0.17	0.05	0.18	0.21	-0.01	0.27	-0.02	-0.01	0.24	0.41	0.39	-0.25	1.15
MOB	-0.34	-0.39	-0.98	-0.27	-0.39	-0.26	0.54	0.06	-0.58	0.84	-0.33	0.32	0.31
CEP	-0.15	0.13	0.33	0.29	-0.13	0.15	-0.48	0.41	0.61	-0.23	0.10	0.62	-0.47
							ABX						
COL	-5.88	-3.18	-1.60	-1.82	-5.62	-4.14	-3.81	-5.60	-1.47	1.92	1.83	5.67	0.21
CRU	-5.14	-4.12	-3.21	0.71	-1.50	-5.15	-5.77	-5.71	3.02	-4.88	-1.22	-3.13	1.67
OIL	-0.19	-0.07	-1.48	1.55	0.11	0.42	-0.39	-1.30	0.00	0.03	-0.69	0.37	1.52
GAS	-6.63	-3.58	-3.08	0.95	0.17	1.28	0.00		2.68	-6.64	0.47	-2.81	0.69
EGW	-3.60	2.85	-0.06		-0.11	-1.22	0.36	0.72	1.46	-0.81	1.14	-2.01	-0.04
IMS	-0.16	0.05	0.18	0.22	0.00	0.27	0.00	0.00	0.27	0.40	0.26	-0.32	1.22
MOB	-0.29	-0.33	-0.94	-0.22	-0.36	-0.20	0.56	0.10	-0.62	0.92	-0.40	0.12	0.39
CEP	-0.12	0.15	0.35	0.33	-0.11	0.17	-0.48	0.44	0.64	-0.18	0.09	0.48	-0.41

The changes relative to BAU in the sectors not reported are negligible
RCA of region R and commodity X = log[(exports of X/imports of X)/(all exports of R/all imports of R)]

Implementing the EU emissions trading directive in Spain: a comparative study of corporate concerns and strategies in different industrial sectors

Pablo del Río González

Universidad de Castilla-La Mancha
Facultad de Ciencias Jurídicas y Sociales
C/cobertizo de S. Pedro Mártir s/n
Toledo 45071, Spain
pablo.rio@uclm.es, pdelrio@jur-to.uclm.es

Abstract

Although emissions trading has proved cost effective at the theoretical level, a very conflicting issue (which may lead to significant administrative costs) is the distribution of allowances between polluting sources. There are winners and losers in this process and, of course, an incentive to get as many allowances as possible. The NAPs are, therefore, a relevant source of conflict between interest groups and their final version reflects the interactions between actors with different interests, strategies and negotiation power. In this context, the Spanish NAP is no exception. By adopting a Public Choice perspective, this paper takes a closer look at the Spanish allocation process, analysing the interests and strategies of the different actors and their interactions. At the individual firm level, the Directive was considered a big threat. At the sector level, distinct sectors put forward different arguments in order to have more allowances allocated to them. Coalitions at different levels were created and lobbying already started very early in the process. The negotiations reflected the tension between the need to control emissions, the arguments put forward by firms and sectors and the minimisation of the negative impact on the overall economy.

Keywords: Emissions trading, allowance allocation, public choice, Spain, national allocation plans.

1 Introduction: aim, scope and methodology

Major legislative steps have been taken in order to implement a CO_2 emissions trading system in the European Union. A Directive creating a CO_2 emissions trading scheme in Europe was formally adopted by the Council on 22 July 2003 and Member States were expected to implement the provisions necessary to comply with the new emissions trading scheme. The European emissions trading scheme (EU ETS) covers carbon dioxide emissions from large stationary sources including power and heat generators, oil refineries, ferrous metals, cement, lime, glass and ceramic materials, and pulp and paper. It is estimated that these sources will emit 46% of the Community's carbon dioxide emissions in 2010.

National authorities issue site-specific greenhouse gas emission permits to installations setting requirements for monitoring and reporting emissions of greenhouse gases. Member States will allocate EU emission allowances to installations, based on a national allocation plan (NAP) developed in accordance with common criteria[1]. Four months after the end of each year, operators are required to hand over, to the national authority, allowances equivalent to the installation's emissions during the preceding year. Operators of installations are free, if they so wish, to buy or sell their allowances. If an operator can reduce emissions, the excess allowances can be sold at a profit.

Countries already carried out their NAPs, which were to be handed in to the Commission by March 31st 2004 at the latest. Sectors were closely watching developments taking place in this regard.

Although emissions trading has proved cost effective at the theoretical level, a very conflicting issue (which may lead to significant administrative costs) is the distribution of allowances between polluting sources. There are winners and losers in this process and, of course, an incentive to get as many allowances as possible. The NAPs are, therefore, a relevant source of conflict between interest groups.

Both formal and informal contacts between the public administrations, the sectors and the NAP teams took place. It can be expected that the final version of the NAP and, therefore, the final allocation will be the result of these interactions between actors with different interests and strategies.

In this context, the Spanish NAP, is no exception. By adopting a Public Choice, this paper takes a closer look at the Spanish allocation process, analysing the interests and strategies of the different actors and their interactions. The focus will be on the sectors covered by the Directive. One indirect but significant source of conflict is the lack of experience in Spain with emissions trading, which causes uncertainties and doubts regarding the use of this instrument.

The paper is, therefore, organised as follows. The next section will provide the theoretical approach. Section 3 will take a look at the strategies and views of the different sectors concerning the EU ETS and allocation. The paper closes with some concluding remarks.

---

[1] In the wording of the CO_2 Directive, "allowances" are the papers that state the authorisation to emit a certain amount of CO_2 (i.e. one ton), while the permit is the general authorisation to take part in trading allowances.

2 The theoretical approach: public choice and climate policy

In the next sections we will discuss what are the stakeholders' preferences and strategies concerning emissions trading and allocation. The theoretical approach followed is one of Public Choice, which will help us structure the information obtained from interviews and documents.

Public Choice can be defined as "the economic study of non-market decision making or, simply, the application of economics to political science" (Mueller 1989, p. 1). The theory of Public Choice is particularly relevant to climate policy since it touches every economic sector and thus matters to all interest groups in the economy (Desai and Michaelowa 2001, p. 328). The array of uncertainties linked to the climate issue allows interest groups to choose divergent positions without being scientifically discredited. Due to the overarching nature of the issue, they can choose from a wide array of instruments.

A major feature of this approach is "lobbyism". Special interest groups lobby for special governmental favours: decisions that create common benefits for their members by redistributing national income to them. Lobbyism can be regarded as an investment. Those who are organising incur costs in order to gain future rents (income). These rents are obtained at the expense of unorganised groups.

The final outcome and design of an environmental policy depends on lobbying by interest groups[2]. Lobbying leads to the formation of coalitions between firms with the same interests. Paradoxically, in order to undertake political action, firms have an incentive to include competitors because the larger the group, the more likely it is to influence government policy, while in the market realm they have an incentive to exclude competitors. This is already visible in the allocation process, with coalitions between sectors and firms being created at different levels (see below).

A Public Choice perspective for analysing environmental policy issues has already been used elsewhere (see Michaelowa 1998; Gerhardsen 1998; Schneider and Volkert 1999; Svendsen 2000; Desai and Michaelowa 2001). However, it has never been applied to emissions trading. This paper aims to make a contribution in this direction. Michaelowa (1998) notes that the question of why economically efficient instruments of climate policy have not been implemented in political reality remains unsolved and that Public Choice methods can help to close this gap.

[2] As stressed by Svendsen (2000, p. 31), the democratic State cannot just pursue the economic interest of the majority. In order to achieve political acceptability, reduce conflict and consequently implement rules of legislation, it must also mediate among the main organised interests.

3 Firms, emissions trading and allocation: the views and strategies of the sectors covered by the EU ETS

3.1 Firms and emissions trading

Traditionally, Spanish firms have been quite reluctant to take climate policy measures. If they were asked what type of regulation they would prefer for tackling climate change, they would probably have the following priorities (in descending order): subsidies, voluntary agreements, emissions standards, grandfathered allowances, auctioned permits and taxes.

The traditional method for environmental regulation has been CAC (Command and Control) policy, mostly based on emission standards. This is preferred to taxes and emissions trading. The reason is that with taxes a high cost for the regulated sources would result (since they would have to pay for all their emissions). Auctioned allowances would have the same effect on firms as a tax (except if the auction is revenue neutral). But even grandfathered permits would be rejected. The reason is that emissions trading allows little negotiation with the regulator and, therefore, "exemptions" or reduced requirements are not possible. CAC allows much discretionary behaviour on the part of policy makers and manipulation by polluting sources. It is probably easier to cheat under CAC regulation than it is under emissions trading because administrative controls are unable to sanction all violations[3]. Sanctions are relatively low[4]. The probability of being caught and sanctioned is generally low as well.

Sectors covered by the Directive differ from one another in many respects: their structure, degree of international openness and the availability of low cost reduction opportunities. However, they all share a significant concern regarding the economic impact of the Directive. Concerns may be a little bit different in the same sector in a different MS. These sectors accounted for 41% of total CO_2 emissions in 2001 (401 Mt CO_2 in that year). Electricity is, by far, the sector with the highest share in these emissions (56.2%), followed by cement (17.2%), refineries (9.5%), steel (8.7%), chemical (4.2%), ceramics (2.3%), pulp and paper (1.9%) and glass (0.1%).

Although Spanish companies are likely to have to buy considerable amounts of CO_2 allowances in the market, they have been for some time relatively unaware of the developments. There was neither a market pull nor an institutional, public policy push motivating them to develop or adopt less carbon-intensive technologies. Of course, this is no longer the case. Suddenly, they realised that the EU ETS would be implemented and that it would have profound implications for the firms. This has been a shock for many firms and sectors.

[3] According to Schneider and Volkert (1999) standards could also be preferred by firms because they can result in market entry barriers leading to large rent-seeking potential for old emitters. However, this could also be true for a grandfathered system.

[4] This is sometimes the result of interaction between the non-compliant firm and the regulator or the judicial authority who are very sensitive to the "unemployment argument" invoked by the firms.

3.2 The allocation process. The views and strategies of firms covered by the EU ETS

3.2.1 A sector by sector overview

3.2.1.1 Electricity generation

Electricity generation is the most important sector in terms of its share in overall CO_2 emissions (see the following table). The Spanish electricity sector is still relatively dependent on coal. At the EU level coal and lignite cover 26% of energy sources in electricity generation while in Spain coal based electricity is more than one third of the total electricity production. This is essential in the context of emissions trading since coal and lignite are responsible for most of the energy sector CO_2 emissions. The share of the second most polluting fuel (oil) is also higher in Spain while gas, which is less CO_2 intensive than coal and oil, is used much less. Non-CO_2 emitting energy sources (nuclear and renewables) together have a lower share in Spain than in the EU. To sum up: the share of polluting fuels is higher in Spain than in the EU, while the share of less polluting (or-non polluting) energy sources is lower in Spain.

Table 1. Generation mix in Spain and Europe in 2000 (% of gross electricity generation).

Energy source	Spain		EU	
	(TWh)	%	(TWh)	%
Nuclear	62.2	27.6	863.9	33.2
Coal	79.1	35.1	673.5	25.9
Oil	22.6	10.0	161.3	6.2
Gas	22.5	10.0	480.3	18.5
Renewables	36.4	16.2	388.2	14.9
Other*	2.3	1.0	31.6	1.2
Total	225.1	100	2598,8	100

Source: Our own elaboration from Eurostat. See European Commision (2002).
*"Other" includes pumped storage and "others" within the conventional thermal generation.

However, a significant reduction in the share of coal and an increase in the shares of natural gas and renewables is expected by the government for the next decade (Table 2).

Table 2. Expected electricity generation share per energy source (2011).

	%
Nuclear	19.4
Coal	15.0
Oil	4.1
Natural gas	33.1
Renewables	28.4
Total	100

Source: Ministry of Economics (2002)

One of the difficulties faced by the whole sector when controlling CO_2 emissions is the growth in electricity demand in the next few years, which is expected to increase by an annual average of 3.5% during the period 2002-2011. The growth in nuclear, large hydro and renewables will not be enough to compensate the increase in electricity demand. Therefore, combined cycles will be very relevant for meeting demand with a lower emissions ratio than conventional electricity[5]. The Ministry of Economics (2002) estimates that combined cycles will save 45 Mton of CO_2 up to 2010. The cost of substituting electricity production from natural gas for electricity generation for coal is considered to be rather low (between 0 and 20 €/tonCO_2[6])[7].

Probably this is the most unique sector in at least two senses. First, contrary to other sectors, there are relatively low-cost reduction opportunities (some renewable energy sources and combined cycles based on natural gas)[8]. The fact that oil and coal still account for almost 50% of power supply leaves some room for improvement (i.e., CO_2 emissions reductions). The EU ETS could be both an opportunity and a threat for the sector. Low cost reduction opportunities provide the opportunity. The threat is related to the uncertainty still existing in the market, both related to the NAP and to the future allowance price. If we add that firms in this sector are the ones that have made serious plans to be engaged in CDM projects, then we may label this as the most carbon proactive sector among those covered by the Directive[9].

Second, in the past electric utilities (grouped in the national association, UNESA) have maintained a homogeneous position concerning compliance with Kyoto targets. The firms had a shared strategy (coalition) to reduce the financial consequences of (not) complying with Kyoto. However, there are now different opinions between one firm and the rest of the sector. This firm is Iberdrola, which accounts for 1/3 of total capacity and 1/4 of total generation, but is only responsible for 8% of the CO_2 emissions from the electricity sector. It has a comparatively lower carbon intensity, even according to European standards (see the tables below). This is related to the use of combined cycles and renewable energy sources for producing electricity (55% of the firm's total installed capacity), and to the use of nuclear and hydro in generation[10]. This is in sharp contrast to other firms, in

[5] Combined cycles have a lower emission ratio (0.36 kg CO_2/kWh) than electricity from coal (0.95 kg CO_2/kWh) and oil (0.75 kg CO_2/kWh).

[6] The range depends on the future prices of coal and natural gas.

[7] Actually the sector argues that it has reduced its "relative" emissions (tCO_2/MWh) by 14% in the 1990-2001 period. The 30% increase in total emissions in the same period is due to a very high increase in demand (a growth of 51% in the period).

[8] Surprisingly, though, simulations made by IPTS with the help of the POLES model show that the marginal abatement-cost curves of this sector do not differ very much in relation to those of the other trading sectors.

[9] Due to a past political/economic crisis in Latin America, some of these projects, which are related to renewable energy sources, never went beyond the feasibility studies.

[10] Also, in September 2003 the firm created the first Spanish green pricing scheme.

which the share of coal in electricity production is much larger[11]. The firm has recently made a significant effort to install combined cycle plants as a normal business practice, although CO_2 reduction may have been an additional factor in the decision[12]. Also, investments in renewable energy have benefited from relatively high feed-in tariffs in the past, covering generation costs and providing a profitable margin. Of course, some firms in the rest of the sector (i.e., Endesa) have also invested in lower carbon intensity energy sources, but not as much as Iberdrola.

Table 3. Generation mix of Spanish power companies (% over total).

Company	Coal	Oil	Hydro
Endesa	44.5	2.6	14.8
Iberdrola	7.2	17.0	49.2
Union Fenosa	53.3	4.8	18.3
Hidrocántrabrico	61	-	16.6

Source: ABC (29/11/2003).

Table 4. Some data on the most important electricity generators in Spain.

Company	Capacity (GW)	Generation TWh)	Emission ($MtCO_2$)	Carbon intensity (kg CO_2/MWh)
Endesa	26	129	73	563
Iberdrola	16	58	9	148
Union Fenosa	6	26	15	559
Hidrocántrabrico	2	13	12	916
Total Spain big-4	50	226	109	482
Total (Europe)	466	2.117	748	353

Source: Our own elaboration based on data provided by PWC-Enerpresse (2002)

Iberdrola regards the CO_2 emission trading market as an opportunity rather than a threat since it is a forerunner in reducing CO_2 emissions. It even tries to use the CO_2 market as an instrument to be more competitive and to gain market quota in the electricity market. Iberdrola claims that compliance with both the EU Directive and the Kyoto Protocol (KP) is possible at reasonable costs, and desirable. This firm, however, disagreed with some of the choices that the National Allocation Plan could take. For example, choosing 1990 as a base year would not be interesting for this firm, since its total emissions have increased since then, but not its relative emissions (emissions per unit of output). It also preferred a production metric based on production projections for 2010[13].

[11] Coal-based generation emits between 879 and 1040 grams of CO_2/kWh, oil-based generation emits 802 grams of CO_2/kWh and combined cycled gas turbines emit 365 grams of CO_2/kWh.

[12] Combined cycles will substitute natural gas for coal as the main input for electricity generation. It is expected to provide 60% of electricity in 2010.

[13] The firm claims that if allocation was based on the data for the year 2002, then all the allowances would go to installations representing 40% of thermal production in 2010.

Iberdrola´s position started a sort of dispute with the rest of the power companies, which claimed that, given the increase in emissions since 1990 (around 38% in 2002) the BSA targets negotiated by the government in 1998 (+15%) have proved insufficient for the smooth functioning of industry and called for a revision of the established quotas[14]. Iberdrola defended by saying that enough time to adapt had been given and that the company had invested heavily in less CO_2 intensive technologies. On the contrary, ENDESA, the fourth largest CO_2 polluter electric utility in Europe and, therefore, an important player in the CO_2 market, had recently increased its investments in coal-based generation in order to maintain its leadership in electricity generation in Spain. This firm and the rest of the sector used the argument that coal is a very important fuel for the security of energy supply of electricity generators and, therefore, this should have been taken into account in the NAP. They also claimed that natural gas also emits CO_2, that its prices are more volatile, that generation of hydro-power should be limited in a country where droughts are a frequent phenomenon and, finally, that renewable energy sources had technical and economic problems and are heavily subsidised. They claimed that the current generation mix should be maintained in the future and that a strategy based on waiting to see what happens with the commitments of other countries in the 2008-2012 period should be followed. These firms prepared reports that informed of the significant financial consequences of compliance and tried to convince the government that they should receive a generous allocation (around 282 $MtCO_2$ in each of the years of the 2005-2007 period).

These firms argued that complying with the targets would increase their costs. This would lead to an increase in electricity prices of around 10€/MWh. This means an increase of around 10-15% in the final price paid by the consumer. However, these prices are regulated by the government which will not allow increases above 2% annually.

Therefore, they showed a more defensive attitude. These firms claimed that, when allocating allowances, an important distinction between the sectors covered and not covered by the Directive should be made and that relatively more allowances should be given to the former. The reasoning was that, since 1990, the increase in emissions from the sectors covered by the Directive had been much slower than the increase from the rest of the sectors (26% and 33%, respectively). They also argued that the sectors not covered had greater reduction possibilities[15]. They proposed an eco-tax on the transport sector with recycling of the revenues to a carbon fund which could be used to buy CERs (from CDM projects). This "let the others reduce" attitude was quite common to all the Spanish sectors and was the general basis for their lobbying activity.

They suggested that allowances should be allocated to covered sectors according to their expected emissions in 2005-2007. The emissions target for the non-

[14] These firms defend by saying that the +15% target does not take into account the impressive economic growth of Spain since 1990 and also the fact that our emissions per capita were lower than in the rest of EU countries.

[15] This is not really true according to the POLES model, which shows that marginal abatement costs in the Spanish non-trading sector are much higher than in the trading sector.

regulated sectors would be calculated as the difference between the total emissions in Spain and the emissions expected in the regulated sectors[16]. The sector also claimed that, contrary to other sectors, it could not pass the additional costs on to the consumers (which they claimed was possible in other sectors) since electricity prices paid by the final consumer are regulated and increases are restricted by regulation.

3.2.1.2 Steel

The steel industry in Spain is one of the most sensitive among those covered by the proposal, since it is both highly energy consuming, highly CO_2 intensive and relatively open to international trade.

This energy intensive industry sector has a limited capability to reduce emissions and sees itself as an allowance buyer in the market. They claimed they have already reduced emissions in the past (as part of normal business practice). Therefore, the aim of the sector was to obtain more allowances than would correspond to them in the initial allocation based on the argument that "early action" was taken. The sector followed a defensive strategy.

The sector was especially worried about the distortions caused by the emissions trading scheme in the international market. ARCELOR, the first world producer and responsible for 2% of European CO_2 emissions even threatened to relocate and start production in non-EU countries. They argued that they already invested heavily in the past to reduce emissions (since 1990 they have been reduced by 17%), that this was not taken into account and that further reduction of emissions would involve high costs[17].

3.2.1.3 Cement

Cement production is an energy and capital intensive process leading to high CO_2 emissions per unit of sales ratio. Energy combustion activities account for 1/3 of CO_2 emissions from cement production. Cement is the third industrial emitting sector in the EU.

The Spanish sector is worried about the effects of the Directive on competitiveness given its high CO_2 emissions per unit of value added. As the steel sector, it points out the risk of eco-dumping (i.e., relocating their installations to countries with lower environmental requirements[18]).

---

[16] The allocation between the regulated sectors should also be made according to the "future needs" in the period 2005-2007.

[17] According to Arcelor, these tariffs should increase the price of imports by 30%. In their opinion, tariffs should be charged on imports of steel from countries that neither ratify the Kyoto Protocol nor comply with its targets.

[18] Firms perceive themselves as net buyers. They fear high prices of allowances which, according to Cembureau (the sector's European association), "would provide an incentive to relocate cement plants outside the EU". The possibility of eco-dumping is usually mentioned by the sector. According to the sector's BREF, transport costs make markets for cement predominantly local. However, international competition is mainly a threat for

This was one of the sectors showing a more defensive strategy and seeing themselves as net buyers in the market[19]. It favoured a “let the others reduce” approach, suggesting additional efforts should be made by the sectors not covered by the Directive but also by the electricity sector itself. They argued that:

- The sector had already made a lot of effort to control emissions: a 30% improvement in energy efficiency had taken place in the 1975-2000 period. This was the only feasible way to reduce emissions, but only an additional 2% improvement was considered possible at moderate costs. The sector demanded that early action (taken as far back as 1988) be taken into account. However, in reality these were BAU measures.
- The sector had “special conditions” which made it hard to further reduce emissions[20].
- Reducing production in order to control emissions was not an option, since the expanding building sector increasingly needed cement[21].
- The costs of buying additional allowances compared to the sector’s value added was one of the highest among the covered sectors (7.2% compared to only 5.4% in electricity sector or 0.2% in the pulp and paper industry, for example)[22]. They claimed it was very difficult to pass this cost on to the final price, since this was a highly internationalised sector in which clinker and cement imports were highly sensitive to price changes. The impact of the Directive on the sectors’ profits would be larger than in other sectors since profit margins (as a percentage of value added) were lower in this sector than in other sectors (i.e., electricity sector)[23].

---

individual plants, and within the EU increasing imports from Eastern Europe do affect local market conditions.

[19] Szabo et al. (2003) show that the marginal abatement cost curve of the Spanish cement sector (compared to 2010 value) would allow lower cost-reductions than in the rest of the MS analysed (France, Italy, Portugal, Germany, Netherlands and Greece). Reductions would only be cheaper in Great Britain. However, the results for the Spanish cement industry show that firms in this sector would be net buyers in the allowance market.

[20] First, 60% of total emissions are from the decarbonization process and can not be reduced since it is a chemical process. Second, substitution of peat-coke with natural gas as a fuel is not economically feasible (production costs would increase by 30%). Third, additional energy efficiency is limited (only 2% improvement in the 40% of emissions not related to decarbonization). Finally, in reality there is no alternative process which allows less carbon-intensive cement production, as they claim is the case in the electricity sector.

[21] According to OFICEMEN, the Spanish association of cement producers, 29 million tons of clinker would be necessary in the next 5 years in order to meet domestic cement demand.

[22] Data provided by OFICEMEN. These were the costs of complying with the Spanish BSA (+15% in 2010). In 1990 emissions by the sector reached 21 $MtCO_2$ and 25.3 Mt in 2001. The target for 2010 would be 24.1 $MtCO_2$. According to CEMEX, emissions in 2010 were projected to reach 27.4 Mt CO_2. Therefore, the reduction in 2010 would be 3.2 $MtCO_2$. On the other hand, an allowance price of 20 € t/CO_2 was considered. Average cost for all sectors covered by the Directive was 2.1%.

[23] This was probably related to the high capital intensive character of the industry. According to the sector’s BREF, the cost of a new cement plant was equivalent to around 3

- Cutting production directly in order to reduce emissions or to buy allowances, thereby increasing the price of the final product, would have a dramatic effect on installation closures and employment levels, because the sector is labour-intensive[24].
- According to both Officemen and Cembureau (the European association), "companies using waste as fuel instead of primary fossil fuels in production processes should be credited with allowances since the use of such alternative fuels reduces overall GHG gas emissions at incinerators and landfills".

Apart from the "let the others reduce" attitude, Oficemen wanted the Spanish government to get 100-150 $MtCO_2e$ in additional allowances per year by investing in CDM/JI projects and by buying AAUs to other MS which would be acquired through additional revenues on consumption taxes.

Concerning the allocation methodology, the sector preferred a benchmarking approach which tied allocation to future production[25].

3.2.1.4 Oil Refining

The most feasible emission-reduction option was improvements in energy efficiency in refineries which had already been done, according to the sector.

Probably influenced by the experience of other firms in the sector (BP and Shell), the Spanish oil industry seemed to be in favour of emissions trading in general and the EU ETS in particular. However, they had some claims regarding the EU ETS:

- The future of CO_2 emissions reductions should take into account demand from the final consumers given the responsibility of consumption activities on CO_2 emissions (transport and domestic sectors) as well as the fact that CO_2 reductions in energy-intensive sectors would require a change in the demand from the consumers.

---

years' turnover, which ranked the cement industry among the most capital intensive industries. The profitability of the cement industry was around 10% as a proportion of turnover (on the basis of pre-tax profits before interest repayments).

[24] According to the sector, if only 22 $MtCO_2$ were allocated to the sector, considering an allowance price of 30 € t/CO_2, this would lead to the closure of 5 installations, a loss of 650 Mt € of net value added and a loss of 3000 jobs (1000 direct and 2000 indirect).

[25] Cembureau developed a method in which the absolute cap was obtained by multiplying a performance standard with a reduction percentage, and with production volume forecasted for the commitment period. The performance standard was equal to the average specific emission of the industry sector in the reference year in a given geographical region (op. cit.). With this system, installations expected to be less damaged than with a pure grandfathering approach based on emissions. This methodology tried to take into account the future growth of the sector and, therefore, to minimize the impact on those installations which made an effort to reduce their CO_2 emissions per unit of output, but which experienced fast growth in production.

- The sector was concerned that if non-EU competitors did not apply the same CO_2 restriction, this would have a negative effect on the competitiveness of EU firms.
- Allocation should be completely free. Auction would lead to an additional burden and an unfavourable competitive situation for the EU industry[26].
- Some actions being taken by the refineries lead to an overall reduction of CO_2 emissions, but they may increase the emissions in the refinery itself[27].

The sector estimated a 36% increase in CO_2 emissions from the refineries in the 1990-2010 period. However cogeneration, higher quality fuels and bio-fuels would reduce overall emissions (in other sectors) by 21%. The sector asked for additional allowances on the basis of these reductions.

3.2.1.5 Pulp and paper

ASPAPEL, the pulp and paper industry association, allegedly supported emissions trading as a means of reducing the cost of emissions reductions. However, it was quite concerned about the implications of the current scheme. In general, firms in the sector followed a defensive strategy, considering the EU ETS as a threat to their business.

The Spanish pulp and paper industry maintained that it was already highly energy efficient since it had taken measures in the past to reduce energy consumption. Thus the potential for further CO_2 emissions reductions was limited[28].

The main concerns of this sector were:

---

[26] The Spanish oil refining industry estimated that allocation by auction would lead to annual costs of 2500 million euros (assuming an allowance price of 20 € t/CO_2).

[27] Among these actions, the following are worth considering: 1) CHP, which is more efficient than conventional electricity generation, has been implemented in many refineries. CHP leads to an increase in the emissions from the refinery, but it reduces overall emissions. 2) Manufacturing of higher quality fuels reduces the energy consumption of vehicles. However, these fuels lead to an increase in energy intensity of refining operations and, thus, to an increase of CO_2 emissions at the refinery level. 3) The introduction of bio-fuels in the market leads to a reduction of CO_2 emissions.

[28] According to the sector, past efforts to install CHP capacity and the increase in paper energy efficiency (which improved by 13% during the last decade) left few possibilities for reducing emissions. Despite claims from the sector, early measures had virtually no CO_2 additionality, but were taken for "normal" business practice reasons. Environmental measures in this sector focused on the reduction of waste discharges to waters, on paper recycling and on the control of emissions of local pollutants (see del Río 2002). These measures were mainly taken as a result of CAC regulation through the implementation of end-of-pipe technologies. Energy efficiency measures (including an increase of CHP and reutilization of organic by-products) were also taken in order to reduce the energy bill. This led to a relative decoupling in the sector (during the 1990-1999 period CO_2 emissions increased at a similar rate to pulp and paper production) although not to an absolute decoupling (CO_2 emissions increased by a cumulative 25% in the same period).

1. Negative effects on competitiveness. Spanish pulp and paper firms were worried that firms in non-Kyoto and non-EU ETS countries would be more competitive because they would not have to reduce their emissions[29].
2. Unfair burden. Firms in this sector claimed that, compared to other sectors, they would bear a disproportionate burden as a result of the scheme.
3. The Spanish pulp and paper sector was expected to experience a 91% cumulative growth in the 1990-2012 period, stimulated by an increase in paper consumption. Firms argued that, unless this increase in production was taken into account, a carbon constrained future would have a large impact on the sector.
4. Treatment of CHP. This was shared by other sectors (see end of 4.2.2).

In the beginning (i.e., when the first draft of the present Directive was approved) firms adopted a "wait and see" attitude towards the EU ETS, being confident that the scheme would not be implemented in 2005-2007. Suddenly, they realised that emissions trading would be a reality and this caused heavy unrest. Some smaller firms, however, were not even aware that an EU ETS system would be implemented and others still do not know the consequences for themselves.

3.2.2 The EU ETS and the allocation process: sectors´ common concerns

**Too fast approval of the Directive.* A common claim of all sectors was that there has not been a transition period to adapt to the Directive. A related argument was that of stranded costs. Most of the firms in these sectors are capital intensive. Investments are planned and recovered only after long periods. The entry by force of the EU ETS would make some of the past investments obsolete since they were made in an unconstrained carbon future. A sufficient transitional period for the implementation of the EU ETS would have allowed firms to plan their investments by allowing the existing equipment to depreciate in a profitable manner.

**Uncertainty.* Firms believed that the initial draft of the was unclear on several decisive points, and thus unsatisfactory. Major concerns were the lack of proper guidelines and the short time available for allocation. There was uncertainty about the way emission allowances would be allocated and about the likely price at which they would be traded. They consider that it was important for the functioning of the market and for investments to know the allocation methods that would be applied across Europe. Investment security and stable framework conditions are needed, and they believed they would not have that until the results of the NAPs were known. Until key elements of the allocation plan were known and uncertainties removed, few participants would be willing to enter the trading arena[30]. Uncertainty about rules and the size of future caps also limited demand, kept trad-

---

[29] For this reason, CEPI, the European association, asked for an extension of the scope of the scheme to increase liquidity and cost efficiency (Point Carbon 10/10/2003). They also asked for a delay in the implementation of the EU ETS scheme to the 2008-2012 period.

[30] An important source of uncertainty in NAPs is the treatment of CHP.

ing prices low and reduced the impact on corporate investment decisions. The Spanish forerunners in the EU ETS were only likely to perform small market interventions in order to acquire specific know-how in this market until NAPs were approved.

**Economy, firm and sector* growth. The Spanish economy is still below the EU average in per capita income (84 % of EU per capita income in 2003). However, the economy shows a high dynamism, fuelled by private demand, which has lead to growth rates high above the EU average in many sectors. This means that, ceteris paribus, CO_2 emissions grow faster in Spain than in other countries. Since Spain has a target of only +15% in the BSA, firms feared that the EU ETS would limit sector and firm growth. Firms asked for an equitable sharing of the burden between sectors of society, between the different industries covered by the Directive and between the companies within the sectors. Many claimed that the allowance allocation should allow for growth. Owing to the expected increase in emission levels, some companies favoured a "forecasting emissions approach" for the NAP. However, a "historical emissions" approach could be preferred for reasons of simplicity. Also, the country's commitment under the KP was questioned by representatives from industry who urged a renegotiation. They argued that high economic growth rates were not considered when those targets were negotiated[31].

**Sectors covered are overburdened.* Sectors covered by the Directive feared there was the political temptation to let the sectors included in the EU ETS carry an unproportionately large burden of national CO_2 reductions due to the expected large increase in emissions from the transport sector. Marginal costs across all sectors should have been estimated, so that the burdens were not inefficiently allocated. The marginal cost curves (MACs) for the sectors included in the Directive are lower than for the sectors not covered[32]. A cost-minimising result should favour reductions by the activities covered by the Directive[33]. The sectors covered carried out reports that showed high compliance costs[34]. The pressure of the covered sectors had an effect on the Spanish government, which was really concerned

[31] There is an apparent 1.4 relationship factor between economic growth and growth in CO_2 emissions.

[32] According to simulations made with the POLES model.

[33] This is in sharp contrast to the statements made by the Spanish Confederation of Business Associations (CEOE) which defends that, in order to minimise the costs for the covered sectors, the largest efforts for emission reductions should be made by the transport and domestic sectors. Lobbying activity from the well organized lobbies of covered sectors have most likely played a role in this.

[34] It can not be said with absolute certainty that the cost to the Spanish economy will be huge. This will depend, among other factors, on the price of allowances. For example, with a price of 5 € t/CO_2, the cost would "only" be 50M€ (around 0.002% of the aggregate value added of the six sectors together). On the other hand, the CEOE estimates that around 100 $MtCO_2$ will have to be bought between 2008-2012. At a price of 50 €/ton this amounts to 5000 M€, around 0.5% of GDP. A report by the CEOE estimates that complying with the KP targets will cost firms around 4000 M€ per year and will lead to increased unemployment, market quota losses and closure of firms. According to CEOE, the source of the problem is the very ambitious target negotiated by the government.

about the impact the Directive and the KP could have on the economy. These sectors maintained that in the future the government should buy credits from emission reduction projects (CDM and JI) in order to reduce the impact of commitments on industry.

**Distortion to competition.* It was feared that non-EU competitors would not experience the same requirements to control emissions. Furthermore, firms were worried about the allocation results of other countries. In order to minimise the negative effects of the NAP on the competitiveness of the Spanish sectors covered by the Directive, these firms even suggested that the Spanish NAP should take into account the methodology of the NAPs of other MS.

**Early measures, high costs and small reduction potential.* Many firms claimed that they made great efforts in reducing emissions in the past and that almost everything that could be done had already been done. Therefore, further opportunities to reduce emissions were either absent or involved very high costs. This CO_2 additionality of early measures can be questioned. They have normally been undertaken on the basis of normal business practice, i.e., to make a profit. Of course, most of these claims were strategic, i.e. made with the aim of achieving a more favourable treatment in the allocation process.

**High compliance costs.* There was a perception that the costs of complying with the Directive would be very high and that firms would be on the buy-side of the market. Firms were quite aware that this was a zero-sum game and that only by lobbying they would increase their allocation. Firms had an incentive to claim that their costs were higher than they really were. The information asymmetry between the public authorities and the polluting sources concerning MACs and the impact of those costs on production and employment was probably being used by some firms to get a more favourable treatment.

**Lobbying activity and forming of coalitions between polluters.* A lot of lobbying activity took place. The firms covered by the EU ETS were precisely those which would be expected to lobby: they were capital intensive firms which had a small-number advantage when organising and lobbying in the political arena. In order to lobby, the firms (grouped at different levels of coalitions) provided information to decision makers[35].

Informal coalitions were formed at different levels between the firms/sectors involved. One of these levels was between the covered sectors. Another was between the energy users (energy-intensive firms) versus the electricity sector. There were also coalitions between firms in a sector.

Transaction costs (bargaining costs) to form coalitions are low because these coalitions at different levels are formed between relatively small groups[36], while

[35] The information from the covered firms had a very important influence on the political arena and in the mass media. An analysis of the news on the EU ETS and the Kyoto Protocol showed that the issue of costs (for the economy or for certain sectors/firms) was the main topic, while it was hardly mentioned before. Firms were particularly successful in giving the general public the impression that emissions reductions would come at a high cost for society.

[36] The small group is more likely to lobby and win the economic struggle in the political arena.

negotiation power can be increased to a point where pressure on public authorities can be very effective (i.e., potential benefits are large). This also makes free riding less likely since the benefits of collective action exceed the costs. The additional costs of forming coalitions are very low, because individual firms would lobby on their own anyway. To sum up: the marginal cost of engaging in coalitions is rather low, while the expected marginal benefits are high.

Coalitions are more profitable at the lower levels (sector level) and the probability increases that they are formed at this level and are stable. Bargaining costs are lower at this level (since the positions of the firms are very clear and more homogeneous and firms know each other rather well by interacting in the industrial association) and the potential benefits for participants are higher (i.e., additional allowances). On the opposite side, the highest level of coalition (between all covered sectors) is much more unstable[37]. This group is rather heterogeneous. The potential benefits of action for each participant are lower, since firms are not sure that the additional allowances given to the covered sectors would be redistributed to them afterwards, although it is also true that the smaller the coalition the smaller the access to policy makers and their influence thereof. The costs of bargaining are also higher at this level since there are more participants and their interests and features are rather different[38]. Therefore, free-riding is also more likely.

Up to the beginning of 2003, there was not too much lobbying activity. The Spanish NAP had not started yet and some firms always believed that at least the first period of the EU ETS would be delayed and it would not enter into force in 2005-2007. In the past there was an almost complete lack of interaction between the business sector and the public authorities on the Kyoto Protocol and, particularly, on how to comply with the targets set in the Burden Sharing Agreements. They believed that Kyoto was distant in time and that something would be figured out at the end that would allow Spain to comply with the targets. But it was no longer possible for public authorities and firms to postpone action.

Energy-intensive industries lobbied to reduce the costs of the emissions trading scheme. Pulp and paper firms, oil refineries, cement and steel producers claimed that, together, they would fall short of 20 million tons annually since 2005[39]. The EU ETS was regarded as a threat to these companies. Their main strategies were:

- Within the national Kyoto commitments, to increase the emissions allowed in the covered sectors versus those in the non-covered sectors.
- Within the emissions allowed to the covered sectors, to increase allowances at the expense of the number of allowances grandfathered to the power companies. They claimed that, in contrast to the electricity sector, they exported

[37] The doubtful effectiveness of lobbying at higher coalition levels has also been noted by Michaelowa (1998) at the EU level.

[38] According to Olson (1965) if the number of involved parties is high, there may still not be a solution because of free-riding. According to Svendsen (2000, p. 19) when the number of partners increases, the incentive for each partner to work for the welfare of the enterprise lessens.

[39] Purchase of allowances would cost them 1000 million euro annually. The expected impact on their bottom line would be dramatic (around 20% of gross benefits).

a high proportion of their production and operated in rather open markets. They could not pass the additional costs they incur on to the consumer. They believed they suffered a double discrimination: against the non-covered sectors and against the electricity sector which is able to pass the costs by increasing the electricity price.

- Within each sector, firms tried to maximise their allocation and influence the NAP with the rules that best fitted their interests.

Another coalition was formed between CHP producers[40]. They argued that the EU ETS had negative effects on CHP. The reason was that new CHP units on sites where there were no previous CHP would increase the emissions from the site but, since CHP was more efficient than producing heat and power separately, total emissions would be reduced. However, the operator of the installation would face additional costs because he would have to buy additional allowances to cover the emissions in excess of his initial free allocation (COGEN Europe 2003)[41].

3.2.3 Responses within sectors: factors affecting different responses by firms to emissions trading and the NAP

Within some sectors, the response by the firms was not homogeneous (i.e., electricity sector) since companies showed significant differences in their climate strategies and other aspects. Depending on the structure and/or main features of the firm, emissions trading might be more or less attractive. Three main aspects of the firm may be relevant in this regard:

* *Firm size.* Although the EU ETS does not cover the smallest installations and firms, some rather small firms will still be covered. Ceteris paribus, the smaller the firm, the less attractive is the emissions trading system. Smaller firms do not usually have the management time and skills to actively participate in such a scheme, and the required efforts are too costly compared to the relatively small reduction achievements.

* *Abatement costs.* The attractiveness of emissions trading also depends on the opportunities to reduce emissions (availability of technologies) and abatement costs, which vary per sector but also per firm. Not all firms within a sector face the same abatement opportunities. Firms with low-cost abatement potentials can make a profit by selling allowances in the market. Therefore, firms with low cost reduction opportunities would probably favour this market while those which are expected to lose the most would be against. In Spain only a couple of firms so far seem ready to embrace emissions trading. The expected losers have been the most influential business voices so far.

---

40 Such as the pulp and paper industry association, the Spanish associations of bricks, tiles, floor tiles and pavements, the association of chemical industries and the association of electricity auto-generators.

41 They proposed that either the CHP installation obtains during initial allocation the allowances required to cover emissions from the separate reference case (separate production of heat and electricity), or that the amount of CO_2 saved by CHP was deducted from the actual annual emissions of the installation.

**Corporate strategies towards climate policy.* Firms differed in their environmental strategy in general and in their climate policy in particular. Strategies differed per sector but, in general, they could be grouped into three categories: reactive, passive and proactive. Proactive firms implemented measures aimed at CO_2 emission reduction and/or had already positioned themselves or carry out a transaction in the emerging CO_2 market (i.e., make offers or bids for emissions reductions)[42]. Reactive firms urgently requested a delay in the implementation of the EU ETS and/or lobby in order to receive a greater allocation. Passive firms adopted a "wait and see" strategy.

Although climate strategies were in their infancy (van der Woerd et al. 2002), a number of firms had already developed strategies aimed at GHG mitigation by implementing energy efficiency measures or by changing to less carbon intensive energy sources. This pro-activity, however, was usually not taken for CO_2 emission reduction objectives alone. Proactive firms are also those that had already taken part in the CO_2 market. Some firms had even shown international leadership in climate change mitigation. According to Dunn (2002), the reasons for taking action in the climate change realm are: the probability of government regulation, demonstration of environmental leadership, shaping the rules of the game, hedging and managing risks, generating revenue, learning by doing and seeking out side benefits[43].

In the Spanish case, proactive firms were the exception and leaders were very hard to find. None had been actively engaged internationally in proactive coalitions (such as the Global Climate Coalition). In general, the firms covered by the emissions trading directive seemed to be quite reactive (i.e., trying to delay the start of the EU ETS system) and others had adopted a "wait and see" attitude (passive strategy). This was due to the short term costs of carrying out action, to the uncertainties involved and to the lack of corporate leadership in climate change matters. Interest in emerging carbon markets remained defensive at that stage (self-protection against a carbon-constrained future predominated, rather than seeing the EU ETS as a new market opportunity)[44]. None of them built a corporate image out of climate change mitigation.

---

[42] According to Dunn (2002, p. 32), "business responses to climate change include a range of internal and external control measures. Internal controls include GHG inventory and management systems, internal GHG reduction targets, internal emissions trading systems and consideration of climate change in outside investments and research and investment into energy efficiency, fuel switching and new technologies (...). Involvement in trading or other flexibility mechanisms is becoming a common external control."

[43] Dunn further considers that the technological, economic and policy dimensions of climate change are essential in order to understand and formulate corporate strategic responses.

[44] According to Point Carbon (7/3/2003) some transactions in the EU allowance market had already taken place. However, at the moment window-shopping seemed to be more common than the real thing, and many firm bids had not been upheld. Still, there were a number of indicative bids around by companies considering taking action in order to hedge the risk of high carbon prices in the future. However, the absence of serious offers was worth noting. Most actors seemed to fear selling at this point, before national alloca-

Initiatives specifically aimed at emissions reductions were virtually lacking, most were carried out for normal business practices (i.e., energy efficiency measures). Firms had neither set internal targets, nor had they created internal emissions trading systems, and measures for positioning in the emerging emissions markets were virtually lacking. They had not considered climate change in outside investments, although electric utilities were increasingly interested (and some even engaged) in CDM projects. Companies covered by the EU ETS perceived emissions trading as a threat to their business. In general, they anticipated that they would be allocated a limited number of CO_2 allowances, and that actively participating in the buy side of the market was necessary in order to maintain or even increase the current production level. Intense lobbying to get a larger allocation was the most common strategy.

4 Concluding remarks

This paper has provided an overview of the distinct views, interests and strategies of different Spanish actors concerning the allocation of allowances in the EU ETS, focusing on the sectors covered by the emissions trading directive.

We can conclude that the emissions trading directive has not been well received by the covered firms. Although emissions trading has proved to be a cost-effective instrument for the reaching of emission targets, it leads to significant impacts for those obliged to abate emissions. They will either incur abatement costs or will have to buy allowances specially taking into account the high growth in production expected by the firms affected. Thus, it is normal that firms complain and devote resources for lobbying to get a larger allocation.

At the individual firm level, the Directive was considered to be a big threat. Spanish firms argued they would be on the buying side of the market. Of course, this could also reflect a strategic positioning in order to be more favourably treated when allowances are allocated by the NAP. At the sector level, we have noticed that distinct sectors put forward different arguments in order to have more allowances allocated to them. Coalitions were created at different levels and lobbying had already started.

The negotiations taking place in the context of the NAPs reflected the tension between the need to control emissions, the arguments put forward by firms and sectors and the minimisation of the negative impact on the overall economy.

References

ABC journal (2003) Various issues from May to December 2003
Carbon Market News (2003) Various dates. http://www.pointcarbon.com

tion plans had been determined and they knew whether they were long or short. The unfinished business over NAPs would be crucial for the future price levels.

Desai S, Michaeowa A (2001) “Burden sharing and cohesion countries in European climate policy: the Portuguese example”. Climate Policy 1: 327-341
Del Rio P (2002) Industry, Tecnological Change and Sustainable Development. Patterns of adoption and diffusion of clean technologies in the pulp and paper industry in Spain. PhD thesis. Department of Structural and Development Economics. Faculty of Economics and Business. Universidad Autónoma de Madrid. July 2002
Dunn S (2002) “Down to Business of Climate Change. An overview of Corporate Strategies”. Greener Management International 39: 27-41
EL PAIS journal (2003) Various issues from May to December 2003
European Commission (2002) EU Energy and Transport in Figures. Statistical Pocketbook 2002, Brussels
European Commission (2003) Directive 2003/…/EC of the European Parliament and of the Council establishing a scheme for greenhouse gas emission allowance trading within the Community and amending Council Directive 96/61/EC
Gerhardsen (1998) Who governs the environmental policy in the EU? A study of the process towards a common climate target. Center for International Climate and Environmental Research, Oslo
Hernandez F, Del Rio P, Gual MA, Caparro’s A (2004) “Energy Sustainability and Global Warming in Spain”. Energy Policy 32: 383-394
Michaelowa A (1998) “Impact of interest groups on EU climate policy”. Eur. Environ. 8 (5): 152-160
Ministry of economics (2002) Plan of Electrical and Gas Infraestructures
Mueller D C (1989) Public Choice II. Cambridge University Press, Cambridge
Olson M (1965) The Logic of Collective Action. Cambridge University Press, Cambridge
Pwc-Energpresse (2002) Climate Change and Power industry. European Carbon Factors. Paris
Schneider F, Volkert J (1999) “No chance for incentive-oriented environmental policies in representative democracies? A Public Choice analysis”. Ecological Economics 31: 123-138
Svendsen G T (2000) Public Choice and Environmental Regulation.Tradable Permit Systems in the United State and CO_2 Taxation in Europe. Edward Elgar. Cheltenham, U.K.
Szabo L, Hidalgo I, Ciscar JC, Soria A (2003) CO_2 Emission Trading in the European Union: the Cement Industry Case. Institute for Prospective Technological Studies. Sevilla
van der Woerd F, Levy D, Begg K (2002) ”Introduction to Corporate Responses to Climate Change”. Greener Management International 39: 22-25

UK's climate change levy and emissions trading scheme: implications for businesses' productivity and economic efficiency

Adarsh Varma*

The University of Hull Business School
Centre for Economic Policy
Hull, HU6 7RX, United Kingdom
a.varma@hull.ac.uk

Abstract

This paper sets out a theoretical model to analyse the effect of UK's climate change levy (CCL) and the emissions trading scheme (ETS) on the productivity and overall economic efficiency of businesses. A microeconomic analysis at plant level enables us to understand the implications of compliance cost for the overall cost structure of the plant. Some of the dynamics of an economic instrument can be lost at an aggregate level. We will use a stochastic translog frontier cost function with a cost inefficiency component to analyse whether the two instruments affect the overall economic efficiency of the firm. The model allows for factor substitution to take into account changes in relative factor prices, which can reduce the perceived adverse effects of the two instruments on productivity and efficiency of plants. The paper also briefly outlines the implications for the cost function due to some of the key developments since their inception in 2001 (CCL) and 2002 (ETS).

Keywords: Energy taxes, emissions trading, cost efficiency, stochastic frontiers

Acknowledgement: I am grateful to my supervisors Chris Hammond and Dr. Jonathan P. Atkins at the University of Hull for their advice and support. A word of thanks to Dr Rüdiger Wurzel, University of Hull for recommending the joint workshop by 'Martin-Luther-University of Halle-Wittenberg', Faculty of Economics and GOR to present my paper.

---

* This paper is part of an on going part-time PhD research at the University of Hull, UK. The author is currently working as an Economist with the Greater London Authority.

1 Introduction

The UK has made headlines all over the world by pioneering the first national emissions trading scheme (ETS) in April 2002. A year earlier it had introduced the climate change levy (CCL) on business use of energy to initiate its climate change programme, which was set up to meet its Kyoto (Grubb et. al 1999) commitments. The first mover advantage of using environmental economic instruments can prove to be significantly worthwhile once the EU emission trading scheme comes into effect in January 2005. This however, does create problems regarding the uncertainties surrounding the scope of the scheme, rules for allocating emission allowances and the likely market price for carbon. Some of these issues will be addressed in the later part of this paper.

The rationale for using economic instruments in environmental policy is provided by the existence of externalities (Pigou 1920), which affect people and the environment. By levying a tax or charge on the production activity, which gives rise to the negative effect, the external cost can be partially or wholly internalised (Pearce 1991; Hjollund and Svensen 2001). A trading programme directly restricts the total amount of carbon (through a "cap" on allowable emissions) and leads to a market price for carbon allowances (i.e. the right to emit a ton of carbon or carbon equivalent) (OECD 1993). Setting a tax rate (P_C) that would lead to a reduction in carbon (Q_C), can also be achieved by setting a level of allowable carbon emissions (Q_C) and allowing trading to result in a carbon allowance price of P_C (Hahn and Noll 1982). The price and quantity would thus be the same with equivalent tax or emissions trading programmes.

The two main economic instruments in question, CCL and ETS, are designed to induce behavioural change and the adoption of more environment friendly technologies. They are implemented as a package containing incentives and opportunities for firms to reduce compliance cost and attain their targets with maximum flexibility. The next section of the paper briefly describes the two instruments. In the third section we present the stochastic translog frontier cost function model, which will be used to analyse the implication of the policy on productivity and the overall economic efficiency of plants in the Yorkshire and Humber Region of UK. In section 4, we will discuss the dynamics of the model. We will briefly present an economic analysis of firms' recent experiences with CCL and ET in section 5. We will present the conclusions in the last section outlining ongoing and future work.

2 Overview of UK's climate change levy (CCL) and emissions trading scheme (ETS)

The CCL (DEFRA, 2001) is a key instrument in the Government's package of measures to meet its Kyoto targets and to reduce CO_2 emissions in the UK. The Government also has a domestic goal of reducing carbon dioxide (CO_2) emissions by 20% from 1990 levels by 2010[1]. The levy has been designed to entail no overall increase in the burden of taxation on business as it is accompanied by a 0.3% cut in employers national insurance contributions (NIC) and support for energy efficiency schemes and renewable energy sources. The CCL is charged only on industrial and commercial use of energy and exempts domestic use of energy, energy used by public transport, energy from "new" renewables and energy in "good quality" Combined Heat and Power (CHP) plants.

The Levy Rates[2] (£/kWh) are:

Electricity	0.43 (0.70 €/kWh)
Gas	0.15 (0.25 €/kWh)
Coal	0.15 (0.25 €/kWh)
LPG	0.07 (0.11 €/kWh)

The design of the levy aims to achieve a balance between its environmental objectives and administrative simplicity (see Varma, 2003 for an economic discussion of the CCL). Due account has been taken of the structure of the energy industries, the need to minimise compliance costs for businesses and to protect the competitive position of UK industry, as well as the need to administer the specific reliefs and exemptions. All firms which are regulated by the EU Integrated Pollution Prevention and Control Directive (IPPC) or are in sectors regulated by the IPPC, but which are themselves too small to be covered by the directive are eligible to sign negotiated agreements under climate change agreements (CCAs) to obtain an 80% discount on the levy. This is based on the condition that the firms will reduce energy use or increase energy efficiency over a 10-year period. The levy thus raised £831 (€ 1350) million (ONS, 2003) instead of the £1 (€ 1.6) billion initially envisaged, due to a number of firms signing CCAs in the course of the year.

Quantitative targets have been derived at sector level by negotiating the potential improvements that could be made in a cost-effective way compared to a base year. Targets may be absolute or relative. Absolute targets are either based on energy use or carbon emissions by a sector and relative targets are based on energy use or carbon emissions per unit output.

The UK emissions trading scheme (ETS) was introduced in April 2002 to provide firms with added incentives and flexibility to reduce emissions and energy use. The Confederation of British Industry (CBI) and the Advisory Committee on Business and the Environment (ACBE) set up the UK emissions trading group in June 1999 to take forward the design of the scheme.

The UK's ETS allows for the following four types of participants (DEFRA, 2003): (1) ***Direct participants*** who are required to take on absolute annual greenhouse gas (GHG)[3] emission reduction targets (on the basis of a 1998-2000 baseline for which the average GHG emissions are calculated) for the years 2002-06. The government made available financial incentives worth £215 (€ 350) million

[1] ACBE Eighth Progress Report. Pg 10 (http://www.dti.gov.uk/acbe8/).
[2] The actual rate reflects the conversion factors of source energy into electricity.
[3] The six GHGs are carbon dioxide (CO_2),methane (CH_4),nitrous oxide (N_2O), halofluorocarbon (HFC), perfluorocarbon (PFC) and SF6.

over five years in order to entice direct participants to take on voluntary absolute reduction targets which operate on a 'cap and trade' basis; (2) ***Participants in climate change agreements (CCAs)*** who have accepted (absolute or relative) GHG emission and/or energy reduction targets by entering into these agreements. Companies which fall under the EU Integrated Pollution Prevention and Control (IPPC) directive can sign (on a company-by-company basis or as an entire sector) negotiated voluntary agreements with the UK government about the reduction of their CO_2 emissions and/or energy use. These companies are entitled to an 80 per cent rebate on the CCL, which explains the high CCA participation. Significantly, CCA participants can make use of emissions trading in order to meet their CCA reduction targets. They can also sell any over-achievements based on a baseline credit approach. A gateway controls the flow of allowances from CCA participants, which are often committed merely to relative (rather than absolute) emission reduction targets, into the general ETS. The gateway aims to prevent the trade in 'hot air' because absolute emission reduction targets put forward by direct participants could be exceeded by relative targets agreed on in CCAs. (3) ***Anyone*** is free to trade allowances on a speculative basis by simply opening a trading account with the emission trading authority (ETA); (4) Any company which has a ***government approved GHG reduction project.***

Both direct and CCA participants have the following three options. First, they can reduce their emissions to the agreed target. Second, they can reduce their emissions below the agreed target while selling or banking their excess allowances. Third, they can leave their emissions above the agreed cap or even increase them if they buy a sufficient number of allowances from other ETS participants so that they can meet their agreed targets

3 The stochastic translog frontier cost function model

We will use a four input and two output stochastic translog frontier cost function with a separate component for change in economic efficiency (Battese and Coelli 1995). This component models the effect of the implementation of the two instruments on the overall economic efficiency (EE) or cost efficiency of plants. Overall EE is a measure given by the combination of allocative efficiency (AE)[4] and technical efficiency (TE)[5]. The two outputs are 'Y' the desired good and 'A*' referred to as pollution abatement activity (PAA), which is a proxy for the undesired good. PAA (A*) is the component of A explained by statistically significant exogenous variables[6] in the regression equation 1. An element of A is capital and it also comprises of ongoing cleaning up costs. Together we will call 'A' as pollution abatement costs (PACs). Using the regression values 'A*' as a measure of regulatory

[4] The right mix of inputs to produce a given quantity of output at minimum cost.
[5] Achieving maximum output from each vector of inputs.
[6] Description of variables in equation 1 given in the appendix.

burden we model the cost function in Equation 2. A* is thus a proxy for environmental improvement i.e. a reduction in the undesired output.

$$A = f(\text{ENGEXP, ENFORCE, TECH1\&2, COMP, INSP, EMISO, POP}) \qquad (1)$$

Since government regulations climate change levy (CCL) and emissions trading scheme (ETS) promote emission reduction, firms will have to adopt techniques for reducing pollution caused by the production of Y and increase energy efficiency to reduce the burden imposed by the levy.

$$\ln C_{it} = \beta_0 + \beta_Y \ln Y_{it} + \beta_A \ln A^*_{it} + \sum_{j=1}^{4} \beta_j \ln p_{jit} + \sum_{j \le k=1}^{4}\sum^{4} \beta_{jk} \ln p_{jit} \ln p_{kit} + \beta_{AA} \ln A^{*2}_{it}$$
$$+ \sum_{j=1}^{4} \beta_{jA} \ln p_{jit} A^*_{it} + \beta_{YY} \ln Y^2_{it} + \sum_{j=1}^{4} \beta_{jY} \ln p_{jit} \ln Y + \beta_{YA} \ln Y_{it} A^*_{it} + v_{it} + u_{it} \qquad (2)$$

note: the cross-product term involving j=k are multiplied by ½ not shown in equation for convenience. Also a time trend can be incorporated to take account of technical progress.

'p_1' is price of capital (K),
'p_2' is price of labour (L),
'p_3' is price of non-energy materials (M),
'p_4' is price of price of energy (E)
v_i accounts for measurement error and other random factors.

The u_{it}s, give the distribution of the overall economic efficiency of the i_{th} firm and are assumed to be independent non-negative truncation of the normal distribution with unknown variance, σ^2, and means δz_{it} (i=1,2,3,...N and t=1,2,3,...T). Thus, the means may be different for different firms and time periods (t) but the variances are assumed to be the same.

$$U_{it} = \delta_0 + \sum_{l=1}^{5} \delta_l z_{lit} + w_{it} \qquad (3)$$

z_1 log of energy spending (inclusive of CCL) divided by the total value of plant's output in that year.

z_2 PACs and energy efficiency expenditures divided by total costs.

z_3 Foreign ownership (1 for FO and 0 for home), assuming that foreign owned firms could relocate production overseas, and reduce the effect on economic efficiency.

z_4 Dummy variable if the plant is under IPPC and has signed a CCA to avail the 80% rebate or is participating in the ETS through the direct entry route.

z_5 log of capital labour ratio (to show capital or labour intensity).

w_{it}s are unobservable random variables, which are assumed to be independently distributed, obtained by truncation of the normal distribution with mean zero and unknown variance σ^2, such that U_{it} is non-negative.

The cost function takes into account relative factor price changes and by assuming fixed factor intensities we can discuss the inefficiency component (U_{it}). The *z* variables may influence economic efficiency (EE) keeping in mind the regulatory factor price manipulation. Changes in relative factor prices may divert resources towards carbon abatement and energy efficiency and away from the actual production of the desired good. A low energy spend (z_1) is unlikely to significantly affect EE and capital/labour intensity will further make the sensitivity to factor price changes more apparent. High pollution abatement costs and energy efficiency expenditures (z_2) means that now due to the regulation more capital is diverted towards pollution abatement and energy efficiency instead of production efficiency[7]. Production efficiency can be interpreted as producing more of the desirable output at lesser cost or improvements to the desirable output at lesser cost. In this case product development may be hampered due reduced inputs in terms of managerial attention, labour, non-energy materials, energy and capital investment. Thus, the input-output ratio of the plant may become higher and the productivity/efficiency of the plant may fall. Foreign ownership (z_3) may allow greater flexibility in resource allocation or enable plants to relocate production to another country where regulation is more favourable. In other words relocate a part or whole of the pollution activity elsewhere to reduce compliance costs. It is also assumed that foreign owned firms are generally more productive and enjoy higher research and development (R&D) spill over effects (Harris 1999). The CCAs and ETS incentive monies (z_4) help to reduce the compliance cost[8] and over all economic inefficiency. Log of capital-labour ratio (z_5) shows capital intensity, so if energy and capital are complements then an increase in energy price will reduce demand for capital and affect economic efficiency for a capital-intensive plant. A labour intensive firm assuming that labour and capital are fairly substitutable will adjust more easily and cost effectively to new factor prices and would thus have a smaller effect (or negative effect) on economic efficiency.

Using FRONTIER (Coelli 1996) we can estimate the cost efficiency estimates for each firm within each industrial sector and cost efficiency change over time[9]. The FRONTIER software also provides an estimate for $\gamma=\sigma_U^2/(\sigma_V^2+\sigma_U^2)$ where σs are the variances of the error term u_i and v_i. This value is bounded between 0 and 1 where, zero indicates that the deviations from the frontier are due to noise, while a value of one would indicate that all deviations are due to cost inefficiency (Coelli et. al 1998). The null hypothesis $\gamma=0$ can be checked by the 'One-sided Generalised Likelihood-Ratio test', the value for which is provided by FRONTIER.

---

[7] Increase in energy efficiency, can also generate 'win-win' solution whereby production efficiency also leads to less pollution (Porter and Linde 1995).

[8] Compliance cost is measured as the plant's average annual Operating costs for pollution abatement divided by the plant's average value of shipments/production over some period.

[9] Since we are using panel data.

4 Dynamics of the model

The model will include data for the time period 1995 to 2004. This will take into account three phases. Since the policy was announced in 1999, the first phase will look at firms' production activity in the pre policy period from 1995 to 1999. The second phase will be the transition period from 1999 to 2001, which is when the policy came into effect. The third phase will be from 2001 to 2004 to infer the implications of the policy on overall economic efficiency.

The regulation policy works through the price mechanism. The incentives provided by the climate change levy agreements help to reduce compliance cost. Essentially, the levy package increases the price of energy and makes labour (and non-energy materials) cheap. Producers can then adjust to the extent implied by the cost function parameters. The two economic instruments (for regulation) will lead to an increase in PAA. The policy can have two main effects - Firstly, it may lead to an increase in PACs and secondly, it will lead to a change in combination of inputs to produce the desired good due to change in factor prices. These two effects will have an impact on the cost function of the plant, a subsequent impact on profits and eventually will affect price and output of the desired good. Allowing for factor substitution due to changes in relative factor prices will reduce the perceived adverse effects of the energy tax on the productivity and efficiency of plants. Barbera and McConnell (1990) have measured the indirect effect of abatement requirements on productivity as a result of change in factor inputs to accommodate the abatement capital. The indirect effect of abatement capital refers to labour and capital diverted from the actual production of the desired good towards the reduction of the undesired good (emissions). We assume that factor inputs are endogenous and factor prices are assumed to be exogenous since the government is manipulating them. We can measure whether these environmental regulations affect the adjustment elasticities of given inputs in addition to labour and pollution abatement capital expenditure. Factor demand elasticities can also determine whether factor intensities can change without affecting efficiency and productivity.

The regulation will also lead to the adoption of new technologies and allocation of R&D resources towards pollution abatement and energy efficiency measures. The cost of producing the desired output with the new or evolved technology will include all pollution abatement and energy efficiency costs. These costs can either be: i) *measurable* - such as end of pipe solutions and other post production measures, ii) *hidden* - these are costs associated with a change in the production process and/or fuel switching, which reduces emission and increases efficiency and iii) *invisible* - costs associated with the new technology i.e. the opportunity cost of the reallocation of R&D resources (DeBoo 1993, 1995). It is very difficult to estimate the total costs of these environmental regulations, as it is difficult to accurately estimate the invisible costs. An estimate of invisible costs can be the benefits from reduction in the levy, receipt of incentives monies and green market image.

The model's aim is to account for post-policy energy price effect as well as the incentives and flexibility provided by CCAs and the ETS for measuring total economic efficiency. The impact on efficiency has to be analysed of the policy as a whole and not just from the point when the price of energy has gone up. The model takes into consideration the price effect and other incentives (e.g. CCAs, ET, etc.) that lead to pollution abatement activity (as we use dummy variables to take these effects into consideration). There is an important linkage between overall economic efficiency and the CCA[10] targets. The CCAs instil greater efficiency and also a reduction in the levy at the 'opportunity' cost of investment diverted from the actual production of the good. Thus, this can be allocative efficient but most likely can also reduce technical efficiency leading to a reduction in overall economic efficiency.

If results show that emissions or SEC[11] has gone down as a result of increase in pollution abatement activity, benefits can arise in the form of rebates on the levy and fall in compliance costs. Overall cost structure may not be significantly affected if the firm achieves AE at the cost of TE. Moreover, emission scenarios of plants (units/establishments) also depend on the growth rates of their usual outputs (and environmental outputs). If a growth rate of the normal output is low, investment opportunities for plant replacement or modernisation will be low. In the cost function output may fall as a result of fall in inputs or relocation of production to other countries or any other reason. Output may fall because of international market issues, cyclical trends affecting one sector more than another and thus the regulation having a greater impact on certain sectors (energy intensive) such as steel, chemicals, automobiles and the engineering sector.

5 The experience so far

How firms have adapted and/or changed their behaviour in the wake of the new environmental regulations determines the effectiveness of our model. The experience so far provides a mixed picture in terms of how successful the two instruments have been in achieving their objectives.

According to DEFRA's Global Atmospheric Division (GAD) the CCAs sector targets add up to a savings of around 2.5 millions tones of carbon (MtC) emissions per year compared to the business as usual scenario. On the other hand the price effect of the levy, assuming the levy is in place with no negotiated agreements and associated discounts, would give rise to a saving of 0.25 MtC per annum (ETSU, AEA 2001).

[10] Climate change agreements provide rebate on the levy for voluntary reduction energy / efficiency targets and incentive monies.

[11] Specific Energy consumption - energy consumed per unit of output.

Table 1. Comparison of Direct Entry UK ETS and CCAs

Comparison of Direct Entry UK ETS and CCAs		
	Climate Change Agreements	Direct Entry - Emissions Trading Scheme
Number of firms	5.500	32
Number of sites	14.000	1.000[a]
Emissions covered ($MtCO_2e$)	100[a]	30[a]
Nominal emission reductions in final year from base year ($MtCO_2e$)	9.2	4.0
Average emission reduction per year ($MtCO_2e/yr$)	0.9	0.8

Note: [a] Enviros Estimate
Source: ENVIROS (2003)

Table 1 shows that about 14.000 sites belonging to 5.500 companies entered CCAs, which covered about 100 million tonnes of CO_2 equivalent ($MtCO_2e$) emission reduction targets. In comparison 34 direct participants with about 1.000 sites were covered for approximately 30 $MtCO_2e$ emission reduction targets. Table 1 also shows that the nominal emission reduction in the final year (compared to the baseline) amounts to about 9.2 $MtCO_2e$ for CCA participants and 4.0 $MtCO_2e$ emissions for direct participants. The CCAs cover significantly more companies and sites than the direct participants and are thus expected to deliver nearly twice as many CO_2e emission reductions. Our cost function can capture this trend to show implications for overall economic efficiency. An increase in companies signing CCAs would lead to an increase in pollution abatement activities and a rebate on the levy (lower energy costs), which are both captured in the model.

However, the price development on the emissions trading front is leading to disequilibrium in the market. The latest closing price for UK allowances was just £2.30 (€ 3.7) per tonne CO_2e for fewer than 2000 allowances[12]. The low price for allowances means that firms unable or unwilling to meet their reduction targets through in-house abatement measures could buy allowances cheaply on the market. On the other hand, the low price of permit and CCA participants being able to ring-fence any over-achievement would restrict permit availability for sale to their competitors. Ring-fenced allowances that are offered for sale first have to be verified which adds costs. This explains why only companies with substantial allowances considered getting their allowances verified. The first year of the ETS has been characterised by a massive surplus of supply over demand with the price for allowances having been determined very much by seller behaviour. Future permit price, which is contingent on permit supply and emission cap will determine strategic behaviour in the current period regarding innovation and output decisions.

The issue which has generated the most rhetoric, is the possibility of firms being allowed to enter "hot air"[13] reductions in the emissions trading market. This is mainly because firms did not behave in a rational way. Instead of developing

[12] Telephone interview with Lucy Mortimer, CO_2e.com on 23.10.2003.

[13] 'Hot air' occurs if a firm adopted an emission limit that was higher than its business-as-usual emissions.

strategies based on marginal abatement costs[14] firms used qualitative risk assessment for rewards at different auction price (DEFRA 2002). It is arguable that loop holes in DEFRA's methodology for arriving at baseline scenarios and/or an extremely wide interpretation of the ETS rules have allowed companies (such as Ineos and DuPont, see Table 2) to trade a significant amount of "hot air" surpluses (ENDS Report 2002, 327, pp. 3-5 and 335, pp.34-5).

Table 2. Top ten successful bidders under the direct entry route for the UK ETS

Winners (34 organisations):	Overall target tCO_2e	Annual target tCO_2e	Annual Incentive Payment	Baseline - tCO_2e	Final Bid - % reduction from baseline	Final Bid - % of reductions secured from ETS
Ineos Fluor Ltd	805.635	161.127	£ 8.599.347 €14.000.000	1.861.863	43.27	20.00
Dupont (U.K.) Ltd	500.000	100.000	£ 5.337.000 € 8.674.760	2.626.226	19.04	12.41
Shell UK Ltd	438.750	87.750	£ 4.683.217 € 7.612.100	3.877.906	11.31	10.89
Rhodia Organique Fine Ltd	430.000	86.000	£ 4.589.820 € 7.460.293	2.012.275	21.37	10.67
UK Coal Mining Ltd	400.000	80.000	£ 4.269.600 € 6.939.807	4.922.175	8.13	9.93
British Petroleum plc	353.500	70.700	£ 3.773.259 € 6.133.055	6.757.799	5.23	8.78
First Hydro Company	285.000	57.000	£ 3.042.090 € 4.944.613	1.370.410	20.80	7.08
Lafarge Cement UK	250.000	50.000	£ 2.668.500 € 4.337.379			6.21
British Airways plc	125.000	25.000	£ 1.334.250 € 2.168.689	1.011.785	12.35	3.10
British Sugar plc	100.000	20.000	£ 1.067,400 € 1.734,951	579.367	17.26	2.48

Source: http://www.defra.gov.uk/environment/climatechange/trading/pdf/trading-progress.pdf

For example, table 2 shows that two chemical companies, Ineos Fluor and DuPont, were overall granted more than £13.8 (€ 22.4) million in fiscal incentives for their reduction targets. However, according to ENDS Report (2002, 335), these two (and some other) companies merely fitted abatement equipment, which had already been required under existing legislation or were part of their investment decisions. Moreover, Ineos' CO_2e baseline scenario became inflated by taking into account all of its volatile organic compound (VOC) emissions for 1998-2000 rather than merely the potent climate change gas HCFCs (which is a VOC) for which it will have received incentive payments of £43 (€ 70) million by 2006.

[14] In theory the auction provides incentive for firms to bid below their business as usual emissions that can be achieved at no cost to capture as much of the incentive money as possible. They would do so up to the point at which marginal abatement costs equal the net benefit of the incentive payments.

Ineos' risk of having taken on a binding CO_2e cap therefore appears non-existent. It is already ahead of its 2006 reduction target for CO_2e emissions (let alone the less demanding annual interim targets). By the end of 2002, Ineos had already achieved a 1 million ton CO_2e emission surplus compared to its first year target.

On the other hand, companies which undertook significant CO_2e reduction measures in the early 1990s (i.e. before the 1998-2000 baseline years) have lost out. One such example is Corus Steelworks (formerly known as British Steel) in Scunthorpe, UK. Corus faces a high marginal abatement cost per unit of emission reduction and further pollution abatement expenditure in addition to its 'early actions' would impose a severe burden in terms of being competitive. It is therefore only prepared to undertake additional reduction measures if rewarded through roughly equivalent fiscal incentives (or required by legislation). Corus is even considering locating most of its CO_2e production activities abroad (Varma 2000). The time period in our model should enable us to infer this phenomenon where companies may have been severely disadvantaged by undertaking abatement and/or energy efficiency measures prior to the introduction of the CCL and ETS in 2001 and 2002 respectively.

However, a crucial question is the links between the UK ETS and the EU emissions trading scheme (EUETS), which came into affect on January 2005. The EU ETS covers 1055 installations in the UK and more than 12,000 across the EU. Some firms under the CCAs and direct entrants of the ETS have been allowed to opt-out from the first phase of the EUETS. The opt-out has been granted if installations can deliver emission reductions equivalent under the EUETS. Currently, there is no link between the UK ETS and the EUETS. The UKETS's approach is to cover all activities on a site which emit CO_2, however the EU scheme is confined to specific activities as decided by the commission. Hence, some activities such as reheat furnaces on iron and steel works and heating units on chemical works, could be excluded from the EUETS (ENDs report, June 2003, 341).

6 Conclusion

On paper both instruments have shown to be successful in reducing emissions. Of the 44 industrial sectors under the CCAs 24 met their target as a whole. The agreements have resulted in a reduction of 4.3 million tonnes of carbon equivalent (FES 2003) from the baseline period. 31 of the 32 Direct Entry participants under the ETS have met their first period emission targets. However, the disequilibrium in the trading market and 'Hot air' has undermined the environmental effectiveness of the ETS. This was a cause for concern for firms seeking to opt-out from the first phase of the EUETS.

The paper's main aim is to investigate micro economic effects at the plant level, which generally gets lost in analyses at the industrial sector level. On the basis of pollution abatement activities (A*) and the parameters in the inefficiency component we should be able to attribute the change in overall economic efficiency due to the two economic instruments. The impact of the entire policy package on firms' behaviour will thus be analysed and not just the price effect of the CCL. The model's ability to capture factor price changes due to the two economic instruments should provide us with a clearer picture of firms' compliance cost and their ability to maintain their competitiveness and overall economic efficiency.

References

Barbera AJ, McConnell VD (1990) 'The Impact of Environmental Regulations on Industry Productivity: Direct and Indirect Effects', Journal of Environmental Economics and Management 18: 50-65

Battese GE, Coelli T (1995) 'A Model for Technical Inefficiency Effects in a Stochastic Frontier Production Function for Panel Data', Empirical Economics 20: 325-332

Business data linking (BDL) (April 2002) An introduction by Matthew Barnes and Ralf Martin, Economic Trends 581 http://www.statistics.gov.uk/articles/economic_trends/business_data_linking.pdf

Coelli T (1996) A Guide to FRONTIER Version 4.1: A Computer Program for Stochastic Frontier Production and Cost Function Estimation, Centre for Efficiency and Productivity Analysis (CEPA) Working Paper, 96/07

Coelli T, Prasada Rao D S, Battese G E (1998) An introduction to efficiency and productivity analysis. Kluwer Academic Publishers, Boston

DeBoo AJ (1993) 'The Costs of Integrated Environmental Control', Statistical Journal of the United Nations Economic Commission for Europe 10(1): 47-64

DeBoo AJ (1995) 'Accounting for Costs of Clean Technologies and Products'. In: Second Meeting of the London Group on Natural Resource and Environmental Accounting: Conference papers, (ed) London Group on Natural Resource and Environmental Accounting, pp. 125-143, Washington, DC: Bureau of Economic Analysis

DEFRA (2001) UK Climate Change Agreements and the Climate Change Levy http://www.defra.gov.uk/environment/ccl/index.htm

DEFRA (October 2002) The UK Emissions Trading Scheme - Auction Analysis and Progress Report.

DEFRA (2003) UK Emission Trading Scheme, http://www.defra.gov.uk/environment/climate change/trading/reia.htm

ENVIROS Consulting Ltd. (2003) A Qualitative Study of the Direct Entry

UK Emissions Trading Scheme. A report prepared by the Climate Change Policy and Strategy Group of Enviros Consulting based on the MSc research carried out by Katharina Kröger at the University of Nottingham

ETSU AEA (2001) Climate change agreements – Sectoral energy efficiency targets http://www.defra.gov.uk/environment/ccl/pdf/etsu-analysis.pdf

Future Energy Solutions (FES) (2003) Climate Change Agreements - Results of the first target period assessment, Version 1.1 - Preliminary Results http://www.defra.gov.uk/environment/ccl/pdf/cca_tp1_prelim.pdf

Grubb M, Vrolijk C, Brack D (1999), The Kyoto Protocol: A Guide and Assessment. The Royal Institute of International Affairs, London UK

Hahn R, Roger N (1982) 'Designing a Market for Tradeable Emission Permits'. In: Magat, Wesley (ed) Reform of Environmental Regulation, Ballinger, Cambridge, pp. 119-146

Harris RID (1999) Productive efficiency in UK manufacturing 1974-1994: Estimates of five leading sectors, University of Portsmouth.

Hjollund L, Svendsen GT (2001) 'Why Green Taxation?' Energy and Environment 12 (1): 29-38

OECD (1993) Taxation and the Environment: Complementary Policies, OECD, Paris.

Office of National Statistics, Annual Abstract of Statistics publication, (2003), http://www.statistics.gov.uk/downloads/theme_compendia/Aa2003/Annual_Abstract_2003.pdf

O'Mahony M (1999) 'Britain's productivity performance 1950-1996: An International Perspective, London: National Institute of Economic and Social Research

Pearce D (July 1991) 'The Role of Carbon Taxes in Adjusting to Global Warming'. The Economic Journal 101(407): 938-948

Pigou A (1920) The Economics of Welfare. London: MacMillan

Porter ME, van der Linde C (1995) 'Toward a New Conception of the Environment-competitiveness relationship'. The Journal of Economic Perspectives, 9 (4): 97-118

Varma A (2000) The UK's Climate Change Levy: Competitiveness and Environmental impacts in the Humber and Yorkshire region. Dissertation submitted for the Degree of MSc (Econ) Public Policy, The University of Hull

Varma A (2003) 'The UK's climate change levy: cost effectiveness, competitiveness and environmental impacts', Energy Policy 31: 51-61

Appendix

Variable	Description	Source
A (Pollution abatement costs)	Annual pollution operating cost for pollution abatement (and energy efficiency i.e. if the system which reduces emissions also increases energy efficiency) divided by the total value of the plant's output in that year[15].	Environment Agency and Questionnaire
ENGEXP	Energy spending (inclusive CCL) divided by the total value of plant's output in that year or percentage of energy costs in total costs.	Questionnaire
ENFORCE	Number of air pollution enforcement actions in a year.	Environment Agency
TECH1, TECH2 (dummy variable)	Dummy variables for each industry to account for plant's technology. It proxies for the difficulty in meeting pollution standards or reducing compliance costs. E.g. oil refineries may use catalytic cracking, steel plants using electric arc technology instead of blast furnaces, etc.	Questionnaire IPC booklet from Environment Agency
COMP (dummy variable)	Compliance status with air pollution regulations (0 if in violation during any period (month) of the year, 1 if not)	Environment Agency
INSP	Number of air pollution inspections in a year.	Environment Agency
EMISO	Total emissions in tonnes per annum divided by the total value of the plant's output (£s) in that year.	Environment Agency and Questionnaire
POP	Population density of the Local Authority District in which the plant is located denominated as people per hectare. This variable proxies for variations in regulatory stringency and the perceived extent of pollution hazard.	Environment Agency and local authorities.
Y	Output of the firm in value terms (£s)	FAME Business database
Price of Capital (K)	Will be a constant value for each plant for a specific sector on the assumption that they will not vary considerably within one region.	O'Mahony (1999)
Price of Labour (L)	Total expenditure on labour costs divided by number of workers.	FAME Business database
Price of Non-energy materials (M)	Total expenditure on non-energy materials divided by quantity of non-energy materials.	Questionnaire
Price of energy (E)	Total expenditure on energy costs divided by quantity of energy.	Questionnaire

[15] To take into consideration plant size. We can also use percentage of costs of investment in pollution control equipment.

The sources of emission reductions: evidence from U.S. SO_2 emissions from 1985 through 2002

A. Denny Ellerman, Florence Dubroeucq[1]

The Massachusetts Institute of Technology
Center for Energy and Environmental Policy Research
77 Massachusetts Avenue, Cambridge, MA 02139-4307, USA
ellerman@mit.edu

Abstract

An enduring issue in environmental regulation is whether to clean up existing "old" plants or in some manner to bring in new "clean" plants to replace the old. In this paper, a unit-level data base of emissions by nearly 2000 electric generating units from 1985 through 2002 is used to analyze the contribution of these two factors in accomplishing the significant reduction of sulphur dioxide emissions from these sources in the United States. The effect on SO_2 emissions of the new natural-gas-fired, combined-cycle capacity that has been introduced since 1998 is also examined. The results indicate that cleaning up the old plants has made by far the greatest contribution to reducing SO_2 emissions, and that this contribution has been especially large since the introduction of the SO_2 cap-and-trade program in 1995. The new natural-gas-fired, combined cycle units have displaced conventional generation that would have emitted about 800.000 tons of SO_2; however, the effect has not been to reduce total SO_2 emissions since the 9.0 million ton cap is unchanged, but to reduce the quantity of abatement required of other units in meeting the cap and thereby the cost of doing so.

Keywords: US SO_2 emission reductions, electricity generation, combined cycle generation, clean aircraft, effectiveness of cleaning up vs. replacing

---

[1] Ellerman is a senior lecturer at the Sloan School of Management at the Massachusetts Institute of Technology (MIT). At the time of writing, Dubroeucq was a candidate for the master's degree in Technology and Policy at MIT. She now works for Électricité de France. Funding for this research from the Environmental Protection Agency (STAR grant #R828630) and from CEEPR is gratefully acknowledged.

1 Introduction

Emissions can be reduced by emission rate reductions at existing plants or by displacing those plants by other plants, frequently new units, with lower emission rates. Accordingly, one of the enduring questions underlying policies aimed at reducing air emissions is the role of these two ways of reducing emissions. A good case study for analyzing their relative contributions is provided by the experience of the United States in reducing sulphur dioxide (SO_2) emissions from the combustion of fossil fuels for the generation of electricity. These emissions have been reduced by about 45%, from a peak of about 18.25 million tons in 1975 when the Clean Air Act Amendments of 1970 became effective to 10.1 million tons in 2002, the last year for which data is available.[2]

Since 1970, SO_2 emissions have been subject to two distinctly different regulatory regimes established respectively by the Clean Air Act Amendments of 1970 and 1990.

The Clean Air Act Amendments of 1970. These amendments instituted a coherent and effective regulatory system for reducing SO_2 emissions whereby

a) *existing facilities* would be subject to emission rate limits imposed by State Implementation Plans that were to ensure attainment of the National Ambient Air Quality Standard for SO_2, and
b) *new plants* would be subject to stringent New Source Performance Standards that would require the adoption of best available control technology.

These provisions had become effective by the mid-1970s when national SO_2 emissions peaked and they have remained in effect to this day.

The Clean Air Act Amendments of 1990. Title IV of these amendments created a nationwide limit on aggregate SO_2 emissions of approximately 9 million tons to be achieved in two phases by an innovative cap-and-trade program that issued allowances in an amount equal to the cap and required all electric utility generating units to surrender allowances equal to the unit's emissions. Since no specific command concerning abatement is given at the unit level, the operators of affected units are free to decide whether they will reduce emissions by lowering the sulphur content of the fuel used to generate electricity (either by switching or retrofitting scrubbers) or by shifting generation to lower emitting units including new units. However, Title IV did not replace the source-specific limits and technology mandates of the earlier 1970 Amendments. The cap and the associated obligation to surrender allowances equal to the tons of SO_2 emitted is an additional requirement imposed on top of the pre-existing structure of prescriptive regulation.[3]

[2] The decrease in SO_2 emissions from all sectors of the economy was slightly larger due to the disappearance of metals processing, mostly copper, within the United States. For the economy as a whole, peak SO_2 emissions were 31.8 million tons in 1973 and they had declined to 15.8 million tons in 2001, or by 50%. (US EPA 2003).

[3] The super-imposing of Title IV on the pre-existing prescriptive rate limits, which are aimed primarily at preventing adverse local health effects means that some plants are not free to increase emissions (and purchase allowances). In practice, these pre-existing con-

The reduction in electric utility SO_2 emissions has been the more remarkable in that the demand for fossil-fuel-fired generation of electricity has grown substantially since 1970 as shown in Figure 1.

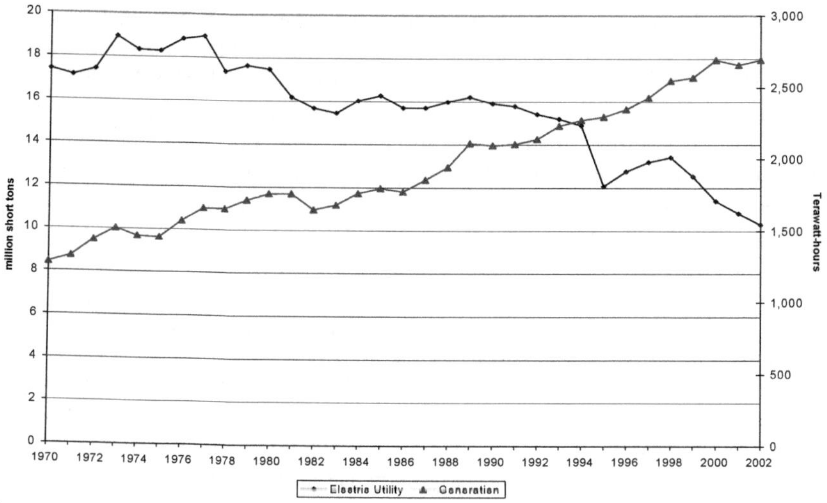

Fig. 1. U.S. Fossil-fuel-fired electricity generation and SO_2 emissions, 1970-2002

In general, generation of electricity from fossil-fuel-fired power plants has increased steadily while SO_2 emissions have regularly declined. Since the year of peak emissions, 1977, fossil-fuel-fired generation has increased at an average annual rate of 2.0% while SO_2 emissions from these sources have decreased at an annual rate of 2.4%.[4] The implied annual rate of reduction in aggregate SO_2 intensity for fossil-fuel-fired generation is 4.3%, from 23 pounds of SO_2 per megawatt-hour in 1977 to 7.76 pounds in 2002. In broad terms, this reduction in aggregate intensity results from two effects: the reduction in emission intensity or rates at individual units and the displacement of higher emitting units by existing sources with lower emission rates or new sources with mandated lower emission rates.

While the trend in SO_2 emissions since the mid-1970s is instructive, the past five years offer an especially good opportunity to examine the effect on emissions of the introduction of low-emitting new generating units. Several factors - the need for new capacity to meet continually growing demand, the availability of more efficient, combined-cycle generating technology, and the expectation of relatively low natural gas prices - coalesced in the late 1990's to create a boom in the

straints have not posed a serious impediment to trading under Title IV since the cap requires a significantly greater reduction of aggregate emissions than what is required to meet the National Ambient Air Quality Standard for SO_2. While generating units can trade only within the prescriptive limits imposed by the 1970 Amendments and these limits have become non-binding for most units.

[4] Over this same 25-year period, total electricity generation, including nuclear, hydro, and renewables, has increased at an annual rate of 2.4%.

construction of new natural-gas-fired generating capacity. Since natural gas emits only trace amounts of SO_2, the deployment of these new units could be expected to reduce SO_2 emissions considerably as pre-existing, higher emitting generating units are displaced in meeting the demand for electricity. As of the end of 2002, the new gas-fired capacity is estimated to be 133 GWe, an approximately 20% increase in generating capacity, and another 56 GWe is under construction and expected to be completed in the next few years, mostly in 2003 (EVA, 2003). About half of this capacity consists of single-cycle combustion turbines that are used mostly for meeting peak demand and offer few if any operating efficiencies compared to existing capacity. The remaining half of the new capacity utilizes combined cycle technology that offers marked operating efficiencies in comparison to existing capacity and new single-cycle units, which would be expected to lead to greater utilization for these units and greater displacement of existing units.[5] Accordingly, we focus mostly on the combined-cycle units.

Our purpose in this paper is to analyze the sources of the reduction in SO_2 emissions and, in particular, to distinguish between the effects of lower emission rates at existing units and the displacement of higher emitting generating units by lower emitting ones, regardless of whether these are new units or existing units with lower emissions. In doing so, we give particular attention to the reduction in SO_2 emissions attributable to the large increase in new natural-gas-fired capacity in the United States since 1998. The methodology we employ in this paper does not discriminate between emission rate reductions and displacements that respond to policy measures and those that would have occurred anyway because of other non-policy-related factors affecting the electric utility generating sector of the economy. Accordingly, the results we report should not be interpreted as being entirely due to regulatory measures, although a large fraction surely is. Where appropriate, mention will be made of these non-regulatory factors.

The next section of the paper explains the data base and methodology that is used to identify the source of observed SO_2 emission reductions. Results are then reported in the next section, and a final section concludes. A technical explanation of the decomposition methodology and the full data results are provided in appendices.

2 Data and methodology

Adoption of the 1990 Clean Air Act Amendments, and specifically the decision to allocate allowances to generating units according to average 1985-87 heat input and the 1985 SO_2 emission rate, required the U.S. Environmental Protection Agency (US EPA) to develop a more detailed and accurate data base than had existed previously. This data base lists annual SO_2 emissions and heat input at the unit level for over 3000 generating units from 1985 on. The availability of data at

[5] For instance, in the third quarter of 2002, combined cycle units constituted 52% of the new capacity and 79% of the generation from the new gas-fired units.

the unit level is particularly important since any given power plant will typically consist of several generating units, usually three to four but sometimes as much as a dozen, usually built in different years and typically subject to differing regulatory requirements. Absent unit-level data, it would be impossible to tell whether an observed change in emissions at a power plant is due to changes in emission rates at all or several units or to the changing utilization of the constituent units with differing emission rates because of different regulatory requirements.

Our analysis is based on this data base from which some 1.000 rarely utilized, old, and small units are excluded. The remaining 1.890 units account for 99% of total SO_2 emissions from the electric utility sector during the years 1985-2002 (US EPA, 2003).[6] Given this concentration of SO_2 emissions in two-thirds of the total generating units (and about 95% of total heat input), any perceptible change in total SO_2 emissions from the electric utility sector as a whole will be determined almost entirely by changes at these 1.890 significant units.

Since annual SO_2 emissions are the product of heat input, measured in million Btus (mmBtu) and the emission rate measured in pounds of SO_2 per mmBtu (#/mmBtu), changes in observed emissions from one year to the next at any given unit can be decomposed into two components: a change in the annual emission rate, which would reflect the use of a higher or lower SO_2-emitting fuel or the installation of emission control equipment, and 2) a change in annual heat input at the unit, which may reflect a change in aggregate demand for electricity or the effect of displacement of one unit by another in meeting any given level of demand. In nearly all cases, both effects operate, often in off-setting directions; however, the relative contributions of each can be identified using analytic techniques explained briefly below and more fully in the appendix.

While the causes of changes in emissions at any individual unit can be decomposed into two effects, changes in observed emissions from any aggregate of generating units must take account of the interaction of all the units in the aggregate. For instance, if one unit is utilized less, as measured by heat input or generation, and the utilization of another unit is increased by the same amount, the effect on total emissions depends on the emission rates at the two units. If the emission rate is lower at the unit increasing utilization than at the other unit, total emissions will decrease without any change in emission rate at either unit. Thus, for any aggregate, changes in total emissions can be broken down into three components - emission rate reductions at individual units, changes in aggregate demand, and changes in the utilization of units with differing emission rates - as represented in the following equation.

---

[6] A unit is included in the data base if it meets one of several criteria developed to determine units that are significant in determining SO_2 emission trends. These criteria are: 1) more than 5 trillion Btu heat input in any year from 1995 through 2001, 2) more than 1 trillion Btu in any two years out of four consecutive years between 1995 and 2001. A 100 MWe unit consuming 1 trillion Btu in a year with a heat rate of 10.000 Btu/kWh would generate 100 GWh of electricity in the year, or 1.000 hours (about 11% of the hours in a year) at full capacity.

$$dE = dEr + dEh_{agg} + dEh_{Disp} \qquad (1)$$

where dEr = the sum of the changes in emissions due to changes in emission rates at individual units,

dEh_{ag} = the change of emissions that can be attributed to changes in aggregate demand,

dEh_{Disp} = the change of emissions that can be attributed to the displacement of some units by others in meeting aggregate demand.

The left-hand-side variable of equation (1) is observed and the first two right-hand-side terms can be easily calculated. The term dEr is the sum of the change in emissions due to changes in the emission rate at all constituent units and the term dEh_{agg} can be found by multiplying the prior year's emissions by the percentage change in heat input for the aggregate. Any difference between the sum of dEr and dEh_{agg} and the observed change in emissions, dE, is due, by definition, to displacement, or the emission effects of the changing shares in heat input of the units composing the aggregate.

The availability of data indicating whether the fuel burned in a generating unit is coal or oil/gas allows us to decompose the displacement effect into a shift between fuels and displacements among the units composing each fuel aggregate, as follows:

$$dEh_{Disp} = dEh_{bet} + dEh_{w/i,Coal} + dEh_{w/i,Oil/Gas} \qquad (2)$$

where dEh_{bet} = the change of emissions that can be attributed to changing shares of generation between coal and oil/gas units,

$dEh_{w/i\ Coal}$ = the change in emissions due to a redistribution of heat input among units using coal, and

$dEh_{w/i\ Oil/Gas}$ = the change in emissions due to a redistribution of heat input among units using oil or natural gas.

One easy way to visualize this decomposition is to recall that the change in emissions due to changing heat input at any individual unit results from the change in aggregate demand for generation, any change in fuel shares, and individual displacements with the two fuel categories. Imagine a situation in which there is no change in the emission rates at individual unit so that all changes in emissions are due to these three demand effects. If a coal-fired unit has increased emissions by 3% while aggregate demand has increased 1% and the demand for aggregate coal-fired generation has increased by 1%, one percentage point of the observed three-percent increase in emissions at this individual unit can be attributed to each of the three effects: dEh_{agg}, dEh_{bet}, *and* dEh_{w/i_Coal}. If observed emissions had not increased at all at this unit while the other conditions applied, then it could be said that this unit experienced a 2% reduction in utilization due to displacement by other coal-fired units. Once these differences are calculated for all units constituting some aggregate, they can then be summed to determine all of the components in equations (1) and (2).

Choosing the appropriate level of aggregation for determining growth in aggregate demand and changes in fuel shares in the United States is not obvious. Fuel shares differ markedly by region, as do the growth rates in the generation of fossil-fuel-fired electricity. Using the national aggregate would not provide an accurate estimate since it would assume that generating units are part of one large integrated national market, which they are not. At the other extreme, a state-level aggregation would be similarly misleading since electricity control areas often encompass several states and electricity flows frequently cross state boundaries even when control areas follow state lines. As a middle ground we have used the nine census regions, the composition of which is given in Table 1 below and for which regional aggregate data is given in Table A1 of the appendix. Accordingly, we calculate dEh_{agg} and the components of dEh_{Disp} on a regional basis and then sum across the nine census regions to obtain national figures.

Table 1. U.S. census regions and constituent states

Region	States
New England	CT, MA, ME, NH, RI, VT
Middle Atlantic	NJ, NY, PA
East North Central	IL, IN, MI, OH, WI
West North Central	IA, KS, MN, MO, ND, NE, SD
South Atlantic	DC, DE, FL, GA, MD, NC, SC, VA, WV
East South Central	AL, KY, MS, TN
West South Central	AR, LA, OK, TX
Mountain	AZ, CO, ID, MT, NM, NV, UT, WY
Pacific	CA, OR, WA

The greater efficiency of the new combined cycle units presents a problem in estimating the SO_2 emission reductions attributable to this new capacity. The heat input used by these new units is fully incorporated into the components of equations (1) and (2), but these units generate more electricity per unit of input than conventional units. Since electricity is the final output, some accounting must be made for this additional displacement and emission reduction, which shows up otherwise erroneously as a reduction in aggregate demand.

This adjustment is made through a three-step process as explained in more detail in the appendix on methodology. First, the heat input savings attributable to the use of combined cycle generating plants is determined. We observe an average heat rate (Btus per kWh) of 7.400 for the combined cycle units and we assume an average heat rate of 10.000 Btus/kWh for the generation being displaced. These figures imply that the heat input displaced by these new combined cycle units is 35% greater than the heat input use observed at those units. The second step is to determine whether the increased generation is displacing coal-fired or oil/gas fired generation, which we do on a regional basis. The last step is to calculate the emission reduction by multiplying the displaced heat input for each type of generation by the respective average regional emission rates.

3 Decomposition results

3.1 National results for the 1985-2002 period

Figure 2 below and Table A2 of the appendix show the national change in SO_2 emissions in tons by year for the 1985-2002 period and by the three components of equation (1), that is, changes in the emission rate, changes in aggregate demand, and changes in dispatch among units from one year to the next.

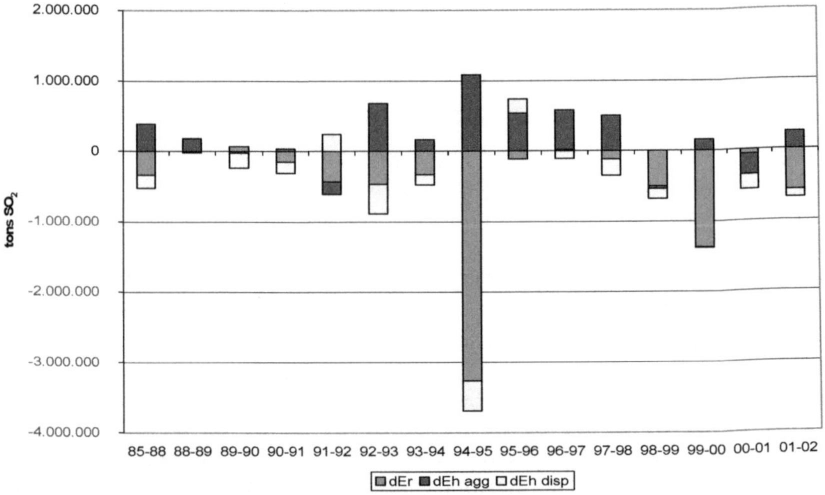

Fig. 2. National change in SO_2 emissions in tons by factor, from 1985 to 2001.

The most salient feature of Figure 2 is the very large reduction in SO_2 emissions in 1995, the first year of Phase I of the Acid Rain Program. This reduction is especially remarkable in that 1) the cap applied only to a sub-set of units in that year (albeit the largest and most highly emitting units), 2) these units reduced emissions far more than was required to meet the cap in that year (or for any year of Phase I), and 3) the much larger set of generating units that did not become subject to the cap until 2000 increased emissions by some 439.000 tons in 1995 compared with 1994.

The second largest annual reduction is in 2000, when all of the other generating units were first subject to the Title IV cap and therefore required to pay the going price of allowances (about $150/short ton in this year) for all SO_2 emissions. The reduction in 2000 occurred despite the large accumulation of banked allowances from the Phase I units (11.6 million tons) that would have easily covered the abated emissions in 2000, had the owners been willing to pay the price of an allowance. That they did not do so suggests that the cost of reducing emissions at these units was less than $150/short ton. The broader point that emerges from the emission reductions observed in these two years is that, when a price must be paid for otherwise permitted emissions, further reductions of emissions can be achieved.

Setting these two years aside, SO_2 emissions typically declined each year (11 out of 15), but by much smaller amounts than were observed in 1995 and 2000. Table 2 summarizes the emission reductions shown on Figure 2 by component and period, pre- and post-Title IV.

Table 2. Emission reductions by component and period, 1985-2002

(000 tons SO_2)	1985-94	1994-2002	1985-2002
dEr	- 2.343	- 6.001	- 8.345
dEh_{agg}	+2.009	+2.748	+4.757
dEh_{disp}	- 1.263	- 1.043	- 2.306
dE	- 1.598	- 4.296	- 5.894

As shown in the lower, right-hand cell, 2002 SO_2 emissions from electric utility generating units had fallen by 5.9 million tons from their level in 1985, and they will fall another million tons in order to meet the Phase 2 cap as the Phase I bank of allowances is drawn down. The decomposition of this change shows that emissions would have increased by 4.7 million tons over this period as a result of increasing generation from fossil-fuel-fired generating units[7], but this effect is more than offset by the combined effect of reductions in emission rates at existing units and the general displacement of generation to lower emitting units. Of these two emission-reducing effects, by far the greater is the effect of emission rate reductions. This effect is also notably larger after 1995 than before.

Moreover, not all of the emission reductions observed over the pre-Title IV period can be attributed to air emission regulations. Ellerman and Montero (1998) estimate that the effect of railroad deregulation in making low-sulphur western coals economically competitive at Midwestern generating units burning local, high-sulphur coals reduced SO_2 emissions by about two million tons between 1985 and 1993. This reduction occurred by switching units burning high sulphur mid-western coal partially or entirely to lower sulphur western coal and by the greater utilization of these units. Their analysis suggests that about half of the 3.6 million ton reduction in SO_2 emissions resulting from emission rate reductions and displacement from 1985 through 1994 was due to reasons other than air emission regulation.[8] Accordingly, the contrast in the magnitude of the emission reductions associated with conventional prescriptive regulation and the cap-and-trade requirements instituted by Title IV is even greater than is suggested by the cumulative amounts in Table 2.

[7] This effect would be larger if it were calculated from some unchanging base year emission rate, such as in 1985, instead of from each succeeding year, which reduces the effect of increasing demand in each year by the emission rate reduction and displacement effects in prior years.

[8] Keohane (2003) shows that the reductions in the delivered price of low-sulphur western coal in the Midwest came to an end in the early 1990s so that the one-year difference in terminal years between the Ellerman-Montero analysis and the analysis in this paper is not likely to be great.

The displacement component in emission reductions observed since 1985 can be further decomposed to reflect the emission effects of shifts in the relative shares of coal and oil/gas and of greater or less use of lower emitting units within each of these fuel types, as shown below by year in Figure 3 and Table 1 of the appendix and cumulatively in Table 3.

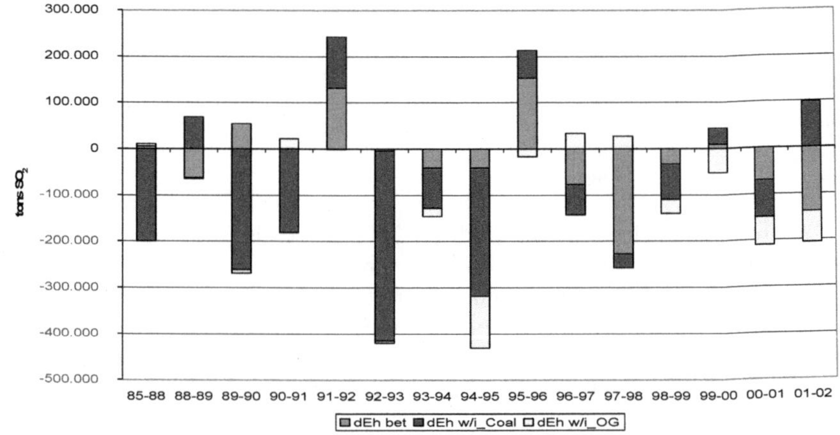

Fig. 3. Decomposition of the displacement effect by year, 1985-2002

Table 3. Cumulative decomposition of the displacement effect

(000 tons SO_2)	1985-94	1994-2002	1985-2002
dEh $_{bet}$	+ 96	- 427	- 330
dEh $_{w/i\_coal}$	- 1.362	- 338	- 1.700
dEh $_{w/i\_OG}$	+ 2	- 278	- 276
dEh $_{disp}$	- 1.263	- 1.043	- 2.306

By far, the largest component of the 2.3 million ton reduction due to displacement of generation among fossil-fuel-fired generating units over the 1985-2002 period is that due to displacement among coal-fired units. This is not surprising since the potential for reduction is large given the range of sulphur content among coals, from as low as 0.5 lbs. SO_2/mmBtu to more than 5 lbs./mmBtu. Most of this reduction occurred in the years before Title IV became effective and it is largely due to the shift to low-sulphur western coal identified by Ellerman and Montero (1998).[9] Once Title IV became effective, the three components of the displacement effect are more balanced and the largest displacement component is a shift to more oil/gas fired generation. This shift is consistent with the abnormally low oil prices experienced in 1998 and the installation of over 150 GWe of new natural gas fired generating capacity including nearly 100 GWe of new combined cycle capacity to which we now turn.

[9] Since units are dispatched on the basis of variable costs, which are largely fuel costs, units switching to lower cost, lower sulphur western coal would tend to be dispatched more.

3.2 The effect of combined cycles on SO_2 emissions

The critical issue in estimating the reduction in SO_2 emission due to the new combined cycle capacity is determining what generation is displaced. This question cannot be answered satisfactorily without a disaggregation to at least the regional level because of the significant differences in the regional distribution of the new combined cycle capacity, differing patterns of displacement by region, and different regional emission rates for coal and oil/gas fired generation. The regional distribution of the new combined cycle capacity is given in Table 4 and additional data used in calculating the effect of the new combined cycle capacity on SO_2 emissions is provided in Table A3 of the appendix.

Table 4. New combined cycle capacity and regional shares of generation

Census Region	CC Capacity 2002 (MW^e)	Regional Share of CC Capacity 2002	Regional Share of US Oil/Gas Generation, 1997	Regional Share of Total Fossil Generation, 1997
New England	6.109	11%	10%	2%
Mid-Atlantic	3.248	6%	13%	8%
East North Central	3.827	7%	2%	20%
West South Central	918	2%	1%	10%
South Atlantic	6.489	11%	19%	20%
East South Central	5.537	10%	2%	11%
West South Central	22.448	39%	40%	16%
Mountain	4.318	8%	2%	10%
Pacific	4.395	8%	11%	2%
USA (lower 48)	57.289	100%	100%	100%

The regional distribution of combined cycle capacity follows the pre-existing distribution of oil and gas generation far more closely than it does the pre-existing generation of fossil-fuel fired generation. Five regions constituting 93% of oil and gas generation in 1997 account for 75% of the combined cycle capacity but only 48% of total fossil generation. The largest share by far of new combined-cycle capacity is in the West South Central census region, encompassing Texas, Oklahoma, Arkansas, and Louisiana, which is also the region with the largest share (and absolute amount) of oil and gas generation. Conversely, regions in which there was little pre-existing oil and gas generation received a smaller share of the new combined cycle capacity.

A solid economic reason explains this pattern. When new, more efficient units compete with existing units using the same fuel, they can be assured of being dispatched first if all other factors are equal. However, when the competing units use a different fuel, displacement depends upon the price difference between natural gas and the other fuel. If the price of the fuel firing the more efficient generation is greater percentage-wise than percent savings in heat input, displacement will not occur. This has been the case for the new combined cycle units when they compete against existing coal-fired units in the U.S., especially since late 2002 when natural gas prices rose to levels that are two to three times the level of coal prices.

There are, of course, other factors concerning location and network dynamics that influence dispatch, but building combined cycle units where reliance on less efficient natural gas generation is already high provides greater assurance of demand for generation from the new capacity, but also less reduction of emissions.

Two distinct patterns of displacement occur, as illustrated by the two charts in Figure 4.

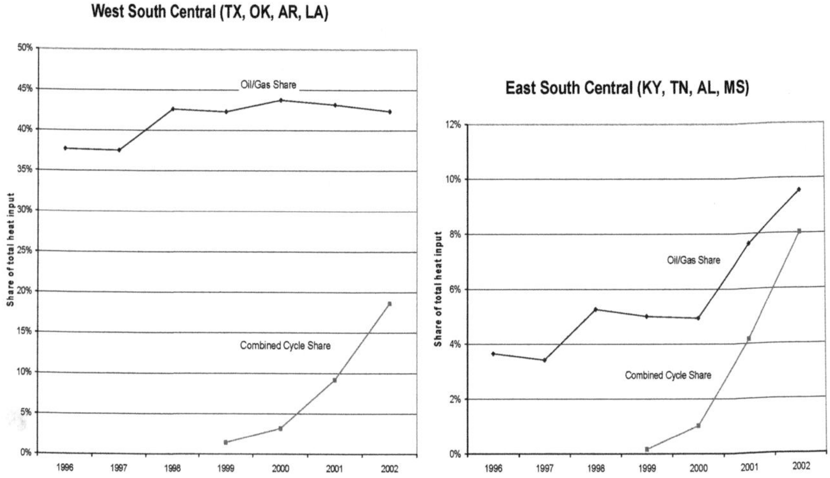

Fig. 4. Combined cycle and oil/gas shares in two census regions

The uppermost line on each chart represents the share of heat input into oil/gas generating units in that region, while the bottom line shows the share of heat input going into combined cycle units. In all regions, the share of combined cycle capacity rises from nearly zero in 1998 to some noticeable positive share by 2002. In cases such as the West South Central census region, the share of combined cycle heat input rose from 1% in 1999 to 19% in 2002. Over the same period, the total oil and gas share of heat input remained relatively constant at 42%-44%. Obviously, the new combined cycle capacity in this region has been displacing existing oil and gas capacity, not coal capacity.

The East South Central region presents a different picture. The 2002 shares of oil/gas and combined cycle heat input are much smaller than in the West South Central region, but the increase in the combined cycle share from zero percent in 1999 to 8% in 2002 causes the oil/gas share of heat input to increase by five percentage points, from 5% in 1999 to 10% in 2002. Accordingly, it can be said that five percentage points of the 8% increase in combined cycle generation displaced coal generation and the remaining three percentage points displaced existing oil/gas generation, which is now 2% instead of 5%.

When displacement is calculated in this manner for all nine regions for each year, the amount of displacement depends not only on the amount of heat input displaced by the new combined cycle units, but also on the emission rate of the Btu's being displaced from coal or other oil and gas-fired units. Figure 5 and Table 5 below provide the year-by-year results for the nine census regions and the nation as a whole.

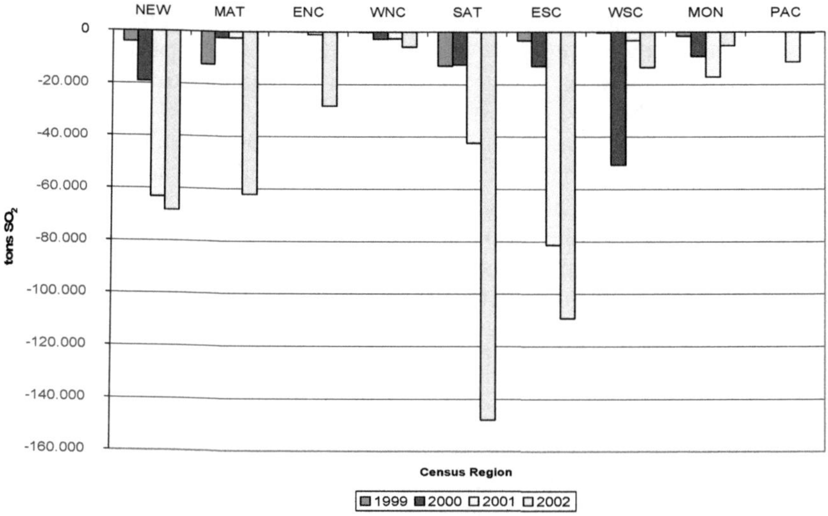

Fig. 5. SO_2 emission reductions due to new combined cycle capacity, by region and year

Table 5. SO_2 emission reductions due to new combined cycle units, by year and region

000 tons SO_2	1999	2000	2001	2002	Cumulative
New England	- 4	- 19	- 63	- 68	-155
Mid Atlantic	-13	- 2	- 2	- 62	- 79
East North Central	0	0	- 1	- 28	- 29
West North Central	0	- 3	- 3	- 6	- 12
South Atlantic	-13	- 12	- 42	-148	-215
East South Central	- 3	- 13	- 81	-109	-207
West South Central	0	- 51	- 3	- 14	- 68
Mountain	- 2	- 9	- 17	- 5	- 33
Pacific	0	0	- 11	0	- 11
Lower 48 States	-35	-110	-225	-441	-810

The two regions with the largest cumulative reduction (the sum of the annual amounts) are the East South Central and South Atlantic census regions. Even though they constitute only a quarter of national combined cycle generation, they account for 53% of the national SO_2 reduction attributable to the new combined cycle capacity. The reason is that the new combined cycle capacity in these regions displaced more coal generation and the emission rate associated with the displaced coal generation is relatively high. In contrast, the much larger displacement of existing generation in the West South Central region reduced SO_2 emissions by considerably less because no coal generation was displaced.

A final observation about the effect of the new combined cycle capacity on SO_2 emissions concerns the interaction between these new units and the Title IV cap.

While the new combined cycle capacity clearly displaced generation that had higher SO_2 emissions, aggregate SO_2 emissions are no lower than they would otherwise be since the SO_2 emissions cap is fixed.[10] The effect of the new capacity is then to reduce the amount of abatement required from the other, mostly coal-fired units. Consequently, the effect of the new combined cycle capacity is not to reduce actual SO_2 emissions, but the emission reduction required of other generators of electricity and therefore the cost of achieving the SO_2 cap.

The extent to which the cost of Title IV has been reduced can be estimated. As shown in Table 5, the cumulative reduction in SO_2 emissions attributable to the combined cycle units as of the end of 2002 is approximately 800.000 tons. The method for calculating the simple counterfactual for 2002 (cf. Ellerman *et al.*, 2000) yields counterfactual emissions that are 6.9 million tons greater than observed emissions of 10.2 million tons; however, this method does not take account of the assumed 35% efficiency gain and greater displacement per unit of heat input associated with the combined cycle units. When this correction is made, counterfactual emissions are 7.1 million tons higher than observed emissions. After subtracting the 800.000 ton emission reduction due to the new combined cycle units, the remaining units reduced SO_2 emissions by only 6.3 million tons or about 11% less than what would have been required to meet the same electricity demand without the new combined cycle units.

Assuming a linear relation between quantity and price for incremental abatement at the current margin, the marginal cost of abatement and the price of allowances is 11% less than it would be absent the introduction of the new combined cycle capacity.[11] The average price of allowances in 2002 was about $150, which would imply marginal costs that would have been $16-$17 higher. Additional combined cycle capacity came on line in 2003, approximately equal in capacity to that added in 2002, so that the ultimate effect might be larger, but this would depend upon the amount of displacement by this new capacity and the data reported so far for 2003 indicates decreasing total oil/gas generation over the past year, probably because of the high natural gas prices that have been observed since the end of 2002. If a round number were to be used for the total effect of the new combined cycle capacity in reducing the marginal cost of abatement, say $20 per ton, the implied annual savings in electricity cost is $180 million when multiplied by the Phase 2 cap of 9 million tons of SO_2 emissions per year.

4 Conclusion

The major source of SO_2 emission reductions in the United States since 1985 has been the reduction of emission rates at existing units. Displacement of higher emitting units by lower emitting ones, whether newly constructed or existing

[10] This effect does not apply for any uncapped emissions, such as NO_x emissions in many states and CO_2 emissions.

[11] This is not to say that allowance prices have fallen as the new combined cycle capacity came on line since its effect of the allowance market would have been anticipated.

units, has also contributed an important share of the total reduction; however, this factor alone has not been sufficient to offset the increase in emissions that would have occurred as a result of continuing growth in aggregate demand. Our analysis also indicates that Title IV has been more effective in reducing emissions during the eight years it has been in effect than the conventional, source-specific, prescriptive regulation had been in reducing emissions in the ten years preceding 1995.

The effect of the new combined cycle capacity is not what might be expected at first sight. These units have clearly displaced more highly emitting generating units, although most often not coal-fired units, but the effect has been to reduce the cost of abatement, not total SO_2 emissions. When emissions are capped, exogenous factors such as the introduction of more efficient combined cycle generation results in less required abatement by other affected units, in this instance, mostly coal-fired units. From the standpoint of the competition among contending fuels, this effect is ironic but it is small and the ultimate beneficiary is the consumer who thereby pays slightly less for electricity without any change in this attribute of environmental quality.

References

Ellerman AD, Joskow PL, Schmalensee R, Montero JP, Bailey E (Ellerman et al. 2000) Markets for Clean Air: The U.S. Acid Rain Program. Cambridge University Press, 2000

Ellerman A D, Montero JP (1998) "The Declining Trend in Sulfur Dioxide Emissions: Implications for Allowance Prices," Journal of Environmental Economics and Management 36: 26-45 (This article is substantively reproduced as chapter 4 of Ellerman et al. (2000)

Energy Ventures Analysis, Inc. (2003) "Tracking the Boom of New Power Plants in the U.S." (Proprietary quarterly report dated September 2003). Arlington, VA

Keohane NO, Busse M (2003) "Pollution control and input markets: The creation and capture of rents from sulfur dioxide regulation." Working paper dated July 28, 2003, available at http://www.som.yale.edu/faculty/nok4/files/papers/sulfur_penalty.pdf

U. S. Environmental Protection Agency (2003) Average annual emissions, all criteria pollutants, years including 1970-2001, Washington, D.C. Available at http://www.epa.gov /ttn/chief/trends/index.html.

Appendix I: Decomposition methodology

Decomposition of changes in emissions at the unit level

The SO_2 emissions produced by a generating unit in the tth year can be described as:

$$e_t = r_t * h_t$$

where h_t is the heat input (i.e. the energy contained in the fuel burnt during year t) and r_t is the emission rate (i.e. the amount of SO_2 emitted per unit of heat input). The change in SO_2 emission between year 0 and t can be described as a function of four observed values, h_0, h_t, r_0, and r_t, such that

$$de_{0,t} = r_t h_t - r_0 h_0 = (r_0 + dr_{0,t})(h_0 + dh_{0,t}) - r_0 h_0$$
$$de_{0,t} = r_0 dh_{0,t} + h_0 dr_{0,t} + dr_{0,t} dh_{0,t}$$

where the $d_{0,t}$'s denote the observed change in e, r or h between year 0 and t.

The change in emissions, $de_{0,t}$, can also be represented in a (h,r) diagram:

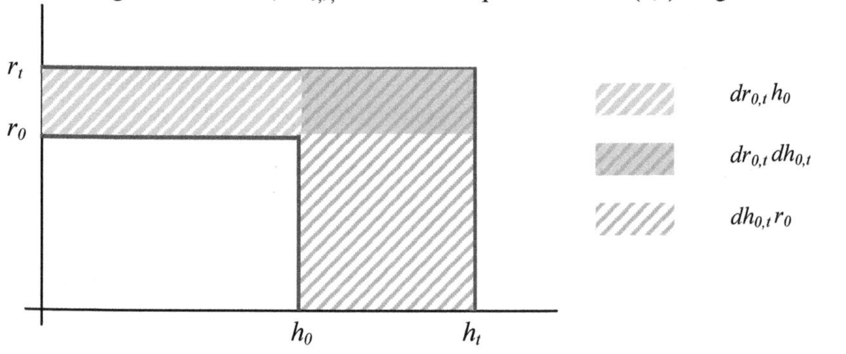

Fig. 6. Representation of the heat input, emission rate and emissions of a generating unit in a (h,r) diagram

In this diagram the surface of the h_0 x r_0 rectangle is equal to the emissions e_0, and the surface of the h_t x r_t rectangle is equal to the emissions e_t. The difference $e_t - e_0$ is represented by the striped areas. The diagram clearly shows that $de_{0,t}$ can be separated into three components:

$\Delta_1 = (r_t - r_0)h_0$ which is created by a change of the unit's emission rate
$\Delta_2 = (h_t - h_0)r_0$ which is created by a change of the unit's heat input
$\Delta_3 = (r_t - r_0)(h_0 - h_t)$ which is created by both changes

We adopt the convention of splitting the third component evenly and attributing each half to the other two components so that we can attribute $\Delta_1 + \Delta_3/2$ to a change of emission rate and $\Delta_2 + \Delta_3/2$ to a change of heat input, which gives us:

- the change in emissions due to a change in heat input

$$de_h = r_0 dh_{0,t} + \frac{dr_{0,t} dh_{0,t}}{2}$$

- the change in emissions due to a change in the emission rate

$$de_r = dr_{0,t} h_0 + \frac{dr_{0,t} dh_{0,t}}{2}$$

When a unit is either shut down or put online (i.e. either h_o or h_t is equal to zero), we set $de_r = 0$ and attribute all the change in emissions to a change in heat input.

Accounting for the interaction of individual units with others in some aggregate

The two components accounting for changes in emissions at the unit level, de_r and de_h, have differing characteristics when the unit is considered as part of some aggregate, such as an electricity grid.[12] A change in emission rate, de_r, such as that resulting from the installation of a scrubber, is a unit specific action that does not imply a change in the emission rate at other units in the aggregate. In contrast, a change in heat input at an individual unit, de_h, will always reflect some change in the aggregate that is shared by other units or is the result of the interaction among the constituent units. For instance, a change in aggregate demand would be expected to affect all units in some measure. Similarly, changes in fuel prices or in conditions on the electricity network, would be expected to change the contribution of constituent units to meeting aggregate demand. While the observed change in heat input at any single unit results from changes in conditions affecting the aggregate, the contributing factors can be analytically separated into three components: the change in aggregate demand, any change in the contribution of coal and oil/gas units viewed as sub-sets, and any change in the utilization of individual units within each fuel share.

More formally, heat input at the *i*th unit can be decomposed into three components.

$$h_t^i - h_0^i = dh_{agg}^{\ i} + dh_{bet}^{\ i} + d\,h_{w/i}^{\ i}\,.$$

where $dh_{agg}^{\ i}$ is the ith unit's share of the change of heat input for the aggregate:

$$dh_{agg}^i = \frac{h_0^i}{H_0}(H_t - H_0)$$

with H_0, H_t being the aggregate heat input for years *0, t*

[12] We use regional aggregates defined along the lines of the U.S. census regions, but the methodology applies for any aggregate.

$dh_{bet}^{\ i}$ reflects what would be the change in the *i*th unit's heat input due to a change in the share of the subset of units constituting "fuel X" in year *t* assuming no change in the shares of the constituent units in that fuel subset:

$$dh_{bet}^{i} = \frac{h_0^i}{H_{0,fuelX}}\left(H_{t,fuelX} - H_{0,fuelX} * \frac{H_t}{H_0}\right) = h_0^i\left(\frac{H_{t,fuelX}}{H_{0,fuelX}} - \frac{H_t}{H_0}\right)$$

with $H_{0,fuelX}$, $H_{t,fuelX}$ the aggregate heat input for 'fuel X' units for years *0, t*.

$dh_{w/i}^{\ i}$ is the remaining part of $h_t - h_0$ which will be equal after substitution and cancellation to:

$$dh_{w/i}^{i} = h_t^i - h_0^i \frac{H_{t,fuelX}}{H_{0,fuelX}}$$

dhw/ii reflects the effect of any change in the role of the ith unit within the fuel X subset after allowing for changes in aggregate demand and for any change in fuel share.

With these definitions, the new decomposition of $de_{0,t}$ in the (h,r) diagram for a unit experiencing an increase in heat input due to all three factors is:

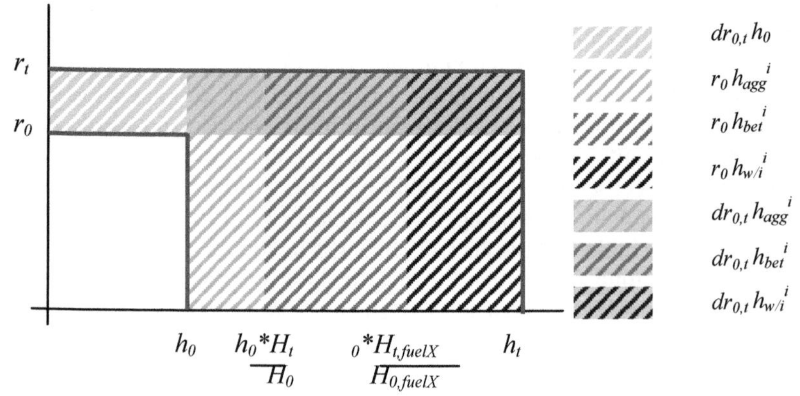

Fig. 7. Representation of the decomposition of the heat input and associated emissions of a generating unit in a (h,r) diagram

Returning now to the formulae for changes in emissions, we can decompose $de_{0,t}$ into four components:

$$de_r^i = dr_{0,t}^i\left(h_0^i + \frac{dh_{0,t}^i}{2}\right)$$

$$de_{h,agg}^{i} = \left(r_0^i + \frac{dr_{0,t}^i}{2}\right) dh_{agg}^{\ i} = \left(r_0^i + \frac{dr_{0,t}^i}{2}\right)\frac{h_0^{\ i}}{H_0}\left(H_t - H_0\right)$$

$$de^i_{h,bet} = \left(r^i_0 + \frac{dr^i_{0,t}}{2}\right) dh_{bet}{}^i = \left(r^i_0 + \frac{dr^i_{0,t}}{2}\right) \frac{h_0{}^i}{H_{0,fuel}} \left(H_{t,fuel} - H_{0,fuel}\frac{H_t}{H_0}\right)$$

$$de^i_{h,w/i} = \left(r^i_0 + \frac{dr^i_{0,t}}{2}\right) dh_{w/i}{}^i = \left(r^i_0 + \frac{dr^i_{0,t}}{2}\right)\left(h_t{}^i - \frac{H_{t,fuel}}{H_{0,fuel}} h^i_0\right)$$

All the above equations are unit-level equations. Aggregate numbers can be obtained by adding up the de^i_x of all the units of the database:

$$dE_X = \sum_{i \in database} de^i_X$$

Special considerations for combined cycle units

Combined cycle units present a problem in accounting for their effect on emissions because they are markedly more efficient in generating electricity than conventional coal and oil/gas generating units. The decomposition methodology presented above is based upon heat input, not electricity, which is the final product. So long as the heat rate, the number of Btu's used to produce a kilowatt-hour of electricity remains relatively constant from year to year and among units that would substitute for one another on the electricity grid, no great distortion results from using heat input as the proxy for electricity output. However, with the recent introduction of a significant amount of combined cycle capacity, this assumption no longer holds and some allowance must be made to recognize that the emission reduction resulting from the displacement of conventional generation by new combined cycle units is greater than would be indicated by a similar displacement among conventional units.

More formally, so long as combined cycles did not play a large role in generation, such as was the case until 1999, it was reasonable to assume that $dEh_{agg} \approx dEh_{elec}$ where the left-hand-side of the equation is defined as the change in emissions due to the observed change in heat input and the right-hand-side, as the change due to the assumed change in demand for electricity from fossil-fuel-fired generating units. With the introduction of a significant amount of combined cycle generation, a new term is required, dEh_{cc}, defined as the change in emissions due to the unobserved heat input savings resulting from the zero-fuel (thus zero-emission) electricity generation by the heat recovery unit of combined cycle facility. Conceptually, this new term can be defined in the following manner:

$$\mathrm{dEh_{cc}} \approx \mathrm{dEh_{agg}} - \mathrm{dEh_{elec}}$$

If the share of combined cycle generation in the aggregate is increasing, then $dEh_{elec} > dEh_{agg}$ and dEh_{cc} will be negative, and vice versa if the share of combined cycle generation is decreasing.

Estimating dEh_{cc} required two analytical tasks to be performed. First, combined cycle units were identified within the subset of oil/gas units. Second, the heat input savings associated with combined cycles was estimated. At first appearance, all of the information required to perform both tasks appeared to be in the quarterly reports whereby emissions are reported to the U.S. Environmental Protection Agency, hereafter called the CEMS (for Continuous Emissions Monitoring System) database, which record not only emissions but also heat input and electricity generation at the unit level, as well as identifiers for the fuel burned and the type of unit. In fact, cross checks with other sources revealed that labels identified as combined cycle in the CEMS database were not always such and that some not so labelled were combined cycle units. Comparison with data reported to the Energy Information Agency and data obtained by web search and direct calls enabled us to identify 276 out of the 948 oil/gas units that could be considered combined cycle units in that these Btu-using generating units had an associated heat recovery unit.

A more serious problem was that the generation reported for combined cycle units in the CEMS database was often only the generation from the gas turbine and not the additional power from the associated heat recovery unit. For instance in 2001, out of the 276 combined cycles, 52 units had an average heat rate above 10.000 Btu/kWh. From discussions with the owners of some of these units, we found that the data reported to the EPA on the CEMS forms sometimes contains only the generation for the turbine (which is the Btu-using and emitting unit) and not the generation from the (non-emitting) recovery unit. Consequently, there is no reliable method within the CEMS data to determine which combined cycle units had complete generation data and which were incomplete. To remedy this problem, we used another database from the Energy Information Administration, EIA Form 906, which reports electricity generation and heat input for all units for the year 2001. We took all the units from the EIA database that were also present in the 276 combined cycles list of the CEMS data and selected a group of 41 units that had been in operation for more than two quarters (thereby avoiding heat rate diminishing start-up problems) and showed steady generation and generally high utilization. This subset of fully operational combined cycles experienced an average heat rate of 7.400 Btu/kWh.

The heat input savings from combined cycle units was then easily calculated using an assumed average heat rate of 10.000 Btu/kWh for conventional generating units. Dividing 10.000 by 7.400 provides the assumed heat input savings of 35% that we use for estimating the emission reductions from the conventional generation displaced by the new combined cycle units.

More formally, it is possible to calculate the total heat input displaced by combined cycles as the sum of the observed heat input of CCs and an estimate of the heat input saved by the recovery unit:

$$h^i_{cc,disp} = h^i_{cc,obs} + h^i_{cc,sav} \quad \textit{for any CC unit I}$$

Since the combined cycle units are present in the database we use, the change in emissions associated with changes in $h^i_{cc,obs}$ are already included in dEh_{bet} and $dEh_{w/i,OG}$. dEh_{cc} is an adjustment, required to account for the emissions savings due to the greater efficiency of combined cycle units, that depends on $h^i_{cc,sav}$, which is related to $h^i_{cc,obs}$ as follows: $h^i_{cc,sav} = 0.35*h^i_{cc,obs}$.

Finally, the savings for the Yth region can be summed across units as:

$$H^Y_{cc,sav} = \sum_{i \in cc, regionY} h^i_{cc,sav}$$

Furthermore, if we assume that the displacement due to the heat input savings, $H_{t,cc,sav}$, is proportional in all respects to the displacement occasioned by the observed heat input at combined cycle units, $H_{t,cc,obs}$, then the heat input savings can be similarly broken down into a component displacing coal generation and another displacing other oil/gas units. Thus:

$$H_{t,cc,sav} = H_{t,cc,sav}{}^{coal} + H_{t,cc,sav}{}^{O/G} = H_{t,cc,sav} * (\%_{coal} + \%_{O/G})$$

The calculation of $\%_{coal}$ and $\%_{O/G}$ can be illustrated taking New England between 1999 and 2001 as an example. Figure 8 represents the heat input shares of coal units, conventional oil/gas units, and combined cycle units.

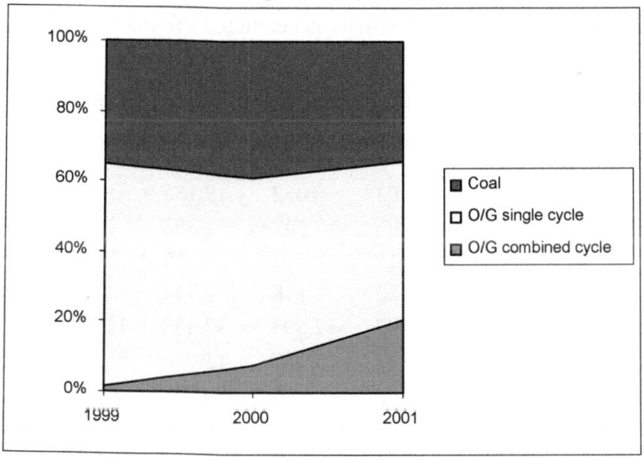

Fig. 8. Shares of Heat Input from coal, oil/gas single cycles and oil/gas combined cycle units in New England between 1999 and 2001.

Three patterns of heat input displacement by combined cycles are possible.

i) If the share of coal heat input either increases or stays equal from one year to the next, then there is no coal displacement and combined cycles exclusively displaced oil/gas units. We then simply calculate the emissions savings by using the average oil/gas emission rate. Accordingly, $\%_{coal} = 0$ and $\%_{O/G} = 1$.

ii) If the share of heat input from coal decreases more than the share of combined cycles increases, then we assume that the combined cycle units have displaced coal units only (and that conventional oil/gas single cycles have displaced coal as well). Thus $\%_{coal} = 1$ and $\%_{O/G} = 0$.

iii) Finally, if there is a decrease in the share of heat input from coal smaller than the increase in the share of combined cycles, then we assume that the combined cycle units have displaced both coal and oil/gas units as follows:

$$\%_{coal} = -\frac{(H_{t,coal} - H_{0,coal})}{(H_{t,cc,obs} - H_{0,cc,obs})} \text{ and } \%_{O/G} = \%_{coal} - 1.$$

Once $\%_{coal}$ and $\%_{O/G}$ have been calculated for each region and each year between 1999 and 2002, the corresponding emissions savings can then be calculated using the average regional (or other aggregate) emission rates of coal r_{coal} and oil/gas $r_{O/G}$, which are observed. Thus,

$$E_{t,cc,sav} = E_{t,cc,sav}^{coal} + E_{t,cc,sav}^{O/G} = H_{t,cc,sav} * (\%_{coal} * r_{coal} + \%_{O/G} * r_{O/G})$$

Table 6 shows the values obtained for dEh_{cc}, as well as the SO_2 savings due to the displacement of generation by the entire combined cycle unit, which is related to the savings by a factor of 1.35/0.35 or about 4)

Table 6. SO_2 emissions savings due to the electricity generation displaced by combined cycles between 1999 and 2002

	SO_2 savings due to CCs' heat recovery unit (dEh_{cc})				SO_2 savings due to CCs as a whole			
	1999	2000	2001	2002	1999	2000	2001	2002
NEW	-1.051	- 4.974	-16.397	- 17.703	- 4.052	- 19.185	- 63.246	- 68.283
MAT	-3.253	- 610	- 632	- 16.096	-12.547	- 2.353	- 2.438	- 62.083
ENC	0	- 9	- 260	- 7.315	0	- 35	- 1.004	- 28.213
WNC	- 36	- 711	- 657	- 1.480	- 138	- 2.744	- 2.532	- 5.709
SAT	-3.343	- 3.228	-10.989	- 38.397	-12.893	- 12.452	- 42.386	-148.102
ESC	- 852	- 3.385	-21.074	- 28.342	- 3.287	- 13.056	- 81.286	-109.319
WSC	- 56	-13.154	- 833	- 3.503	- 216	- 50.738	- 3.212	- 13.511
MON	- 402	- 2.402	- 4.436	- 1.343	- 1.552	- 9.264	- 17.110	- 5.179
PAC	0	0	- 2.951	- 74	0	0	- 11.382	- 285
USA	-8.992	-28.473	-58.229	-114.252	-34.685	-109.825	-224.596	-440.685
USA Cumul.		-37.466	-95.694	-209.946		-144.510	-369.106	-809.792

Appendix II: Data tables

Table A1. Heat input, SO_2 emissions, and average emission rates by region

	1985	1988	1989	1990	1991	1992	1993	1994
New England								
Heat input	484	527	556	485	461	413	327	331
SO_2 emissions	380	406	409	354	337	306	234	212
Emission rate	1.568	1.540	1.470	1.461	1.459	1.483	1.430	1.283
Middle Atlantic								
Heat input	1.818	1.962	2.031	1.923	1.843	1.737	1.657	1.627
SO_2 emissions	1.645	1.694	1.696	1.670	1.623	1.557	1.465	1.429
Emission rate	1.809	1.727	1.670	1.737	1.762	1.793	1.767	1.756
East North Central								
Heat input	3.456	3.601	3.662	3.741	3.808	3.713	3.906	4.069
SO_2 emissions	5.435	5.149	5.232	5.167	5.091	4.784	4.673	4.686
Emission rate	3.145	2.860	2.857	2.762	2.673	2.577	2.393	2.303
West North Central								
Heat input	1.518	1.756	1.754	1.800	1.825	1.754	1.829	1.933
SO_2 emissions	1.582	1.355	1.317	1.311	1.306	1.169	974	1.088
Emission rate	2.085	1.543	1.502	1.457	1.431	1.334	1.065	1.126
South Atlantic								
Heat input	3.341	3.661	3.817	3.640	3.634	3.678	3.873	3.867
SO_2 emissions	3.372	3.559	3.527	3.469	3.420	3.457	3.444	3.306
Emission rate	2.018	1.944	1.848	1.906	1.882	1.880	1.778	1.710
East South Central								
Heat input	1.795	1.919	1.797	1.910	1.932	1.969	2.235	2.157
SO_2 emissions	2.234	2.245	2.310	2.354	2.267	2.342	2.556	2.354
Emission rate	2.489	2.340	2.571	2.464	2.347	2.379	2.287	2.182
West South Central								
Heat input	3.340	3.331	3.324	3.335	3.344	3.311	3.503	3.456
SO_2 emissions	802	677	715	732	750	774	838	763
Emission rate	0.480	0.406	0.431	0.439	0.448	0.468	0.478	0.442
Mountain								
Heat input	1.643	1.974	2.038	2.043	2.009	2.138	2.097	2.201
SO_2 emissions	480	428	466	454	442	466	457	484
Emission rate	0.585	0.434	0.457	0.444	0.440	0.436	0.436	0.440
Pacific Contiguous								
Heat input	707	646	639	569	543	681	590	728
SO_2 emissions	75	76	78	71	71	84	87	86
Emission rate	0.212	0.235	0.245	0.248	0.263	0.248	0.295	0.235
National								
Heat input	18.102	19.378	19.619	19.446	19.398	19.394	20.018	20.369
SO_2 emissions	16.006	15.591	15.751	15.581	15.306	14.939	14.727	14.408
Emission rate	1.768	1.609	1.606	1.602	1.578	1.541	1.471	1.415

Note. Heat input in trillion Btus, emissions in thousand short tons, emission rate in lb. SO_2/mmBtus.

Table A1. (con't) Heat input, SO_2 emissions, and average emission rates by region

	1995	1996	1997	1998	1999	2000	2001	2002
New England								
Heat input	369	413	558	550	505	470	511	512
SO2 emissions	204	195	265	274	243	211	192	128
Emission rate	1.105	0.943	0.950	0.996	0.961	0.897	0.750	0.500
Middle Atlantic								
Heat input	1.695	1.699	1.839	1.961	1.858	1.870	1.827	1.917
SO2 emissions	1.324	1.292	1.369	1.420	1.283	1.270	1.241	1.170
Emission rate	1.562	1.521	1.490	1.448	1.381	1.359	1.359	1.220
East North Central								
Heat input	4.250	4.579	4.707	4.875	4.791	4.832	4.619	5.450
SO2 emissions	3.258	3.667	3.804	3.762	3.489	3.015	2.761	2.730
Emission rate	1.533	1.602	1.616	1.543	1.457	1.248	1.195	1.002
West North Central								
Heat input	2.211	2.259	2.320	2.417	2.422	2.472	2.484	2.190
SO2 emissions	996	959	918	939	896	790	813	974
Emission rate	0.901	0.849	0.791	0.777	0.740	0.639	0.655	0.890
South Atlantic								
Heat input	4.253	4.579	4.792	5.056	5.073	5.059	4.870	4.888
SO2 emissions	2.750	2.954	3.086	3.269	3.148	2.840	2.713	2.746
Emission rate	1.293	1.290	1.288	1.293	1.241	1.123	1.114	1.124
East South Central								
Heat input	2.534	2.513	2.620	2.570	2.676	2.762	2.724	2.406
SO2 emissions	1.781	1.807	1.866	1.823	1.767	1.651	1.496	1.099
Emission rate	1.406	1.438	1.424	1.418	1.320	1.195	1.098	0.914
West South Central								
Heat input	3.775	3.830	3.880	4.142	4.266	4.361	4.142	4.205
SO2 emissions	924	96	994	971	984	836	834	841
Emission rate	0.490	0.506	0.513	0.469	0.461	0.384	0.403	0.400
Mountain								
Heat input	2.200	2.302	2.378	2.499	2.507	2.601	2.650	2.587
SO2 emissions	503	489	507	484	434	408	418	391
Emission rate	0.457	0.425	0.426	0.388	0.347	0.314	0.315	0.302
Pacific Contiguous								
Heat input	467	454	504	594	643	866	1.003	708
SO2 emissions	60	83	70	88	104	98	87	32
Emission rate	0.255	0.367	0.278	0.297	0.325	0.227	0.173	0.089
National								
Heat input	21.753	22.629	23.598	24.663	24.740	25.292	24.829	24.863
SO2 emissions	11.799	12.415	12.880	13.030	12.349	11.119	10.554	10.112
Emission rate	1.085	1.097	1.092	1.057	0.998	0.879	0.850	0.813

Note. Heat input in trillion Btus, emissions in thousand short tons, emission rate in lb. SO_2/mmBtus.

Table A2 . Decomposition of US SO_2 emission changes from 1985 to 2002 (tons SO_2)

	85-88	88-89	89-90	90-91	91-92	92-93	93-94	94-95
dEr	(1.006.068)	(21.095)	68.099	(151.580)	(431.379)	(467.854)	(333.388)	(3.262.739)
dEh agg	1.158.880	177.574	(24.002)	36.217	(176.880)	676.632	160.129	1.085.564
dEh disp	(567.998)	4.032	(214.656)	(159.150)	241.372	(420.647)	(146.430)	(431.280)
dEh bet	*17.003*	*(61.896)*	*54.290*	*972*	*131.061*	*(4.149)*	*(40.983)*	*(40.992)*
dEh w/i_Coal	*(585.000)*	*65.927*	*(268.946)*	*(160.123)*	*110.311*	*(416.498)*	*(105.447)*	*(390.288)*
dEh w/i_OG	*14.837*	*(2.671)*	*(7.520)*	*21.081*	*(885)*	*(4.928)*	*(17.920)*	*(112.684)*
dE	(415.185)	160.51	(170.559)	(274.513)	(366.887)	(211.869)	(319.689)	(2.608.455)

	95-96	96-97	97-98	98-99	99-00	00-01	01-02
dEr	(116.650)	18.020	(118.570)	(504.606)	(1.375.826)	(64.047)	(576.993)
dEh agg	535.560	556.647	499.392	(35.830)	154.141	(289.325)	242.235
dEh disp	196.802	(109.935)	(230.446)	(140.263)	(8.622)	(211.068)	(107.741)
dEh bet	*152.300*	*(76.646)*	*(227.473)*	*(33.027)*	*9.076*	*(70.293)*	*(139.518)*
dEh w/i_Coal	*44.502*	*(33.289)*	*(2.973)*	*(107.236)*	*(17.699)*	*(140.775)*	*31.777*
dEh w/i_OG	*(16.480)*	*33.494*	*27.133*	*(30.322)*	*(52.333)*	*(60.269)*	*(66.537)*
dE	615.712	464.732	150.376	(680.698)	(1.230.307)	(564.440)	(442.498)

Table A3. Data for calculating combined cycle SO_2 emission reductions, 1998-2002

	1998	1999	2000	2001	2002
Heat input into combined cycle-units (trillion Btu)					
New England	-	7	37	105	191
Middle Atlantic	-	10	11	10	58
East North Central	-	0	0	7	41
West North Central	-	1	6	15	19
South Atlantic	-	14	24	52	253
East South Central	-	4	29	113	193
West South Central	-	61	139	381	783
Mountain	-	6	41	74	162
Pacific Contiguous	-	0	0	21	95
National	-	104	287	778	1.796
Combined cycle share of fossil-fuel heat input					
New England	-	1.4%	7.9%	20.6%	37.4%
Middle Atlantic	-	0.5%	0.6%	0.5%	3.0%
East North Central	-	0.0%	0.0%	0.1%	0.8%
West North Central	-	0.0%	0.3%	0.6%	0.9%
South Atlantic	-	0.3%	0.5%	1.1%	5.2%
East South Central	-	0.2%	1.0%	4.2%	8.0%
West South Central	-	1.4%	3.2%	9.2%	18.6%
Mountain	-	0.3%	0.5%	1.1%	5.2%
Pacific Contiguous	-	0.0%	0.0%	2.1%	13.5%
National	-	0.4%	1.1%	3.1%	7.2%
Oil/gas share of fossil-fuel heat input					
New England	68%	65%	61%	66%	68%
Middle Atlantic	29%	31%	28%	29%	32%
East North Central	2%	2%	2%	2%	6%
West North Central	2%	2%	2%	2%	2%
South Atlantic	18%	18%	16%	18%	20%
East South Central	5%	5%	5%	8%	10%
West South Central	43%	42%	44%	43%	42%
Mountain	4%	5%	7%	9%	9%
Pacific Contiguous	75%	77%	84%	85%	80%
National	18%	18%	19%	20%	20%
Average emission rate for coal units (lbs. SO_2 per million Btu)					
New England	1.31	1.32	1.32	1.32	1.00
Middle Atlantic	1.88	1.86	1.75	1.77	1.60
East North Central	1.58	1.49	1.27	1.21	1.02
West North Central	0.79	0.75	0.65	0.66	0.90
South Atlantic	1.39	1.35	1.22	1.21	1.29
East South Central	1.44	1.35	1.23	1.13	1.01
West South Central	0.81	0.79	0.67	0.69	0.69
Mountain	0.41	0.36	0.34	0.34	0.33
Pacific Contiguous	1.18	1.43	1.40	1.15	0.44
National	1.22	1.16	1.03	1.00	0.97

Policy-business interaction in emissions trading between multiple regions

Jürgen Scheffran[I], Marian Leimbach[II]

[I] ACDIS, University of Illinois at Urbana-Campaign
505 East Armory Ave., Champaign, IL 61820, USA
scheffran@uiuc.edu

[II] Potsdam Institute for Climate Impact Research
Telegrafenberg A31, 14473 Potsdam, Germany
leimbach@pik-potsdam.de

Abstract

Emissions trading helps to bring about emission reductions in regions and business sectors where they are least costly. The cost level highly depends on the absolute level of emission reduction required and the capacity of economies to shift to a carbon-free mode of production. As a framework for understanding this transition and the policy-business interaction, a multi-region framework is used which integrates political decisions on global and regional emission targets, equity considerations, technical choice and the market dynamics of permits. Welfare maximization for each region, subject to production, investment, damage functions and emissions trading costs, results in regional permit threshold prices and a subsequent demand-supply adjustment to an average price. Based on this price, the optimal emission reductions for each region serve as target levels in reaction functions, leading to an iterative dynamic game and Nash equilibria of future emission permits allocation. The model is applied with stylized data for 11 world regions and four cases (business as usual, equal per capita, 10% reductions of baseline, stabilization).

Keywords: Emissions trading, emission reduction, fair allocation, multi-regional dynamic game, Nash equilibrium, price adjustment, reaction function, policy-business interaction, technical choice

1 Emissions trading on multiple levels

International climate negotiations (Kyoto and beyond) aim at finding agreed limits on greenhouse gas (GHG) emission trajectories and associated emission reductions compared to baseline emissions (Philibert and Pershing 2001, Ipsen et al. 2001). One of the unresolved questions is to find mechanisms to allocate emission rights and limits from global levels down to regional, national and local levels, including individual firms and consumers (Morthorst 2003). While at global levels only a few actors are involved (the groups of industrialized and developing countries; a dozen major regions, about 200 States), with local levels the number of actors increases to millions, if not billions. It is not possible to organize a negotiation process to find accordance with so many actors. A key issue is to find a fair allocation among actors, based on equity principles such as equal per capita emissions (see Leimbach 2003 and the given references there). Market mechanisms such as emissions trading are usually assumed to provide an efficient and cost-effective allocation. A variety of models have been applied in the field of emissions trading (see the survey in Springer 2003), some of which combine general equilibrium models with the selection of policy instruments (Jensen and Rasmussen 2000, Springer and Varilek 2004). The underlying micro-macro link is a challenge for the theory of emissions trading as well as for its practical implementation. This paper is seen as a contribution to further development of the framework for emissions trading analysis on multiple levels, supported by data-based computation.

1.1 Emission baselines, targets and emission reductions

The task is to ensure that total flow of emissions $G(t)$ at time t, accumulated by n actors, each with emissions $G_i(t)$ ($i = 1,\dots, n$), should not exceed an agreed total limit $G^*(t)$ which serves as a target:[1]

$$G(t) = \sum_{i=1}^{n} G_i(t) \le G^*(t) \,.$$

The target trajectory $G^*(t)$ is based on value judgements on the meaning of dangerous climate change, but also on estimates of what is actually achievable. One way to construct these limits is to define a tolerable level of climate change and then to derive in an inverse manner the admissible set of emission trajectories that stay within the target set, excluding those that are perceived as dangerous and thus intolerable (in particular with regard to singular and extreme weather events).[2] Within the definition of what is tolerable or dangerous, there can be a range of views (Ott et al. 2003), leaving the borders fuzzy.

---

[1] Emissions are a flow variable, representing the amount of gases emitted in a particular time period, usually years.

[2] On the tolerable windows approach see WGBU (1995), Petschel-Held et al. 1999, Bruckner et al. 2003.

Assuming that there is an aggregate cap on emissions, it is a challenge to find institutional mechanisms to ensure that a large number of actors jointly pursue a particular emission limit G^*, avoiding collective actor problems (free-riding, prisoners' dilemma, tragedy of the commons) by cooperative and regulative measures. The usual approach is to define baseline trajectories $\overline{G}_i(t)$ for each actor, assuming that these trajectories would be pursued without regulative mechanisms. Altogether this leads to a joint baseline trajectory

$$\overline{G}(t) = \sum_{i=1}^{n} \overline{G}_i(t)$$

which usually deviates from the set of target trajectories $G_i^*(t)$ and the actual emissions $G_i(t)$.[3] To adapt the baseline path to the target path, actors would take efforts to reduce emissions

$$R(t) = r(t)\overline{G}(t) = \sum_i R_i(t)$$

where $R_i(t) = r_i(t)\overline{G}_i(t)$ are reductions for each actor i and $r_i(t)$ is the percentage of reductions with regard to the baseline emissions. Altogether we have the condition for emissions staying below the target

$$G(t) = \overline{G}(t) - R(t) = \overline{G}(t)(1 - r(t)) = \sum_i \overline{G}_i(t)(1 - r_i(t)) = \sum_i \overline{G}_i(t) - \sum_i R_i(t) \leq G^*(t)$$

Seeking $G(t) = G^*(t)$ translates into emission reduction requirements

$$R^*(t) = \sum_i R_i^*(t) = \overline{G}(t) - G^*(t).$$

The question is how to distribute reduction requirements $R^*(t)$ to the individual actors i, depending on global targets $G^*(t)$ and baseline emissions $\overline{G}_i(t)$. Here equity considerations come into play.

Defining the emission baselines, measuring the differences between baseline $\overline{G}(t)$, actual $G(t)$ and target emission trajectories $G^*(t)$, and estimating the costs of emission reduction is not an easy task, and so far no agreement has been achieved in this regard. This is further complicated by the fact that emission reduction cannot be achieved just by building a new, technically more advanced facility which would always add more emissions. Emission reduction in the true sense is accompanied with closing down previous "emitters" (or pursuing the possibility of carbon capturing and sequestration). Thus, investing into new and often more costly low-emission technologies and facilities is accompanied by shutting down old ones to provide a net emission reduction which implies a loss of production and

[3] In general, there is a set of admissible paths from which we select here a particular one for analytical purposes.

invested money (sunk costs). This can become significant not only on the macro-economic level, but also on the micro levels of firms and specific products and technologies.

1.2 Allocation of permits from macro to micro levels

We now study how a collective emission target G^* can be allocated to emission permits G_i^* for individual actors *i*. Different allocation mechanisms could be considered, including equal emissions per capita, auctions and size- or efficiency-dependent mechanisms (Jensen and Rasmussen 2000). As a baseline we suggest the following procedure:

1. For allocation of global emission permits G^* to regions $I = 1,\dots, N$ we assume that each region receives a share G_I^* proportionate to its population.
2. Regional emission permits are further allocated to permits of individual producers $i = 1,\dots, n$ in the region proportionate to its amount currently emitted

$$G_{Ii}^* = \kappa_I \cdot G_{Ii} \ .$$

where κ_I is the required percentage reduction or increase of emissions in region *I*.

The latter expresses the fact that larger emitters demand more emissions and thus have the right to a bigger share. With the need for emission reduction, big emitters would have to take a greater share of reduction and thus would be more severely affected if global permits shrink (similar statements apply for other size-dependent mechanisms). Previous efforts to decrease emission intensity are not included.

The factor κ_I is common for all producers in a region, in order to ensure that the sum of individual permits is equal to regional permits:

$$\sum_i G_{Ii}^* = \sum_i \kappa_I G_{Ii} = G_I^*$$

leading to

$$\kappa_I = \frac{G_I^*}{\sum_i G_{Ii}} \ .$$

Thus, whether a producer is actually in demand or in supply of permits depends on the relative "excess" $G_{Ii} - G_{Ii}^* = (1 - \kappa_I) G_{Ii}$ which generates the price on the market.

2 Links between economic growth, emission reductions and mitigation costs

2.1 Basic factors

In the following we describe the interactions between the key variables in emissions trading, each for a particular time period and actor, who can be a region, state or firm:[4]

- Human population N
- Energy consumption E
- Greenhouse gas emissions G
- Economic output Q
- Investment (costs) C
- Utility U generated from production
- Damage D induced by climate change
- Costs and gains Π for buying or selling emission permits

We further use the following ratios which play an important role in the debate (more can be added):

- $g^e = G/E$: Emission intensity of energy
- $q^e = Q/E$: Energy productivity (inverse of energy intensity e^q)
- $q^n = Q/N$: Output per capita
- $q^g = Q/G$: Emission productivity (inverse of g^q)
- $c^g = C/G$: Cost per emission unit, with emissions per cost unit g^c as inverse (accordingly c^e for energy and c^q for production)
- $u^q = U/Q$: Utility per production unit
- $d^g = D/G$: Damage per emission unit[5]
- $\pi^g = \Pi/G$: Price per emission unit (in the following always $\pi \equiv \pi^g$)

Thus with the basic variables and their ratios, we can use numerous functional relationships, such as $Q = q^g \cdot G = q^g g^c \cdot C = U/u^q$, $G = g^q \cdot Q = g^c \cdot C = D/d^g$, and so on. Perhaps the most famous of these relationships is the Kaya formula, which connects two variables with three ratio factors in a row (Edenhofer 2001):

$$G = g^e \times e^q \times q^n \times N \quad .$$

To avoid misunderstanding, the ratio factors are in general not constant, only for linear relationships between the respective variables (as explained later). They can be functions of the variables themselves (e.g. representing learning, escalation

[4] To facilitate notation, we neglect the indices t and i as long as they are not needed.

[5] Every actor contributes to the climate damage for any other actor which represents an important coupling factor. For incremental changes actors only take their own contribution into consideration, leading to the collective action problem.

or satisfying behaviour of actors). The impact of non-linearities will be studied later. These ratios are important indicators in the climate debate and they are measurable as long as the basic variables themselves are measurable.

These factors have developed quite differently in various regions of the world. In North America both population N and annual emissions G have increased by one third between the early 1970s and early 1990s, leaving emissions per capita constant. In the same period both output per energy unit q^e and per capita q^n have been growing by about 40% which became possible through a decline in carbon intensity by about 10%. In Europe the increase in the output-to-energy ratio was half of that in North America (about 20%), while the decline in carbon intensity was more than twice as much. Emissions remained almost the same. Different is the development in China, where total emissions increased by a factor of three between 1972 and 1994, much more than the population (about +50%). While primary energy consumption increased comparable to emissions, both emission intensity and energy-specific output increased only a little.

With these variables we can define an overall value function for each actor, taking into account the utility and costs associated with additional production, as well as the damage and cost associated with additional emissions (for different reference variables):

$$\begin{aligned} V = U - C - D - \Pi &= (u^q - c^q)\cdot Q - (d^g + \pi)G \\ &= (u^q - c^q - d^q - \pi g^q)Q \\ &= (u^g - c^g - d^g - \pi)G \\ &= (u^c - 1 - d^c - \pi g^c)C \ . \end{aligned}$$

In emissions trading schemes the target emissions G^* would be free, while only deviations $G - G^*$ are taken into account for emissions trading, with the gains and costs of selling and buying emission permits given as

$$\Pi = \pi \cdot (G - G^*) .$$

For $G < G^*$ actors tend to supply permits by staying below their emission target. A variation in emissions $\Delta G = -R$ induces a value change

$$\Delta V = -(v^g - \pi)R - c^r R$$

where v^g is the value gain per additional emission unit (or the value loss in case of reduction due to production losses). In the linear case we have $v^g = u^g - c^g - d^g$. $C^R = c^r R$ are the costs for investment in emission reduction measures with c^r being the costs per reduction unit. For emission increases $R < 0$ we have no direct reduction costs ($c^r = 0$). To ensure positive value gains $\Delta V > 0$, for increasing emissions, $\Delta G > 0$ the price should stay below the threshold

$$\pi < v^g + c^r \equiv \pi^* .$$

If the market price exceeds this threshold ($\pi > \pi^*$), then it is profitable to sell emission permits ($\Delta G < 0$). The permit price adjusts to this demand-supply interaction. It is interesting to note that, according to the last relationship, the threshold price increases with productivity q^g and decreases with cost c^g per emission unit. There is also an impact of utility u^q per production unit which may decline due to saturation effects. In the following we set $u^q = 1$, thus using the amount of production in monetary units as the basis in which value is measured, avoiding further discussion on utility here.

2.2 The price mechanisms for multiple actors

We now distinguish $i = 1,\ldots, n$ actors, spending investment C_i and producing economic output Q_i, emissions G_i and damage D_i. Then the condition for positive net value change $\Delta V_i \geq 0$ leads to an actor-specific threshold price

$$\pi \leq v_i^g + c_i^r \equiv \pi_i^* .$$

Since these threshold prices vary among the actors, for a given permit price π there are some actors above it ($\pi_i^* > \pi$) who demand further permits $\Delta G_i > 0$, and other actors below it ($\pi_i^* < \pi$) who become suppliers $\Delta G_i < 0$. The price adapts until no further production is feasible for anyone and the price change stops.

We assume that for each actor permit demand increases the further the actual price is below the threshold, while supply increases above the threshold (because it becomes more attractive to sell). Positive $\pi_i^* - \pi$ is an incentive to buy and, if negative, to sell permits. Thus, it makes sense to use this difference to define individual demand and supply functions around one's own threshold price π_i^*:

- Demand $\Delta G_i^+ = D_i(\pi, \pi_i^*) = a_i^+(\pi_i^* - \pi) \geq 0 \quad (for\ \pi \leq \pi_i^*)$

- Supply $\Delta G_i^- = -S_i(\pi, \pi_i^*) = a_i^-(\pi_i^* - \pi) \leq 0 \quad (for\ \pi \geq \pi_i^*)$.

Parameters a_i^+ and a_i^- indicate the "reactivity" of demand and supply to the price difference. Summing up all demands and supplies, leads to the well-known linear relationships between price, demand and supply

$$\pi = \frac{\sum_{i^+} a_i^+ \pi_i^* - \sum_{i^+} \Delta G_i^+}{\sum_{i^+} a_i^+}$$

$$\pi = \frac{\sum_{i^-} a_i^- \pi_i^* - \sum_{i^-} \Delta G_i^-}{\sum_{i^-} a_i^-}$$

where i^+ and i^- are the index sets of actors for demand and supply. To identify the equilibrium market price the total demand and supply should be equal:

$$\sum_{i^+} \Delta G_i^+(\pi) = -\sum_{i^-} \Delta G_i^-(\pi) .$$

Demand and supply reactivity can be constant or size-dependent (e.g. be proportionate to GDP or actual emissions). If they are identical for one actor ($a_i^+ = a_i^- \equiv a_i$) we have one common reaction function $\Delta G_i = a_i (\pi_i^* - \pi)$. The market balance condition (supply=demand) $\sum_i \Delta G_i = \sum_i a_i (\pi_i^* - \pi) = 0$ finally leads to the market price

$$\pi = \frac{\sum_i a_i \pi_i^*}{\sum_i a_i} .$$

For homogenous reactions $a_i = a$ for all actors $i = 1,\ldots, n$, the price is the average of all value gains per emission unit

$$\pi = \sum_i \pi_i^* / n = \sum_i \frac{v_i^g + c_i^r}{n} .$$

This function does not depend on individual reactivity, only on the net value gains and costs of all actors, depending on the input-output ratios of production, damage, costs and emissions which are important in the climate debate. For instance, the permit price increases with productivity of investment q_i^c and decreases with emission intensities g_i^q. Thus, with countries like China or Russia on the market the threshold price would be low, while with the EU or the US it would increase.

3 Choice between technical options

We now discuss the possibility to invest into different production paths $k = 1,\ldots,m$, based on different technologies. Then production, damage and costs induced by emission G are the sum over all paths,

$$Q = \sum_{k=1}^{m} Q^k = \sum_k q^k G^k = G \cdot \sum_k p^k q^k$$

$$D = \sum_{k=1}^{m} D^k = \sum_k d^k G^k = D \cdot \sum_k p^k d^k$$

$$C = \sum_{k=1}^{m} C^k = \sum_k c^k G^k = C \cdot \sum_k p^k c^k$$

where Q^k, G^k, D^k, C^k define the production, emission, damage and cost on the respective path k and p^k is the share (allocation percentage) of emissions on this path, with $\sum_k p^k = 1$. Since all ratios refer to emissions as a basis, we neglect the upper index g for reasons of convenience here. Then value change for additional emissions ΔG depends on the allocation vector $\vec{p} = (p^1, \ldots, p^m)$

$$\Delta V(\vec{p}) = (\sum_k p^k v^k - \pi)\Delta G$$

where we used the unit value, $v^k \equiv u^k - c^k - d^k - c^{rk}$ with c^{rk} being the cost per unit of emission reduction for path k (which is zero for emission increase). Then for $\Delta G > 0$, the condition $\Delta V \geq 0$ leads to the threshold price

$$\pi \leq \sum_k p^k v^k = \pi^*(\vec{p})$$

which depends on allocation. For n actors, the market price

$$\pi(\vec{p}) = \frac{1}{n} \sum_i \sum_k p_i^k v_i^k$$

also depends on the allocation chosen by all actors. Each actor chooses the allocation vector $\vec{p}$ that maximizes its value gain. Which path k is to prefer, depends on the allocation dependent market price. Thus, there is an implicit interdependence between price π and allocation $\vec{p}$ among all actors which can be seen by introducing the price function into value change:

$$\begin{aligned}\Delta V_i(\vec{p}) &= (\sum_k p_i^k v_i^k - \sum_j \sum_k p_j^k v_j^k / n)\Delta G \\ &= [\sum_k (p_i^k v_i^k 1 - 1/n) - \sum_{j \neq i} p_j^k v_j^k / n)]\Delta G \quad i,j = 1, \ldots, n\end{aligned}$$

If all actors switch gradually $\Delta\, p_i^k$ to the production path k^* generating the highest unit value, this leads to a dynamic interaction between actors and technologies, including the phenomenon of collective switching.

4 Non-linear ratios

4.1 General case

Skipping the linearity assumption, we now treat the case of non-linear ratios between G, Q, C. Then emission change can result from two causes - changing production/costs and changing specific emissions:

$$\Delta G = g^q \cdot \Delta Q + Q \cdot \Delta g^q .$$

The first term represents mitigation of emissions, resulting in reduced production $\Delta Q < 0$ (which is perceived as a loss and thus a cost), the second term represents induced technical progress for emission reduction. An emission increase due to production growth $\Delta Q > 0$ can be compensated by a reduction in specific emissions $\Delta g^q < 0$, associated with additional investment in technology. The condition for declining emissions is then

$$\frac{\Delta g^q}{g^q} \leq -\frac{\Delta Q}{Q} \leq 0 .$$

Thus the pace of technical progress should keep up with the pace of economic growth to avoid emission growth.

In the following, we include non-linearities to represent increasing production efficiency per emission unit, increasing marginal costs for high emission reduction R and disproportionate damages for high emissions:

$Q \;= q^0 \cdot G^\alpha \;,\; \alpha > 1$ Increasing emission productivity

$C^R = c^0 \cdot R^\gamma \;,\; \gamma > 1$ Increasing abatement costs

$D \;= d^0 \cdot G^\delta \;,\; \delta > 1$ Increasing vulnerability to damage

The factors q^0, c^0, d^0 represent the respective output for the first unit of input, thus the outcome on the left side of the equation if the variable on the right side is set to 1, whatever the chosen unit (ignoring upper indices g and r). Note that for $R < 0$ (emission increase) the emission reduction costs C^R are zero ($c^0 = 0$). Then we have a non-linear value function in emissions:

$$V = Q - C - D - \Pi = q^0 \cdot G^\alpha - c^g \cdot G \; - d^0 \cdot G^\delta - \pi(G - G^*)$$

where the state-dependent output-input ratios are given as $q^g = Q/G = q^0 G^{\alpha-1}$, $c^g = C/G = const.$, $d^g = D/G = d^0 G^{\delta-1}$, $c^r = C^R/R = c^0 R^{\gamma-1}$ which are connected to the respective differential partial derivatives via the product with the exponents.

For an incremental change in emissions $R = -dG$, the differential value change can be expressed in terms of the partial derivatives, taking into account the incremental cost for emission reduction C^R and neglecting the individual impact on price $\partial\pi/\partial G$:

$$dV = \frac{\partial V}{\partial G} dG - C^R$$

$$= -(q^0 \alpha G^{\alpha-1} - c^g - d^0 \delta G^{\delta-1} + c^0 R^{\gamma-1} - \pi) R = -(v^g(G) + c^0 R^{\gamma-1} - \pi) R .$$

Here $v^g(G) \equiv q^0 \alpha G^{\alpha-1} - c^g - d^0 \delta G^{\delta-1}$. Then for $i = 1, \ldots, n$ actors, $dV_i > 0$ for $R_i > 0$ leads to the threshold price as a function of emissions and emission reductions

$$\pi > v_i^g(G_i) + c_i^0 R_i^{\gamma-1} = \pi_i^* .$$

The average price

$$\pi = \sum_i \pi_i^*(\vec{G}, \vec{R}) / n = \sum_i \frac{v_i^g(G_i) + c_i^0 R_i^{\gamma_i - 1}}{n}$$

is a nonlinear function of emission vector $\vec{G} = (G_1, \ldots, G_n)$ and incremental reduction vector $\vec{R} = (R_1, \ldots, R_n)$ including the individual contributions of all actors.

To identify the individually optimum emission reductions for a given price π we use the partial derivative for a single actor

$$\frac{\partial dV_i}{\partial R_i} = \pi - v_i^g(G_i) - c_i^0 \gamma_i R_i^{\gamma_i - 1} + R_i \cdot \frac{\partial \pi}{\partial R_i} = 0 .$$

with $\partial \pi / \partial R_i = c_i^0 (\gamma_i - 1) R_i^{\gamma_i - 2} / n$, we obtain the optimal reduction

$$R_i^* = \left(\frac{\pi - v_i^g(G_i)}{c_i^0 [\gamma_i (1 - 1/n) + 1/n]} \right)^{\frac{1}{\gamma_i - 1}} .$$

In particular, this shows that optimal emission reductions are only positive (decreasing emissions) for a price $\pi > v_i^g(G_i)$. Thus, for increasing emissions G_i further emission reductions are more difficult to achieve, except for high permit prices. For larger reductions R_i, the differential value change only holds approximately, and has to be expanded by higher terms in the Taylor series.

Inserting the price function $\pi(\vec{G}, \vec{R})$ into the previous equation, it can be resolved for

$$R_i^* = \left(\frac{\sum_{j \neq i} (v_j^g(G_j) + c_j^0 R_j^{\gamma_j - 1}) - v_i^g(G_i)(n-1)}{c_i^0 (n\gamma_i - 1)} \right)^{\frac{1}{\gamma_i - 1}}$$

Thus, optimal reduction of actor i depends on the reduction vector of all other actors R_{-i}. The adjustment of all actors towards their optima R_i^* shifts the price and thus the optima. For each time step this corresponds to a dynamic game in which an actor i seeks the best emission reduction R_i^* for given reductions of all others. The Nash-Cournot tatonnement adjustment is a mechanism to achieve a Nash equilibrium, if stability conditions are satisfied. A dynamic process that represents this interaction is given by

$$\Delta R_i(t) = R_i(t+1) - R_i(t) = p_i (R_i^*(t) - R_i(t))$$

with adjustment reactivity p_i (see Scheffran 2002).

4.2 Special case for square functions

To better understand the qualitative properties of the model, we analyze the special case for $\alpha = \gamma = \delta = 2$. Then we obtain a linear-square value function

$$V = (v^q \cdot q^0 - d^0)G^2 - c^g \cdot G - \pi(G - G^*)$$

where the state-dependent output-input ratios are given as $q^g = Q/G = g^0G$, $d^g = D/G = d^0G$, $c^r = C^R/G = c^0R$. The incremental value change becomes

$$dV = \frac{\partial V}{\partial G} dG - C^R$$

$$= -(q^0 \alpha G - c^0 - d^g \delta G + c^0 R - \pi)R = -(v^g(G) + c^0 R - \pi)R$$

where $v^g(G) \equiv q^0G - c^g - d^gG$ is linear in G. Then $dV > 0$ for $R > 0$ is satisfied for the threshold price as a function of emissions and emission reductions:

$$\pi > v^g(G) + c^0 R = \pi^*$$

To identify the individually optimum emission reductions for a given price π we use the partial derivative for a single actor

$$\frac{\partial dV}{\partial R} = \pi - v^g(G) - 2c^0 R + R \cdot \frac{\partial \pi}{\partial R} = 0$$

For $i = 1,\ldots, n$ actors the price is given as $\pi = \sum_i \left[(q_i^0 - d_i^0)G_i - c_i^g + c_i^0 R_i\right]/n$. Inserting this into R_i^* we obtain a linear reaction line

$$R_i^* = \frac{\sum_{j \neq i}(v_j^g(G_j) + v_j^0 R_j) - v_i^g(n-1)}{c_i^0(2n-1)}$$

In particular, this shows that optimal emission reductions would become negative (increasing emissions) for a price $\pi < v^g(G)$. Thus, for increasing emissions G, further emission reductions are more difficult to achieve for a single actor, except for high permit prices.

5 Model-based data and computations for the world's regions and selected cases

In the following we feed the outlined model with stylized data for 11 world regions and apply some of the explained methodology to determine the threshold and permit prices and the optimal emission reduction for three selected cases. The regions are the same as in Leimbach 2003:

- AFR Sub-Saharan Africa
- CPA China, Mongolia, Vietnam, Cambodia, Laos
- EEU Eastern Europe
- FSU Former Soviet Union
- LAM Latin America and the Caribbean
- MEA Middle East and North Africa
- NAM North America
- PAO Pacific OECD (Japan, Australia, New Zealand)
- PAS Other Pacific Asia
- SAS South Asia (mainly India)
- WEU Western Europe.

All computations are based on Gross Domestic Product (GDP) and emission data of the year 2005 as applied in the reference scenario of the ICLIPS model (see Leimbach and Toth 2003). Net economic growth (production minus costs) is assumed to be 3% and damages are assumed to be 1.5% of GDP for all regions. Moreover, we assume for the non-linear case that the exponents in the cost and production functions are between a linear and square functional relationship, thus $\alpha = \gamma = 1.5$, while for damage we refer to the often used square function $\delta = 2$. The mitigation cost function is calibrated on the data point assuming that 5% incremental reduction leads to 2% GDP loss.

All figures show results from this non-linear approach, whereby the final prices and optimal reduction values are based on an iterative procedure of price adjustment. The chosen cases differ with respect to the allocation of emission permits which are implicitly assumed to correspond to the levels of actual emissions:

- BAU - Business as usual (this refers to the emission baseline of the ICLIPS model)
- epc - equal per capita
- 10% red. - 10% reduction of 2005 baseline level (only for industrial countries)
- stabilis. - stabilisation of emissions on 1990 levels.

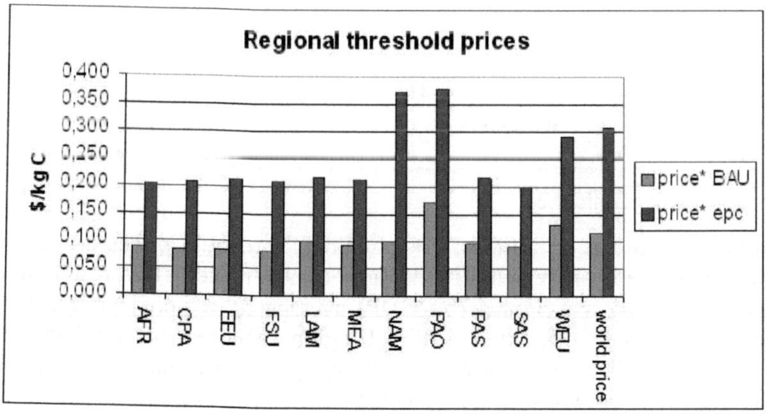

Fig. 1. Regional threshold prices

The difference between the regional and the world prices represent the bias of regions to sell or to buy emission permits. The comparatively high world price is due to higher weights of industrialised regions.

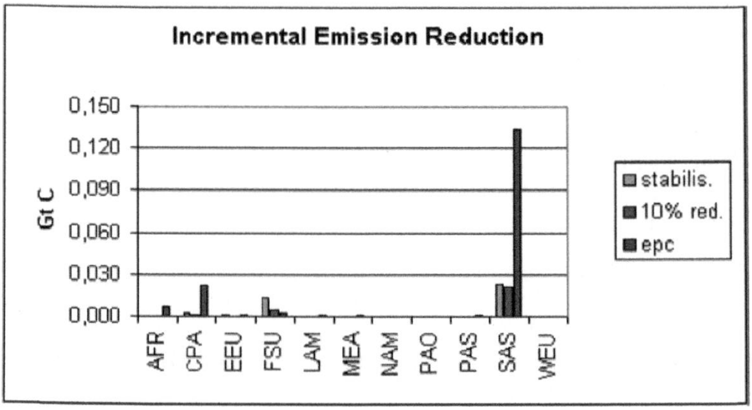

Fig. 2. Incremental Emission Reduction

Simulations for region SAS result in an emission level far above its baseline. Hence, emission reduction is cheap, and profits from selling permits are welcomed. In the case of stabilisation, region FSU is allowed to emit above its baseline and to sell permits ("hot air").

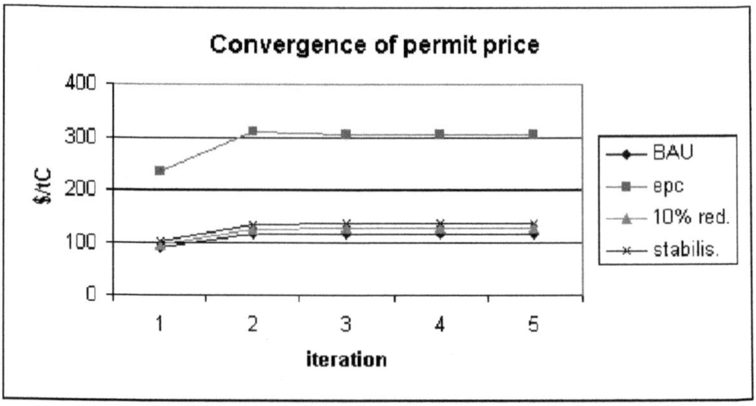

Fig. 3. Convergence of permit price

The iteration procedure is as follows: First, regional threshold prices are determined by assuming zero (incremental) reduction in each region. This results in a world price by creating a weighted (by GDP) average of regional prices. This world price can be used to determine optimal regional (incremental) reductions. Again new prices result, and so forth. Presumably, prices are especially high in cases where high trade volumes can be expected (as in the epc case).

References

Bruckner T et al. (1999) Climate change decision-support and the tolerable windows approach, Environmental Modeling and Assessment, 4: 217-234

Edenhofer O (2001) The Sustainability Transition and Technological Change. A Research Agenda. Working Paper. Potsdam, PIK

Ipsen D, Rösch R, Scheffran J (2001) Cooperation in Global Climate Policy: Potentialities and Limitations, Energy Policy, 29(4): 315-326

Jensen J, Rasmussen TN (2000) Allocation of CO_2 Emissions Permits: A General Equilibrium Analysis of Policy Instruments, Journal of Environmental Economics and Management 40: 111-136

Leimbach M (2003) Equity and carbon emissions trading: a model analysis, Energy Policy, 31: 1033-1044

Leimbach M, Toth FL (2003) Economic development and emission control over the long term: the ICLIPS aggregated economic model. Climatic Change, 56: 139-165

Morthorst PE (2003) National environmental targets and international emission reduction instruments, Energy Policy, 31(1): 73-83

Ott K et al. (2004) Reasoning Goals of Climate Protection - Specification of Article 2 UNFCCC, Report for the Federal Environmental Agency, Berlin

Petschel-Held G et al. (1999) The tolerable windows approach: Theoretical and methodological foundations, Climatic Change, 41: 303- 331

Philibert C, Pershing J (2001) Considering the options: climate targets for all countries, Climate Policy, 1(2): 211-227

Scheffran J (2002) Economic Growth, Emission Reduction and the Choice of Energy Technology in a Dynamic-Game Framework. In: P. Chamoni et al., Operations Resarch Proceedings 2001, Heidelberg, Springer, 329-336

Springer U (2003) The market for tradable GHG permits under the Kyoto Protocol: a survey of model studies, Energy Economics 25 (2003): 527-551

Springer U, Varilek M (2004) Estimating the price of tradable permits for greenhouse gas emissions in 2008-12, Energy Policy, 32(5): 611-621

WBGU (1995) Scenario for the derivation of global CO_2 reduction targets and implementation strategies, German Advisory Council on Global Change, Bremerhaven

The changing role of the project mechanisms in emissions trading

Katherine G. Begg

University of Edinburgh
Centre for Environmental Change and Sustainability
Crew Building,West Mains rd, Edinburgh EH9 3JN, UK
katie@sun-shine.org.uk

Abstract

The perception of the project-based mechanisms has varied since their inception under the name of Activities Implemented Jointly (AIJ) to their form under the Kyoto Protocol of Joint Implementation (JI) and the Clean Development Mechanism (CDM). AIJ projects of course still exist under the Marrakech Accords. The controversy surrounding the concept of emissions trading and the need for purely domestic reduction actions has not really abated. The role of JI and the CDM and of AIJ is therefore interesting in the way these mechanisms have alternately been propelled along by their role as market mechanisms in achieving a low carbon future 'at least cost to society' and then reined in by those concerned that the priority should be domestic action. New arguments have appeared with time. The environmental integrity of the reductions achieved by project based mechanisms is now seen as a key selling point in an international market beset by 'hot air'. Thus the popularity of the project mechanisms is now being propelled along by concerns on environmental integrity but then limited by problems with increasing transaction costs due to the complexity of the system to ensure that integrity.

The pressure to reduce transaction costs for the project mechanisms has led to investigation of possible simplified and standardised procedures for baselines for the project mechanisms. This paper therefore traces the evolution of the project mechanisms and some of the latest developments in carbon accounting for reductions for large and small-scale projects as well as looking at the implications for projects under the EU and UK emissions trading schemes.

Keywords: Kyoto protocol, Clean Development Mechanism, emissions trading, Joint Implementation, baselines, small scale projects

1 Introduction

Recent studies (Stainforth et al. 2005) have highlighted that the world is running a higher risk of dangerous climate change than previously thought and that in the next 10 years we have to make some very serious changes to our emission rates if we are to avoid serious risks. So far it has been convenient to play down climate change as an issue. After all any effects attributed to climate change don't seem too bad so far and some areas would actually benefit. However as the adoption by the UK government of the target of 60% reduction in greenhouse gases (GHGs) by 2050 has shown, there are very good reasons why we need to take action. The risks we run are to the whole of human society. Increasing climate change will lead to mounting pressures from climate refugees and disputes over natural resource access that will feed into the existing inequality in the world. This in turn can only worsen the security and peace of all nations.

Actions under the Kyoto Protocol (KP) offer one way forward. The Kyoto Protocol though subject to the vagaries of international negotiations has developed mechanisms and processes as a starting point to engage the major actors so that we can decrease our GHG emissions and try to avoid the worst of the predicted changes. The ratification of Russia in November 2004 meant that the Protocol came into force in February 16th 2005 and the emission reduction commitments agreed under the Protocol are legally binding. The market mechanisms introduced under Kyoto are already becoming a reality with the onset of the EU emissions trading scheme in 2005 with its linking directive with the Clean Development Mechanism (CDM) and trades in the carbon market. The activities of the Executive Board in developing the modalities and approvals process for the CDM have meant that projects have cleared the validation and official registration phase and are ready for implementation though this has been a protracted process. In the following sections we discuss these mechanisms in more detail and trace some of this process of development.

2 Development of the project mechanisms

Economic theory suggests that emissions reduction targets for greenhouse gases can be achieved 'at least cost to society' through emissions trading where targets can be met through trading 'permits'. Where abatement can be made at low cost a firm may overachieve its target and sell excess reduction permits to those who find it more difficult or expensive to abate. Trading equalises abatement costs across all the actors and results in 'least cost to society' for a given target reduction to be achieved. Emissions trading has the advantages of flexibility for each player and specific environmental targets can be reached. The US experience with their SO_2 emissions trading scheme showed that such a system can achieve deep cuts at less cost than anticipated. Emissions trading is also efficient compared to command and control measures which can also be costly and fail to produce the required reductions.

The other attractions are that market mechanisms allow markets to work so that they allow more flexibility and decreased regulation on business. Compared to a pollution tax they represent a much more positive 'carrot' approach rather than a punitive approach. However there are still financial flows involved but these will only be between businesses and countries.

Under the UNFCCC, pre-1995, emissions trading was initially discussed using the language of 'Joint Implementation' (Begg et al. 2001a). Most of the arguments against the project mechanisms were made at this time and still have to be taken into account in developing equitable systems. Trading was seen by its opponents as a way for the developed world to avoid changing its ways by not taking domestic action. It was also considered that such a system could compromise the sovereignty of host nations and their ability to harness indigenous resources and develop their own markets. An increase in transaction costs in achieving emission reductions was also foreseen.

In 1995, the UNFCCC agreed to institute a pilot phase for these 'Activities Implemented Jointly (AIJ)' to gain experience. Specific rules were drawn up and projects were undertaken. In 2004 at COP10 it was reported that the AIJ Pilot phase would continue. There were 157 projects under the AIJ pilot phase with the next report being due in June 2006. These projects were initially not entitled to any credits though after COP10, projects built after 2000 can now be submitted as JI or CDM projects. With the advent of the Marrakech Accords where the details of the CDM were agreed it would seem unnecessary to continue with the AIJ projects. However it does provide a means for the US to continue gaining project experience without Kyoto and from that point of view AIJ is a valuable link.

At Kyoto in 1997 the AIJ project mechanisms were developed into two separate mechanisms called Joint Implementation (JI) and the Clean Development Mechanism (CDM) taking into account whether the host country already had targets (JI) or whether it was a developing country without targets (CDM).

2.1 Project mechanisms and complexity under Kyoto

One of the main issues of contention is the complexity of the JI and CDM processes under Kyoto. This reflects the tension between environmental integrity in delivering 'real' reductions and the necessary quality assurance processes that must then be involved which inevitably increase transaction costs for participation. The current models for the project mechanisms are developing with time. In the following sections existing developments and new forms of the project mechanisms are described.

Joint Implementation (JI)

JI projects cannot generate credits in advance of the first commitment period (2008-2012) but nevertheless projects need to be planned and built to be operational in this period. Interest in JI projects has come from governments who are finding difficulty in finding cheap domestic emission reduction options such as the

Netherlands government with their ERUPT programme and from the Prototype Carbon Fund (PCF) of the World Bank set up to facilitate experience in this area. Further recent government developments have been the Japanese Carbon Fund, the Danish JI programme in the Russian Federation and the New Zealand (JI) projects mechanism (JIQ 2004a).

There are two main forms of Joint Implementation. Track 1 and Track 2 JI. Track 1 JI can take place between countries who have emission reduction targets and who have ratified the Protocol. These countries are subject to a series of conditions in terms of emission inventories and reporting which involve quite a high commitment but which are necessary for any government serious about reducing GHG emissions. Track 1 JI is subject to almost no controls relying on governments and their agents to deal honestly. The environment is protected only by the overall need to meet the national quantified emission limitation and reduction targets. In addition, for projects in accession countries being traded through the linking directive into the EU emissions trading scheme (EU ETS) the reductions generated must be demonstrated to be below a baseline incorporating actions under the Acquis Communitaire. Installations covered by the directive can also not host JI projects. This restricts the availability of projects.

In general though, if host countries go out of compliance then, depending on the extent of under compliance, carbon credits from Track 1 JI projects may not be able to be honoured and there will be no carbon credits for the investors.

As a form of insurance against this, Track 2 JI is available where the validated reductions are guaranteed for the investor. Track 2 JI will follow similar procedures to the CDM in that external validation will be required but the Supervisory Committee for JI will be set up at the first COP/MOP and have yet to finalise procedures. We therefore discuss the CDM in the next section on the understanding that many of the features of CDM are expected to be mirrored in Track 2 JI.

The Clean Development Mechanism (CDM)

The CDM has a complex project cycle of validation, registration, verification, certification and issuance of credits which is described in the Marrakech Accords (UNFCCC 2002). Within the process project participants must prepare a Project Design Document (PDD). This document concerns itself with the issues of Additionality[1], Baselines and Monitoring Plan. The process for large projects (small projects are discussed separately below) deals with the host country approvals, the external consultation and feedback and the environmental and social impacts as well as the independent verification of reductions. Thus both the process and the PDD are focussed on producing reductions that would not have taken place in the absence of the project (additionality) and ensuring that the quantification of the reductions is not overestimated. The main reason for this focus on environmental integrity is that there are no targets and no incentives for host countries not to game play to maximise carbon credits claimed. However all these checks do in-

[1] A project is deemed additional if the emission reductions produced would not have occurred in the absence of the project i.e. under business as usual.

crease transaction costs for the project so that projects approximately over 50000t/y are considered necessary before transaction costs are not a determining factor (Michaelova and Stronzic 2002). These broad statements on size of project of course rely on assumptions about the behaviour of the carbon price and also of the validating and verifying organisations or operational entities (OEs).

The 9th Conference of the Parties (COP 9) saw decisions on modalities for afforestation and reforestation projects being taken which can produce temporary or long term certified emission reductions (tCERs and lCERs) with a fixed crediting period or one which can be renewed. Positive aspects are the inclusion of socio-economic and environmental criteria within the PDD and the use of IPCC good practice guidance. It is not yet clear whether these credits have high environmental integrity.

This integrity rests on the CDM process with its independent validation and verification as well as on the baseline methodology and the additionality test mentioned above. The Executive Board for the CDM was set up as the oversight body for the mechanism and has the authority to register projects and issue certified emission reductions on the basis of reports from an independent operational entity (OE). A recent development has been the adoption by the Executive Board of consolidated tools for demonstrating additionality which requires five steps to explore the policy additionality of the project. This involves determination of whether it is a legal requirement already in the country (policy additionality), its investment additionality either through financial tests or a barrier analysis or both and the role of CDM status in enabling the project. The project has also to be set in the context of the sector in which it will operate to ensure that it is not a business as usual project. These are additional hurdles in preparing the project design document but are a real attempt at ensuring environmental integrity. The additionality question and how the risks of non additionality may be minimised are discussed in later sections.

Small scale projects

Small scale projects are recognised as important in delivering direct sustainability benefits particularly to the poor (Begg et al. 2001b, 2003). The Executive Board for the CDM addressed the problem of small scale projects through an expert group. Their principal aim was to reduce the transaction costs associated with small scale projects by streamlining procedures for the determination of additionality, baselines and monitoring. They have produced guidelines for baseline methodologies and monitoring that should be applied and have recommended the barriers test as the additionality test for small-scale projects. Such projects have originally been defined in terms of renewable energy projects, energy efficiency projects and other projects such as landfill methane[2] and are prioritised under the

---

[2] Small scale projects are defined in one of the three categories as follows: renewable energy projects <15MW output, Energy efficiency projects reducing energy consumption by 15GWh/y and other activities reducing emissions and which emit less than

Accords. A new category of small scale afforestation and reforestation sink projects has been added to these with methodologies elaborated at COP 10 (UNFCCC 2004) and a decision to adopt simplified modalities in the first commitment period. However there is still much to be done in this area before very small-scale projects can become viable. One area which needs clarification is the overlap between some energy projects and sink project boundary issues as well as bundling small projects to reduce transaction costs and leakage issues.

Green investments

'Hot air' was generated in the system when the reduction targets compared to 1990 levels were agreed at Kyoto for Russia and Ukraine as well as the accession countries. As the economies of these countries had collapsed since 1990 and have since recovered more slowly than expected, the actual emissions levels are well below the emissions target. This has given rise to a large amount of emission permits or Assigned Amount Units (AAUs)[3] which do not represent any form of mitigation action and are dubbed 'hot air'. The amount of Russian hot air was more than enough to offset the US target at that time. However recent estimates for Russia suggest that there will be a surplus of 800 $MtCO_2$ in 2000 (JIQ 2003a) possibly falling to a minimum of 275$MtCO_2$ in the commitment period depending on the assumptions. This is much less hot air than anticipated.

Such a lax target has proved problematic to Russia. Facing a buyers market with the withdrawal of the USA, they have sought to overcome the lack of environmental integrity of the AAUs produced by linking the revenues from emissions trading with environmentally friendly energy efficiency projects called 'green investments' (Tangen et al. 2002). These are not JI but would be linked to action within the country. This initiative has met with some scepticism and some of the problems are discussed in detail by Korpoo (2003) but nevertheless this could still be a way forward. Denmark is working with Russia on greening AAUs as well as on its programme on JI.

Green emissions trading

In December 2002 a new trading arrangement was agreed whereby the Slovak Republic will sell some of its assigned amount to a Japanese trading house. Reductions generated by small scale projects will be aggregated and buyers are provided with government guaranteed Kyoto compliance Assigned Amount Units (AAUs). The process avoids the track 1 or 2 JI paperwork and the delays inherent in normal project mechanisms but there are no environmental safeguards other than trust in

---

15ktCO_2equivalent/y. Removal by sinks less than 8ktCO_2/y was added as a category at COP 9.

[3] AAUs are the emission permits available for trading under an emission trading regime at international level between governments or their legal entities. The amount of AAUs available to trade depends on the national target and the current level of national emissions relative to 1990 emissions.

the host and donor countries and the fact that the AAUs are linked to concrete projects rather than national emission inventories.

Gold standard projects

The WWF in 2002 launched their 'Gold standard' for JI and CDM projects. The project types are limited to demand side energy efficiency and renewable energy projects. Additionality tests are used in conjunction with a range of other criteria including an assessment of the impact of the project on the local, regional and global environment, the social sustainability and development impacts, and the economic and technological development impacts. The project should also not use Official Development Assistance (ODA), should have two participation sessions with local stakeholders and a separate environmental assessment. These requirements mirror the existing CDM system but in an extended and more rigorous way. These are expected to be received in the market at a premium because of environmental integrity and their association with sustainability benefits.

2.2 Progress on the CDM

Given the complexity described above, the first batch of 14 projects submitted for desk review by the secretariat for approval of new methodologies in 2003 were rejected by the Executive Board (EB); most of them on the grounds of their (lack of) additionality or problems with the baseline methodology. This caused a storm of protest (JIQ 2003B) which led to resubmission and re-evaluation and finally approval of some of the projects. The EB have since engaged in an effort to make transparent and efficient the approach for registration of CDM projects.

Various criticisms have been made of the CDM process and the role of the Executive Board but the main concern seems to have been that it is more important to engage rather than discourage investors at this early stage in the CDM. No doubt a process of revision and learning will continue for some time until a portfolio of experience has been accumulated and more and more streamlining can occur without compromising environmental integrity. With increasing experience, the Executive Board and the methodological Panel are processing more projects and decisions on registration are beginning to come through. Up to end of December there were 1189 proposed projects for JI and the CDM and 272 of these have submitted Project Design Documents (Point Carbon 2004). Despite this, the difficulty of the process may cause disenchantment with the CDM by private investors which could cause problems for its viability.

However it is also obvious that markets are nervous of hot air. Too much leeway on integrity is not a good thing either. This was also shown in a recent laboratory study by Casa (2003) which indicated that some verification process is needed to deal with the problem of hot air. The introduction of the WWF gold standard on projects in 2002 showed that not only the NGOs but also the large corporations are concerned that if they make investments, then these should hold their value and not be arbitrarily devalued along with less well documented credits. Companies such as BP have given their backing to this initiative. Of course extra safeguards like this prompt a return to the argument on costs. Most economists would say that the extra transaction costs will not be paid by investors when there are cheaper alternatives on the market. Woerdmann (2000, 2001) argues that transaction costs for JI and the CDM are not necessarily higher than for permit transfers under the International Emissions Trading (IET) system and has suggested (2000) that the project mechanisms will be more acceptable as they involve real investment, could use standardised approaches and early crediting and they avoid the allocation issues on property or user rights associated with the IET. The interplay between the supply of hot air and CDM transaction costs examined by Sager (2003) supports the development of a CDM market share when hot air supply is limited. However the risk of devaluation of 'poor' credits and the reward of a premium on credits from projects shown to produce not only well documented and verifiable GHG reductions but also delivering sustainability benefits seems to be sufficient to drive gold standard projects forward.

Grubb (2003) considered the development of the emissions trading markets in terms of the development of the price of carbon credits but also in terms of the environmental integrity of the traded carbon. He suggested that there could be differentiation in the market on credits depending on their association with real projects and real reductions, their crediting from 2000 and ability to deliver sustainability benefits to the poor. Gold standard Certified Emission Reductions (CERs) would be most preferred over Small scale CERs which in turn would be more preferred than Emission Reduction Units (ERUs) especially from Track 1 JI.

Market demand

The World Bank report on the state of the carbon market (2003) showed that the traded volume for 2003 wais more than double that of 2002 for CO_2e and most of these trades were in project based credits. Private investor participation had increased to 40% of the carbon finance sector. All of these factors are encouraging and point to the importance of the project mechanisms.

Ratification and the EUETS appear to have had an accelerating effect on interest in the project mechanisms. Besides the developments documented above, the compulsory nature of the EU ETS market and the realisation by many countries that there has to be some reliance on the project mechanisms to meet Kyoto targets has led to the announcement of a series of initiatives on JI and the CDM.

In addition to the early initiatives on ERUPT and C-ERUPT from the Netherlands government, there is the latest development of a partnership with the EBRD to form a new Carbon fund with the EBRD (32m euros) in a facilitating role for JI projects from 2004-6. Denmark and Belgium have declared a new programme on JI and CDM while China is gearing up to provide CDM projects for investors stimulated by the PCF. A new consortium in Spain has created a Carbon Fund for the CDM and a similar initiative has been announced in Japan with a budget of 75m euros. Transition economies such as Latvia, Estonia, Slovakia, Romania and Bulgaria are also gearing up to act as hosts to JI while Austria, Finland and Sweden continue to enlarge their JI and CDM programmes. The emergence of a series of national emission trading initiatives in Switzerland, Japan, and Canada as well as in some parts of the US has also increased activity and awareness and adds to the pressure being generated under the EU ETS.

For the project mechanisms therefore this background context suggests that interest and activities in this area are increasing though several issues still divide and possibly discourage some investors from investing in JI or CDM projects.

These issues include the complexity and lack of clarity in the current requirements especially for the CDM as well as the complexity in terms of the number of different market instruments available. Another problem is the dilemma in carrying out measures to reduce emissions to deliver both good quality GHG reductions and sustainability benefits without incurring excessive costs which is particularly acute for small scale projects. With small scale projects carbon revenues can be small while transaction costs can be a relatively high proportion of total costs.

Some of these process problems have been addressed under a number of studies and the rest of this paper discusses some of the recent work with which the author has been associated.

3 What can be done to reduce costs in the CDM and where is progress being made?

3.1 Developments in accounting for GHG reductions: the EU PROBASE project

At the moment the services of consultants are frequently used to prepare the Project Design Document (PDD) for a CDM project and even then there are uncertainties in getting it validated and approved. A major activity within the PDD is the derivation of baselines for the GHG accounting and the associated costs and uncertainties. One way of minimising the risks and the costs is to have a system which uses Standardised baselines for large projects along the lines already in use for small scale projects. Recently the EU project PROBASE reported on their investigations into this area (van der Gaast et al. 2003) and this work will be discussed in more detail below. Other studies such as Kartha et al. (2004) have also followed this line on the electric power sector.

Multi-project or standardised baselines

The PROBASE team addressed the problem of streamlining procedures especially for selecting appropriate baselines for projects. They developed an initial version of an 'enabling tool' which provides the basis for a manual and a decision tree to aid the choice of baseline for project developers. In this case we are talking about multi-project or standardised baselines using benchmarks or other types of baseline rather than purely project specific baselines derived for that single project only. Having made that distinction the boundaries are in fact blurred as a project specific baseline may involve a benchmark and could be generalised in some circumstances to other projects of the same type and context.

Specific questions that have been addressed within the Probase study concern the environmental integrity of such baselines. How can they be selected and how can the uncertainties be managed to avoid inflation and gaming? What about additionality? Can these baselines be used to address that problem as well? Some key results to answer these questions in Begg and van der Horst (2004) are considered here.

The problem of selecting a suitable standardised baseline for a particular project such as this wind farm in Poland can be easily seen in Figure 1.

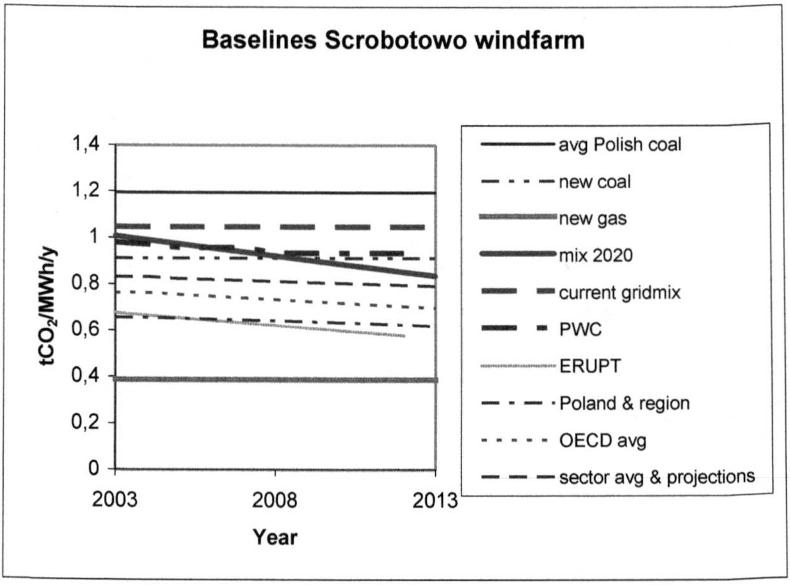

Fig. 1. Skrobotowo Wind Farm in Poland scenario (mix 2020) and standardised baselines

Given the possible range of baselines here where arguments can be made for replacement of emissions from new coal average polish coal, new gas the current grid mix as well as baselines derived from other studies on this project such as ERUPT and PWC and standardized baselines such OECD, the Poland and region baseline which is lowered by a weighting factor as well as sector baselines, how can an appropriate standardized baseline be chosen from such an array of possibilities?

Environmental integrity for large project baselines

The PROBASE study concentrated on the heat and electricity sectors in a range of host countries. They generated a range of multi-project baselines using a range of benchmarks and using an energy economic model Perseus. In this analysis the environmental integrity was explored by comparing these benchmarks against the 'envelope of uncertainty' generated by a project specific set of scenario baselines for the project.

Figure 2 shows the results for the uncertainty analysis (Begg, van der Horst 2004) on the Combined country and Regional benchmark (Anagnostopoulos et al. 2004). This is a regional level benchmark lowered by a weighting to discriminate against high carbon options. This graph was produced by examining the difference between the most conservative project specific scenario and this benchmark over a range of projects in different countries.

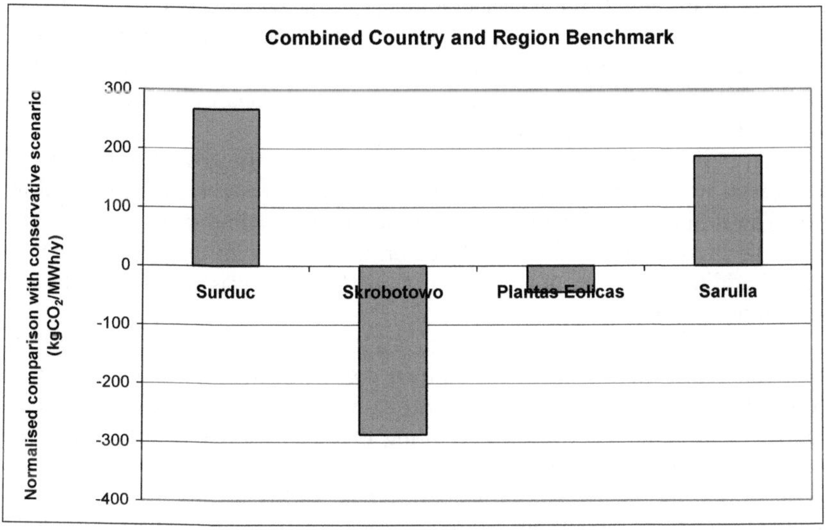

Fig. 2. Comparison of the Combined Country and Regional Benchmark (CCB) with the scenario baselines for 4 grid-electricity projects (- = more conservative; + = less conservative).

Similar comparisons for other benchmarks (e.g. OECD) showed that conditions varied from country to country so that a multi project baseline at the regional level could give one country an advantage over another.

Therefore the selection of an appropriate multi-project baseline has to be related to the country context. For an electricity supply project it must also relate to the project type and whether it is contributing to meeting new demand or replacing existing demand. The baseline chosen must also be conservative to minimise the risk of overcrediting.

The scenario baseline analysis showed that the uncertainties associated with the estimations were found to be quite high (±12% to ±57%) and lowering the conservative baseline or benchmark with a weighting factor should be considered to manage these uncertainties. This is in line with previous studies (Begg and Parkinson 2001). Such a step may fulfil several functions while reducing transaction costs; in dealing with additionality issues, managing uncertainties and minimising inflation and gaming. The additionality aspect is discussed in more detail in the next section.

Additionality and environmental integrity

Additionality is another aspect of ensuring environmental integrity. In the AIJ Pilot Phase the original aim of using additionality as a test for projects was to minimise the use of the mechanism by 'free riders'. This means that projects which would have been done anyway as part of a business as usual situation should not be credited. Of course it is extremely difficult in practice to ensure this. The recent problems with the approval of CDM projects by the Executive Board highlight the confusion surrounding the issue. Some assert that the wording in the Marrakech Accords means only that the project must 'beat' the baseline to be additional while others maintain the original aim of the test which was to ensure that the project would not have taken place anyway. This is a problem for project developers as the uncertainty surrounding the interpretation of additionality can cause delays and unexpected upsets in project schedules.

What is important in all this debate is to focus on the original objective of the test which was to ensure 'real' reductions and thus preserve environmental integrity. The initial test for additionality is therefore very important. To this end the EB have tried to clarify their guidance on this issue with the development of an agreed additionality test process from COP10 which was detailed in the earlier section on the CDM.

However for the CDM there are now many other safeguards in place besides the additionality test such as the limited crediting lifetimes and the third party verification process to help ensure environmental integrity in the context of a host country with no targets. Limited crediting lifetimes are one of the most effective ways of minimising this risk and preserving the integrity of the credits produced so that it is unfortunate that it only applies to the CDM and not to JI.

What was done in the Probase project was to explore some of the consequences of non additionality in terms of the contribution to the uncertainty of the emission reductions to see what difference it makes.

3.2 Additionality and crediting lifetime: Implications for environmental integrity

There is a relationship between the consequences of non additionality and the crediting lifetime used. Obviously where there are no restrictions on crediting lifetime and there is no separate additionality test then the only factor limiting the creation of hot air is the level at which the baseline is set. Depending on whether additionality is interpreted just as beating the baseline this means that for track 1 JI there are real risks of producing more hot air into the system. The risk is lessened with track 2 JI with third party verification and a possible separate additionality test but no restrictions on crediting lifetime is a serious risk to integrity.

In the case of JI there is therefore a very strong case for JI with a separate additionality test, a stringent baseline and third party verification (track 2) as far as possible to help ensure environmental integrity in the reductions achieved. A limited crediting lifetime inclusion would be best of all.

Where there is a large CDM project with third party verification, an initial separate additionality test and a 3 times 7 year crediting lifetime, there will be revisions in the baseline every seven years which will mitigate against the need for concern on the initial additionality test compared to the original AIJ and JI situation. However conservative assumptions on the baseline and a reasonable additionality test are still required because of the large amount of greenhouse gas reductions involved.

However it is suggested that the main risk to the integrity of reductions from projects is the case where the crediting lifetime is 10 years.

CDM projects with 10 year crediting lifetime

The contribution to the uncertainty in the reductions if the project becomes non additional in the period from 5 years after commissioning to the 10 year limit was examined. The assumption was that an additionality test should be able to be reasonably confident for at least the first 5 years of the project implementation. However from 5 years onwards the additionality uncertainty is significant and contributes from ±6 to ±35% to the relative uncertainty of the reductions.

This uncertainty from non additionality at the later project stage is not addressed by the original test. To cope with the uncertainty there is therefore a case for having a stringent benchmark lowered by some weighting factor. A project which beats this stringent benchmark could be considered to have a higher probability of being additional as it minimises the uncertainties in the reductions. This approach is therefore rooted in the analysis of uncertainties and the management of environmental integrity in a direct way while simplifying the approach. It is not the same as just 'beating the baseline'.

Sathaye et al. (2001) suggested a stringent benchmark based on performance criteria as an additionality test in the electricity and cement sectors. However this was based on technological additionality rather than the arguments above. The problem with the approach is that these benchmarks are limited to specific project types and require performance data in the sector which this proposal does not.

Thus for CDM projects with a 10 year crediting lifetime and for JI projects such a stringent baseline lowered by a weighting factor would be a simple way of managing some of the uncertainties inherent in the system.

Managing the risks: The way forward

From the study therefore a possible process was suggested which could include: an initial additionality test such as a combined barriers test, the selection of an appropriate standardized baseline as proposed above followed by weighting of the baseline to produce a stringent benchmark to address the uncertainties and any residual additionality problems. This could be applied to all projects but the weighted stringent baseline would apply particularly to JI and CDM projects with a 10 year crediting lifetime and would fit with the existing suggested framework. In addition for JI which have no restrictions on crediting lifetime this would provide some compensation for the likelihood of non additionality over the lifetime.

3.3 The initial additionality test

A study was carried out on the possible project entry into the UK Emissions Trading Scheme (UKETS) which focussed on the electricity sector, Combined Heat and Power (CHP), transport projects, built environment projects and on methane capture projects. Baseline methodologies and additionality approaches were explored for these projects and are reported in a final report (Begg et al. 2002). Unfortunately the EU ETS poses a severe problem of incompatibility with the UK system which makes the electricity, CHP and built environment project entry problematic due to issues such as double counting of reductions upstream and downstream. However there is scope in some sectors, especially as the EU system expands, to include other gases and sectors such as transport. Most of the projects examined were small scale projects which would have to be aggregated to minimise transaction costs. Project additionality methods covering a range of approaches and policy additionality were analysed and recommendations made. Additionality tests which seem easy to apply in terms of a simple yes/no assessment, such as performance benchmarks, in fact require large amounts of data and preparation. The most versatile test was found to be a combined barriers test which looked at the barriers and how they would be overcome by the project. Though it was qualitative in nature it provided more insights than the use of investment criteria which can be manipulated. It is therefore interesting that both the investment and the barriers tests are being promoted in the latest recommendations for the CDM. However the introduction by the Executive Board of the CDM of such a delineated process must help to standardise procedures and in the end lead to a reduction in transaction costs.

3.4 Small scale projects

Small scale mitigation projects at first sight may seem to be less important in delivering reductions than the large scale projects. However small scale projects are usually aggregated to minimise their individual transaction costs and large numbers of small scale projects can deliver significant reductions. In a developing country or transition economy host country context (Slovakia green emission trading for example) they can represent a significant share of the opportunities available.

As small scale projects are usually at the household or the community level, these projects can deliver direct sustainability benefits if, and only if, they are implemented well. The delivery of these sustainability benefits has been examined through a number of studies (Thomas et al. 2000; Troni et al. 2002; Sutter 2003; Begg et al. 2003). Some of the issues under the Gold Standard address this but it does not go far enough.

The sustainability benefits which are supposed to be delivered by CDM projects have been subsumed to the responsibility of the host countries. However Begg et al. (2003) suggest that if the project is not designed to deliver these benefits the long term viability of the reductions is called into question. This accords with the recent discussions at COP10.

In order to minimise transaction costs small scale projects should usually be aggregated or bundled and the bundled project is than treated as one project with the costs then shared among many small projects. However, there has been insufficient work done on how best to achieve the aggregation needed for small scale projects within a CDM framework. Costa Rica led the way with their 'Umbrella model' for forestry projects (Hambleton et al. 1999). There, the government set up an intermediary office where investors could come and buy credits and the government office managed the interface with the thousands of small farmers who were involved. The government also underwrote the risk associated with the Certified Tradable Options issued. Currently there are few institutional structures set up in developing countries to deal with small scale CDM projects though the idea of a one stop shop could be a viable option. The need for new structures rather than existing systems also needs to be considered.

There is currently no guidance under the methodological guidance from the UNFCCC/EB on bundling for small scale projects. On the contrary some advice actually militates against this. For example the project boundary for a Solar Home System is set as the household itself.

How the projects should be bundled in terms of baselines and monitoring with the relevant sampling and averaging of information is not explored at all in current advice from the EB; nor is what should or could be bundled as project types. Some of these issues have recently been addressed by Green et al. (2002) and Begg et al. (2003) who suggest that there a number of different options which could be used depending on the baseline situation and the energy needs and possible project types in the communities involved. They also point out that the current official guidance is also not yet sufficient to cope with the actual complexity of some baseline situations nor are other issues such as equivalence of service between the project and the baseline adequately or explicitly addressed in all areas. A more holistic approach would be beneficial for host countries.

4 Conclusions

The controversies over the project mechanisms have not really abated. Until more domestic action is seen to be taken and until solutions are found to maintain environmental integrity and ensure delivery of sustainability benefits without incurring high transaction costs, there will be scope for criticism of the project mechanisms. However recent developments within the EB in formulating the additionality modality and in progressing projects are encouraging as is market value being assigned to sustainability benefits from projects whether they are JI or CDM. This will be crucial for the 'ownership' of projects to ensure the delivery of long term emission reductions. It is nevertheless early days of 'learning and doing' where new methodologies are evolving. Some practical studies carried out to address the issues of complexity and transaction costs while maintaining integrity do exist to standardise and streamline approaches for carbon accounting and additionality. However these are not necessarily being used by developers or the EB. Particularly in the key area of small scale projects little progress is being made on bundling projects which is key to their operationalisation on the large scale needed to deliver the equity embedded in the Convention and Protocol. New initiatives under the EB are required on these issues. Fast moving developments under the UK and EU emissions trading schemes mean that perceptions are changing with much more interest in the project mechanisms for meeting Kyoto targets with a range of country initiatives being announced where previously the Netherlands government was the sole player. The market is already generating new ideas with new variations on the mechanisms. Environmental integrity and sustainability of the credits being traded appears to be an aspect of value which is being reflected in relative preferences for 'gold standard Certified Emission Reductions (CERs)' and ordinary CERs compared to other carbon units. Given that the complexity of the mechanisms is being reflected in these new trading markets, it may be expected that in the future, the market preferences will in their turn influence the further development of the project mechanisms.

References

Anagnostopoulos K, Flamos A, Askounis D, Psarras J (2004) The Multiple Benchmark System Application to Indonesia, Russia and Panama, Mitigation and Adaptation Strategies for Global Change 9(2): 147-180

Begg K, Jackson T, Parkinson S (2001a) Beyond Joint Implementation - Designing Flexibility into Global Climate Policy, Energy Policy 29(1): 17-27

Begg K, Parkinson S, Wilkinson R (2001b) Maximising GHG emissions and sustainable development aspects in the Clean Development Mechanism, World Resources Review 13(3): 315-334

Begg K, Jackson T, van der Gaast W, Jepma C, Sorrell S, Smith A, van der Horst D, Bandsma J, ten Hoopen M (2002) Guidance for UK emissions trading projects: Advice to Policymakers, (Final report Policy Document for UKDTI)

Begg K, Wilkinson R, Parkinson S, Amissah-Arthur H, Atugba S, Brew-Hammond A, Theuri D, Mathenge M, Gitonga S, Meena H, Mwakifwamba S, Mwakifsonde S and van der Horst D (2003) Encouraging CDM energy projects to aid poverty alleviation, (Summary Report to UK DFID).

Begg K, van der Horst D (2004) Preserving Environmental Integrity in Standardised Baselines: The Role of Additionality and Uncertainty, Mitigation and Adaptation Strategies for Global Change 9: 181-200

Cason TN (2003) Buyer liability and voluntary inspections in international greenhouse gas emissions trading: A laboratory Study, Environmental & Resource Economics, 25(1): 101-127

Green J, Mariyappan J, Plastow J, Mensah-Brown H, Agbey S, Brew-Hammond A, (2002) 'Bundling and small scale projects', Report for FCO CCCF

Grubb M (2003) The Economics of the Kyoto protocol, World Economics 4(3): 143-189

Hambleton A, Figueres C and Chatterjee K (1999) Do AIJ Projects Support Sustainable Development Goals in the Host Country? Chapter 8 in The UNFCCC AIJ Pilot: Experiences and Lessons Learned, pp. 167-181

JIQ (2003a) Joint Implementation Quarterly JIN Foundation 9, 2: 2-4 (July 2003)

JIQ (2003b) Joint Implementation Quarterly JIN Foundation 9, 4: 1 (December 2003)

JIQ, (2004a) Joint Implementation Quarterly JIN Foundation 10, 4 (December 2004)

JIQ (2004b) Joint Implementation Quarterly JIN Foundation 10, 1: 14 (March Issue)

Kartha S, Lazarus M, Bosi M, (2004) Baseline Recommendations for GHG mitigation projects in the electric power sector, Energy Policy, 32(4): 545-566

Korppoo A (2003) Forging alliance with Russia: The example of the green investment scheme, Climate Policy 3(1): 67-76

Michaelova A, Stronzik M (2002) Transaction costs of the Kyoto Mechanisms, HWWA discussion paper 175

Pointcarbon (2004) CDM and JI monitor, Dec 2004

Sager J (2003) An analysis with the CERT model of the FSU market power in the carbon emissions trading market, Environmnetal Modelling & Assessment 8(3): 219-238

Sathaye J, Price L, Worrell E, Ruth M, (2001) Multi project baselines for evaluation and Industrial Energy efficiency and Electric Power projects, LBNL-48242, (LBL report for the EPA)

Stainforth DA, Alna T, Christenesen C, Collins M, Faull N, Frame DJ, Kettleborough JA, Knight S, Martin A, Murphy JM, Plant C, Sexton D, Smith LA, Spicer RA, Thorpe AJ, Allen MR (2005) Uncertainty in the prediction of climate responses to rising levels of greenhouse gases, Nature, 433-436 (January 11th 2005)

Sutter C (2003) Sustainability Check up for CDM projects, Wissenschaftl. Verlag, Berlin

Tangen K, Korppoo A, Berdin V, Sugijama T, Drexhage J, Egenhofer C, Grubb M, Legge T, Moe S, Stern J, Yamaguchi K, (2002) A Russian Green Investment Scheme-securing environmental benefits from international emissions trading, Climate Strategies Report, Nov 2002, http://www.climate-strategies.org

Thomas F, Ullrich S, Mazzoni O, (2000) Developing a Methodology to Evaluate CDM Projects According to National Sustainable Development Criteria. SUSAC project for UK FCO and EU, http://www.esd.co.uk/Tindex.htm?, http://www.esd.co.uk/html/CDM_Susac.htm~tmain

Troni J, Ageby S, Costa PM, Haque N, Hession M, Gunaratne L, Rodriguez H, Sharma A (2002) Moving Toward Environmentally Neutral Development (MEND), (Final Report to UK DFID)

UNFCCC (2002) FCCC/CP/2002/7/Add.3 Simplified modalities and procedures for small scale projects

van der Gaast W, Jepma C, Begg K, Michaelova A, Butzengeiger S, Flamos A, Brodmann U (2003) Procedures for Accounting and Baselines for JI and CDM projects (PROBASE), Final report to EU FPV

Woerdman E (2000) Implementing the Kyoto protocol: why JI and the CDM show more promise than international emissions trading, Energy Policy 28(1): 29-38

Woerdman E (2001) Emissions Trading and Transaction Costs: analysing the flaws in the discussion, Ecological Economics 38, 2: 293-304

World Bank (2003) State of the Carbon Market (December 2003)

Prevailing technologies and locations of CDM projects: the current situation compared with expectations

Jobert Winkel

EcoSecurities
Mauritskade 25, 2514 HD The Hague, Netherlands
jobertwinkel@lycos.nl

Abstract

The Clean Development Mechanism is one of the flexible mechanisms in the Kyoto Protocol. In this paper the prevailing technology types and project locations of CDM projects are analyzed. The prevailing technology types are hydropower (24% of all projects), followed by bio- and landfill gas (16%). The distribution can be generally explained by the values for the Internal Rates of Return (IRRs), including the revenues from CERs (Certified Emission Reductions), of all of these technology types, except for hydropower. Besides the fact that landfill CDM projects generally have an attractive IRR, an additional advantage of this type is the fact that methane emissions are reduced, resulting in higher revenues from CERs.

The most popular CDM countries are Brazil, India, Costa Rica and Panama. One of the indicators for a high suitability of a country is a high country-specific Carbon Emission Factor. An analysis shows that more than 50% of all electricity CDM projects are located in a country whose CEF is among the 40% lowest. It can be concluded that other factors must play a more important role in those cases.

Also the sensitivity of the IRR has been tested in a case project by varying the CER price, the MWh price for electricity, and the total amount of waste used. All three parameters are approximately linearly related to the IRR. It can be concluded that all three determinants of the IRR that are analyzed have a significant influence.

Keywords: CDM, technology, internal rate of return, CER price, MWh price

Acknowledgements: The author wishes to thank Jan-Willem Martens and Michiel ten Hoopen for many useful comments.

1 Introduction

The Kyoto Protocol enforces industrialized countries to decrease their greenhouse gas emissions. A part of this commitment can be realized through the so-called 'flexible mechanisms' by purchasing emission reductions that are realized in other countries. These emission reductions are identified in Joint Implementation (JI) or Clean Development Mechanism (CDM) projects. The revenues from the sale of the identified emission reductions (or carbon credits) increase the financial feasibility of the project and thus stimulate the market penetration of new innovative technologies in emerging market economies. This impact differs significantly per type of project.

The objective of this paper is to provide explanations for the current mix of technology types and locations of CDM projects, and to show the impact of principal market and project parameters on the profitability of a case study.

First, an overview of the current mix of CDM projects is presented, including two principal characteristics for the projects that are developed: the types of technology used and the project locations. Factors that may have a substantial impact on the popularity of certain technologies or project locations are examined, such as the internal rate of return of a project.

Secondly, a case study will be analyzed, in order to show the variability of the project's internal rate of return.

Note: At present, no CDM projects have passed the entire approval process until registration as an official CDM project. However, projects for which Project Design Documents (PDD) have been prepared, are denoted as 'CDM projects' in this paper. It should be born in mind that they have not yet been officially approved.

2 Overview of CDM projects at present

Until October 2003, Project Design Documents (PDD) have been prepared according to the rules of the UNFCCC for 63 CDM projects. The preparation of a PDD is one of the CDM requirements a CDM project has to meet in order to generate Certified Emission Reductions (CERs). For comparison: there are now 38 JI projects for which similar project documentation has been prepared and buyers of the credits (ERUs) have been found.

In the next sections, an analysis of the technology types and of the project location is performed.

2.1 Technology types

Figure 1 shows the shares of different technologies in all CDM projects.

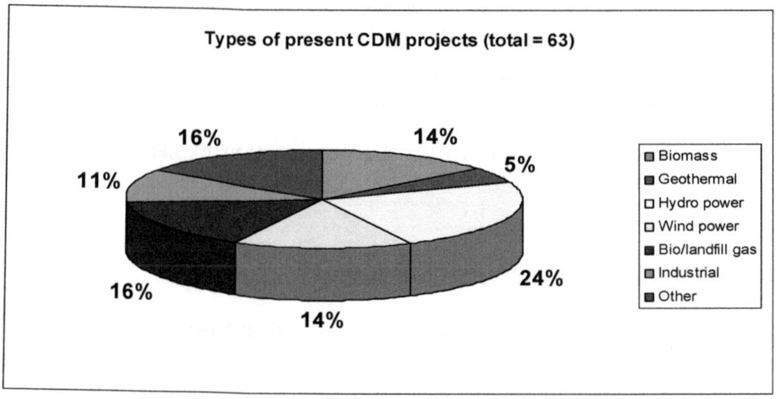

Fig. 1. Total *number* of CDM projects divided per technology category (not the total *emission reductions* of the projects).

Figure 1 clearly demonstrates that most CDM projects are renewable energy projects. At present, the share of renewable energy including 'bio/landfill gas' is 73%. Hydro power is the most dominant category CDM projects. Most projects reduce CO_2 emissions, except the projects in the bio/landfill gas category (16%), which reduce CH_4 emissions, and one project in the industrial sector, which reduces HFC-gases. Regarding the renewable energy mix, it is apparent that there are no solar energy projects and only a few geothermal energy projects (5%).

Of all projects, those that have their baseline and monitoring methodology submitted to the EB of the CDM or even have their methodology approved, are closest to being registered as an official CDM project. Therefore, these 'most serious' projects have been considered separately (see Figure 2).

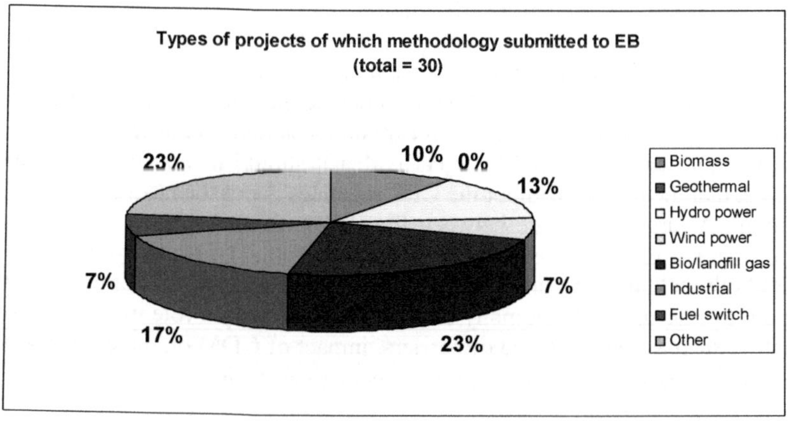

Fig. 2. Total *number* of CDM projects divided per technology category (not the total *emission reductions* of the projects).

It is apparent that compared to the total mix of CDM projects, projects of which the methodology has been submitted to the EB:

- include much less renewable energy projects (in particular hydropower) than can be expected from their share in the total,
- have been mostly landfill gas to energy projects,
- include relatively a large share of industrial projects.

Now the actual situation is known, in the next section explanations for the composition of this mix are sought.

2.2 Explanations for prevailing technologies

Projects that can be expected to be abundant among CDM projects are likely to possess the following characteristics:

- Generating a substantial amount of emission reductions
- Having a high financial feasibility when the sale of CERs is included, compared to that of the project excluding the sale of CERs

In order to explain why some technologies are more popular than others among CDM projects, these criteria are discussed in relation to the prevailing technology types.

The first mentioned characteristic applies for many projects that reduce greenhouse gas emissions, since the revenues of the CERs can easily be several times higher than all costs involved in transforming the project into a CDM project.

The second characteristic concerns the financial feasibility. The Internal Rate of Return (IRR) can be used as an indicator for the financial feasibility. Technology types face different risks and therefore require different rates of return. As a consequence, the value of the IRR required by a project developer will not be the same for each technology (and project location).

A project is feasible if the project IRR is higher than the required IRR, which results in a positive net present value. Although other factors will play a role in the feasibility of a CDM project as well, first it is considered whether typical values of the IRR for different technology types can explain the actual mix of technologies.

Another requirement of a CDM project is that it should demonstrate its additionality, i.e. that the project without the CER revenues is not the most likely baseline scenario. A large difference between the IRR with and without the sale of CERs will make it easy to prove this additionality. Table 1 lists technology specific IRRs both including the sale of CERs as well as excluding these revenues. It is important to bear in mind that many other factors also play a role in the issue of additionality, such as the existence of barriers, impact of CDM on the diversity of cash flow, and impact of CDM on improved debt service ratio.

Table 1. Average IRR of projects of different technologies

Technology	Av. IRR including sale of CERs	Av. IRR excluding sale of CERs	Nr. of projects on which data is based
Landfill	19	10	3
Energy efficiency	19	17	1
Wind power	15	14	3
Biomass	14	12	5
Geothermal	13	12	1
Hydropower	11	10	2

Based on: World Bank Prototype Carbon Fund, 2002 (preliminary data)

Note that these figures are based on data from only a few projects. In addition, the IRR values for wind power and biomass vary significantly per specific project. These figures are meant to be illustrative and no firm conclusions should be drawn from the average IRR rates mentioned in this table.

Although the IRR required by project developers depends on many factors (such as country and technology), for most technologies it can be assumed to range between 10% and 20%. Landfill projects and 'energy efficiency' projects use relatively mature technology and are, therefore, expected to require a somewhat lower IRR. Solar photovoltaic projects could require a relatively high IRR.

Bearing these requirements in mind, one would expect the largest share of CDM projects to be landfill projects, especially since the difference in IRR is very large. The other technologies as ranked in Table 1 are expected to have subsequent lower shares.

The next issue to analyze is if the technology specific IRRs can explain the current shares of technologies in the total mix of CDM projects (see Table 2 below).

Table 2. Comparison of actual and expected occurrence of technologies based on impact on IRRs

Technology	Ranking (actual shares of technologies among CDM projects)	Expected ranking based on average IRR incl. sale of CERs
Hydropower	1 (24%)	5
Bio/landfill gas	2 (16%)	1
Wind power	3 (14%)	3
Biomass	3 (14%)	4
Industrial / energy efficiency	5 (11%)	2

Note: The two categories 'industrial' and 'energy efficiency' were identified as separate categories, while they overlap partially. For comparison purposes, they have been considered one group here.

Landfill projects are very popular, as is expected. An additional advantage for landfill projects is the type of emission reductions that is generated. Methane emissions are reduced instead of carbon dioxide, and methane has a much higher global warming potential. As a result, reduction of methane accounts for much more CO_2-equivalent emission reductions. The development of a landfill project can, therefore, be financed to a large extent through carbon revenues.

Biomass and wind power are quite popular, which was to be expected based on their average IRR given in Table 1. Solar photovoltaic projects do not occur, which is in line with expectations since this type of project has in general a low IRR and small project size. Only hydropower does not satisfy this rationale: it has the biggest actual share while the CDM impact on hydropower project was assumed to be small and it has on average a lower IRR (see Table 2). On the other hand, Table 2 is only illustrative, so no firm statements can be drawn from this figure. The popularity of hydropower may be explained by the fact that this technology is relatively mature and, therefore, less risky. Moreover, compared to other renewable energy sources, hydropower is a much less intermittent source.

As stated in Table 1, energy efficiency projects have a relatively high IRR, both including as well as excluding the sale of CERs. As mentioned above, it is generally perceived that energy efficiency is relatively mature and would therefore require a lower IRR by the project developer. Hence, one would expect many energy efficiency projects in the CDM. One reason why this is not the case could be that a part of the energy efficiency projects is already financially feasible and does not need to be developed as CDM projects. Another explanation is that, due to specific barriers for energy efficiency, the required IRR for energy efficiency needs to be much higher than for power supply projects before implementation is considered.

It should be noted that Table 2 is built up by *average* values of the project IRR, while for individual projects this may vary widely, due to several other project parameters. In Chapter 3, an analysis is presented of some key parameters that influence the IRR.

It is chosen to consider IRR values in this comparison, instead of the difference in IRR values between inclusion and exclusion of CER revenues (second column of Table 1). This is because the absolute IRR can explain the actual situation better than the difference in IRR values.

2.3 Project locations

Besides the most popular technologies among CDM projects, a second interesting characteristic of CDM projects is the project location. The success of a CDM project can be greatly influenced by the country in which it is located. Important country-specific factors are:

- The intention of the government to ratify the Kyoto Protocol
- Institutional organisation and willingness to endorse the transfer of CERs
- In case of electricity generation: the average amount of greenhouse gas emissions of the existing electricity park, which influences the amount of emission reductions a CDM project can generate
- Investment climate

First, an overview of the most popular countries is provided. After this, an explanation for these results is sought, based on the country-specific criteria mentioned above.

Table 3 provides an actual overview of the most popular countries, sorted per technology. The two most popular countries are mentioned per technology (in case there is more than one project).

Table 3. Most popular CDM countries

Technology types	Total number of projects per type	Most popular countries
Biomass	9	India (5), Brazil (2)
Geothermal	3	Different countries
Hydropower	15	Costa Rica (3), Panama (3)
Wind	9	India (4), Costa Rica (2)
Bio/landfill gas	10	Brazil (3)
Industrial	7	Costa Rica (2)
Fuel switch	3	Brazil (2)

It is obvious that India and Brazil are very popular locations for CDM projects in general.

In the case of electricity generation projects, a country with a high level of greenhouse gas emissions per unit electricity is favourable for a CDM project. This level can be denoted by the Carbon Emission Factor (CEF), expressed in tonnes CO_2 equivalents per MWh electricity.

The popularity of the countries listed in Table 3 might therefore be explained if they have a relatively high country CEF, which is examined in the next section.

2.4 Explanations for prevailing project locations

Before we look at a specific explanation for the popularity of the countries mentioned in Table 3, first an overview is provided for all countries in which electricity CDM projects are developed. The popularity of countries is compared with the country-specific CEF. (Only *electricity generation* projects are considered since emission reductions that result from other project types do not depend on the CEF of the existing electricity park.)

In order to generally indicate the level of a country CEF, all developing countries have been ranked from low to high CEF. From the study of Martens et al. (2001) the CEF for almost 100 Non-Annex I countries have been obtained. These countries have been split into five groups of equal sizes, i.e. all groups contain the same number of countries.

Figure 3 shows the prevalence of electricity CDM projects in countries that are grouped by their country CEF. One would expect many electricity CDM projects in countries with a high CEF (i.e. group 5) and few projects in countries with a low CEF (i.e. group 1). It is obvious that this does not correspond with reality.

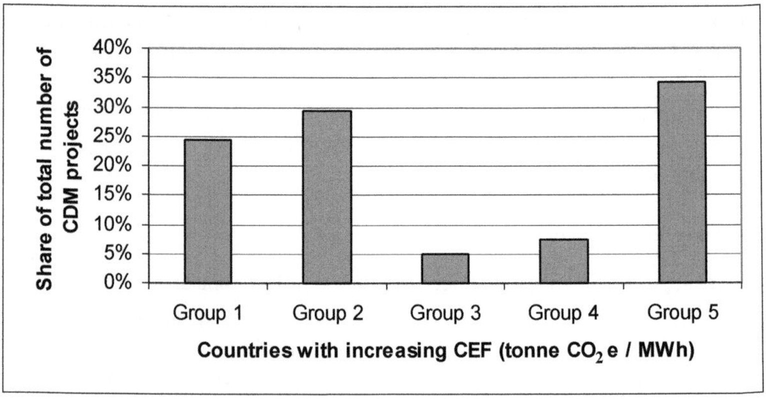

Fig. 3. The prevalence of electricity CDM projects (41 in total) in countries that are classified into groups depending on the level of their country CEF

N.B.: the CEF ranges of the groups are (in tonne CO_2-e / MWh):

Group 1: <0.611
Group 2: 0.611 to 0.743
Group 3: 0.743 to 0.778
Group 4: 0.778 to 0.924
Group 5: $>=0.924$

Although many projects are actually developed in countries with a relatively high CEF (group 5), the same applies for low CEF countries in group 1 and 2.

Possible explanations are:

- In most projects, it is not the *national* electricity park that is used for the calculation of the emission reductions of projects. Instead, a relevant *part* of the electricity park is considered.
- Instead of using *present* data in a number of projects, data of the past or projections of the future are used.
- The CEF values of the middle groups do not vary widely to make a large difference.

However, it is very likely that other criteria for a good project location than the country CEF (see the beginning of this chapter) will play an important role for countries in group 1 and 2. A recommendation for future research is to examine the potential influence of these criteria.

After this general analysis, the specific cases of the most popular countries as presented in Table 3 are examined (see Table 4 below).

Table 4. Country CEF of popular CDM countries

Country	CEF (tonne CO_2e / MWh_{el})[1]	Group nr. (5 = highest CEF)	Total number of CDM projects at present
India	1.055	5	13
Panama	0.688	2	3
Brazil	0.642	2	10
Costa Rica	0.128	1	8

[1]Source: Martens et al., 2001

India has a high country CEF and as a consequence will reduce a relatively high quantity of emissions compared to most other countries (in case of electricity projects). This can explain why India is more attractive for CDM projects than other countries. On the contrary, it is surprising that Costa Rica has a very low CEF, but hosts many CDM projects (eight in total). An explanation needs to be sought in other country characteristics. In the case of Costa Rica, it is known that the government is stimulating the development of CDM projects and the institutional organisation to facilitate this process is in place.

All three projects in Panama are hydropower projects. The disadvantage of the country's relatively low CEF is probably outweighed by the fact that Panama has excellent possibilities for hydropower.

Brazil belongs to the same group in terms of its CEF. However, it hosts many projects. The specific explanation for Brazil could be the country's favourable investment climate.

A joint explanation for the popularity of (the latter three) countries with a relatively low CEF could be the fact that all are already familiar with renewable energy projects.

3 Determinants of the IRR of a landfill project

We have seen that the distribution of technologies in the present mix of CDM projects can be explained partially by technology specific IRR values. In order to learn more about the determinants of a project's IRR, the influence of selected parameters on the IRR is examined in this chapter.

One of the CDM projects that EcoSecurities is developing is used as a case study in order to perform a sensitivity analysis. This sensitivity analysis focuses on a number of parameters that have an influence on the IRR of the project. (These parameters will have different values in other projects.) The chosen parameters are:

- CER price (USD / tonne CO_2-e)
- price per MWh of generated electricity
- Amount of waste present and added per year

These parameters have been chosen since they are important determinants of the IRR and they can vary substantially between different technology types or countries. Besides investment costs and operational costs, other parameters that are of influence are:

- Annual electricity production of the project
- Tax levels of the country in which the project is located

3.1 Description of the case

The case CDM project chosen is a landfill gas to energy project. This is because landfill gas projects have a significant share in the total mix, and three methodologies have been approved for this technology. For the IRR analysis the technical aspects of an EcoSecurities project have been used.

The case project concerns two adjacent landfill sites that will be covered and landfill gas will be flared and combusted for electricity production. One landfill site has been used for several years and has been closed, since it has reached its maximum capacity; the other landfill site has just been opened and will be used for the next decades.

The emission reductions that are claimed through the CDM are only those that originate from the transformation of methane into carbon dioxide (which has a much lower effect on global warming). The emission reductions that result from the production of clean electricity instead of 'dirty' electricity have not been claimed.

The project excluding revenues from the sale of carbon credits (emission reductions) is financially infeasible. The IRR for this situation is around 5%. However, when the project is developed as a CDM project, the IRR is raised to 18%.

The main parameters used in this calculation are listed in Table 5:

Table 5. Main parameters influencing the IRR

Parameter (unit)	Value
Amount of waste in place at start of project	2 M tonnes
Anticipated future daily waste disposal for next 10 years	Average of 1.500 tonnes per day
Electricity tariff	30 USD / MWh
CER price	4 USD / tonne CO_2e
Total emission reductions over 10 years	2.5 million CERs

In the next sections the influence of these parameters is examined.

3.2 Influence of CER price

Since each project developer demands different values of a project's IRR, for a number of IRR values the required change in CER price has been calculated, assuming all other parameters remain constant (see Figure 4). In this way, a project developer can see at what CER price a similar project as the case project would become attractive.

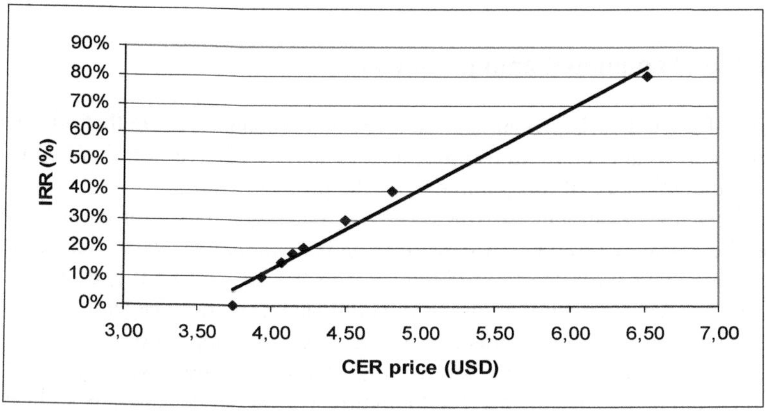

Fig. 4. Required CER price in order to achieve a preferred project IRR

The relation between CER price and IRR is almost linear. Deviations from this line are due to varying waste input over time (a low waste input in the end of the project lifetime will decrease the influence of a high CER price on the IRR). Because of this linearity, it can be stated that for this landfill project, an increase in CER price of one dollar results in an increase in IRR of almost 30 percentage points. Since the CER price is highly variable and is expected to alter significantly within the next years, it is clearly a very important determinant of the project IRR.

3.3 Influence of MWh price

For a number of IRR values the required change in MWh price has been calculated, assuming all other parameters remain constant (see Figure 5).

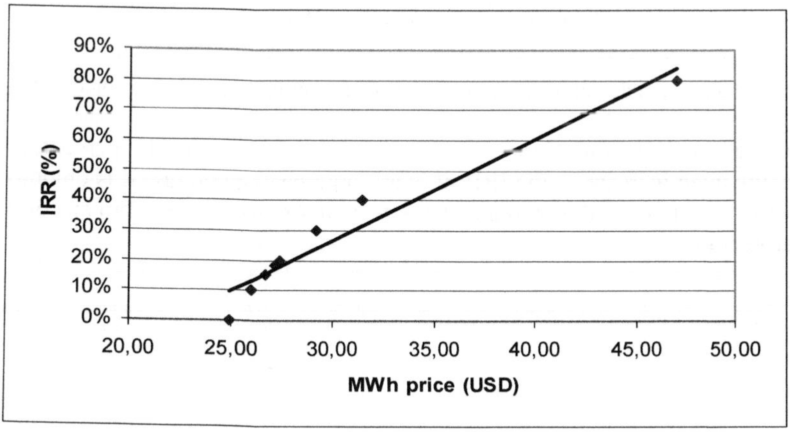

Fig. 5. Required MWh price in order to achieve a preferred project IRR

It can be stated that for this landfill project, an increase in MWh price of one dollar results in an increase in IRR of around almost 4 percentage points.

3.4 Influence of estimated amount of waste

The amount of waste that is present and the expected amount of waste that will be disposed in future years is hard to estimate adequately. For this reason it is important to know the influence of this parameter on the project IRR.

Substantially higher amounts of waste will require higher costs for equipment to process the waste. Therefore, in case of a higher amount of waste than estimated, this parameter will also be different (not in the analysis of this case project). However, in case the amount of waste would also turn out to be smaller than estimated, it is useful to see how this affects the IRR, assuming all other parameters will remain constant. Figure 6 shows this relation for the case project. (A value of, for example, 110% means that the amount of waste in place, as well as the rate of waste that will be deposited in the future, are 10% higher than expected.)

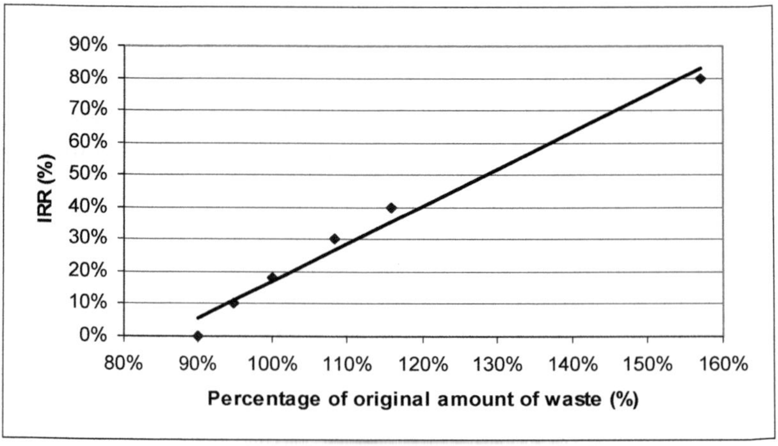

Fig. 6. The influence of a deviation from the expected amount of waste on the project IRR

On average, a deviation of the expected amount of waste of 10 percentage points results in an increase in the IRR of over 10 percentage points. It is obvious that also (a deviation in) the amount of waste is a significant determinant of a landfill project IRR.

4 Conclusions

The present mix of CDM projects (in the process of approval) consists of mostly renewable energy projects. However, a different trend of most promising technology types for the near future has recently been set by the Executive Board (EB) for the CDM. The EB approved the first methodologies, which are applicable for landfill projects, fuel switch projects and HFC reducing projects.

The prevailing technology types among current CDM projects are hydropower (24% of all projects), followed by bio- and landfill gas (16%), and biomass and wind (both 14%). This distribution can be generally explained by the values for the IRRs of all of these technology types (including the revenues from CERs), except for hydropower. Hydropower projects have a relatively low IRR. However, the popularity of hydropower could be explained by the fact that this technology is most mature and therefore less risky, requiring a lower IRR. (Note that IRR values have been obtained for a small sample of projects only.)

The most popular CDM countries are Brazil, India, Costa Rica and Panama. One of the indicators for a high suitability of a country is a high country-specific CEF.

Among the most popular project locations, in particular the very low CEF of Costa Rica does not fit this rationale.

An analysis shows that more than 50% of all electricity CDM projects are located in a country with a relatively low CEF. More precisely, they are located in countries that are of group 1 or 2 out of 5 groups (among the 40% lowest country CEFs). It can be concluded that other factors must play a role (which is recommended to examine in further research).

In order to learn more about the determinants of a project's IRR, the sensitivity of the IRR has been tested in a case project by varying the CER price, the MWh price for electricity, and the total amount of waste used. Such a sensitivity analysis is useful background analysis for investors who need to judge investment decisions in the emerging CDM market.

A landfill gas project has been chosen as a case study. All three parameters are approximately linearly related to the IRR. Therefore, the linear relation can be stated that an increase in CER price of one dollar will result in an increase in IRR of almost 30 percentage points. An increase in MWh price of one dollar results in an increase in IRR of around 4 percentage points. A deviation in the total amount of waste results roughly in an equal deviation of the IRR in percentage points. It can be concluded that all three analyzed determinants of the IRR have a significant influence.

Proposed new research questions:

- A more comprehensive comparison of the actual prevalence of technology types of CDM projects and their expected prevalence based on the technology specific IRR. The comparison can be improved by identifying a uniform typology for technology types. For example, in this paper the types 'industrial' and 'energy efficiency' have been used from different sources, while they overlap partially.

- Since a high country CEF is no guarantee for favourable CDM project circumstances, other factors play an important role. The influence on project location of these other factors, as mentioned in this paper, could be analyzed.
- Further analysis of key determinants of CDM projects regarding the profitability of CDM projects to inform decision-makers regarding investments.
- Small-scale projects in general are financially less attractive. Further recommendations for simplification of rules for small-scale projects would reduce transaction costs and increase their financial feasibility.

References

World Bank Prototype Carbon Fund (2002) Impact of Carbon Finance on Project Financing; PowerPoint presentation

Martens JW, van Rooijen SNM, Bovée V, Wijnants HJ (2001) Standardized Baselines for Small-Scale CDM Activities - A proposal for the CDM programme of the Netherlands, ECN C-01-122

GPSR Compliance

The European Union's (EU) General Product Safety Regulation (GPSR) is a set of rules that requires consumer products to be safe and our obligations to ensure this.

If you have any concerns about our products, you can contact us on ProductSafety@springernature.com

In case Publisher is established outside the EU, the EU authorized representative is:

Springer Nature Customer Service Center GmbH
Europaplatz 3
69115 Heidelberg, Germany

Batch number: 10725800

Printed by Printforce, the Netherlands